ENFORCING INTERNATIONAL MARITIME LEGISLATION ON AIR POLLUTION THROUGH UNCLOS

This book explores the scope and applicability of the United Nations Convention on the Law of the Sea (UNCLOS), relating to the enforcement of international maritime legislation on air pollution. It focuses on enforcement of Annex VI of the International Convention for the Prevention of Pollution from Ships (MARPOL Annex VI) and the strengthened global sulphur limit which comes into force in 2020. The first chapters in Part I provide an overall introduction to relevant regulations of MARPOL Annex VI, UNCLOS, Port State Control (PSC), the EU Sulphur Directive and basic jurisdictional principles of international law. Part II analyses the amplified enforcement and notifying obligations of UNCLOS chapter XII placed on flag States and the broadened jurisdictions for port and coastal States to enforce. This includes extraterritorial enforcement by port States on the high seas and how overlapping jurisdictions are resolved. These theoretical discussions on jurisdiction are tied to practical applications pertaining to PSC and sanctioning. Part III builds upon the conclusions of Part II in relation to the enforcement of other legislation adopted by the International Maritime Organization (IMO), including regulations on Greenhouse Gases (GHG) which the IMO is set to adopt through its GHG Strategy. Finally, with the increased environmental challenges relating to global warming, and given the special legal status of ships, Part IV offers an analysis of whether specific IMO regulations on GHG could, in the future, be considered peremptory norms of a 'jus cogens' character, and addresses the potential legal implications.

Enforcing International Maritime Legislation on Air Pollution through UNCLOS

Jesper Jarl Fanø

•HART•
OXFORD • LONDON • NEW YORK • NEW DELHI • SYDNEY

HART PUBLISHING

Bloomsbury Publishing Plc

Kemp House, Chawley Park, Cumnor Hill, Oxford, OX2 9PH, UK

HART PUBLISHING, the Hart/Stag logo, BLOOMSBURY and the Diana logo are trademarks of Bloomsbury Publishing Plc

First published in Great Britain 2019

A catalogue record for this book is available from the British Library.

Library of Congress Cataloging-in-Publication data

Names: Fanø, Jesper Jarl, author.

Title: Enforcing international maritime legislation on air pollution through UNCLOS / Jesper Jarl Fanø.

Description: Oxford ; New York : Hart, 2019. | Includes bibliographical references and index.

Identifiers: LCCN 2019042221 (print) | LCCN 2019042222 (ebook) | ISBN 9781509927760 (hardback) | ISBN 9781509927784 (Epub)

Subjects: LCSH: Marine pollution—Law and legislation. | Air—Pollution—Law and legislation. | Maritime law—Environmental aspects. | United Nations Convention on the Law of the Sea (1982 December 10) | International Convention for the Prevention of Pollution from Ships (1973 November 2)

Classification: LCC K3591.2 .F36 2019 (print) | LCC K3591.2 (ebook) | DDC 344.04/6342—dc23

LC record available at https://lccn.loc.gov/2019042221

LC ebook record available at https://lccn.loc.gov/2019042222

ISBN: HB: 978-1-50992-776-0
ePDF: 978-1-50992-777-7
ePub: 978-1-50992-778-4

Typeset by Compuscript Ltd, Shannon

For Esther and Arendse

PREFACE

This book has been many years in the making. Seven years to be exact.

I was working at the Danish Environmental Protection Agency (EPA) in 2012, as a lawyer (legal clerk), when I first heard of MARPOL[1] Annex VI and the EU Sulphur Directive.

That same year I started working part-time as an assistant teacher at the Faculty of Law at Copenhagen University, teaching International Law, which includes the Law of the Sea (UNCLOS), the law on treaties, basic jurisdictional principles, sources of international law, State responsibility, etc.

It was through my work at the Danish EPA, incorporating Annex VI and the Sulphur Directive into Danish law, that I learned about the challenges with enforcing these regulatory measures. A sulphur-enforcement taskforce was established, consisting of the Danish EPA, the Danish Maritime Authority (DMA), the Danish Attorney-General, the police and public prosecutors.[2] The Attorney-General stated, through talks with the Justice Department, that since Denmark did not have jurisdiction outside its EEZ over such emissions violations, it would prevent the Danish authorities (EPA and public prosecutors) from imposing fines that would wipe out savings achieved outside the EEZ. For example, if a ship(owner) saved \$500,000 by not complying with the sulphur regulations under MARPOL Annex VI, but only \$10,000 thereof was achieved when sailing in Danish waters, then the fine should be \$10,000 dollars plus a punitive element, perhaps resulting in a total fine of \$25,000–\$50,000, but still far from adversely affecting the \$500,000 saving. It was deemed to be up to the flag State to penalise the part of the violation that occurred on the high seas.

This was a troubling conclusion in terms of enforcing the 0.1% sulphur limit that came into force in 2016 in certain defined (SECA) areas. But, more importantly, it was a conclusion of *apocalyptic proportions* pertaining to enforcement of the strengthened 0.5% global sulphur limit that would come into force on 1 January 2020. Not complying with this provision can yield enormous savings for shipowners, but it also negates the environmental and human health benefits of the regulatory measures, including preventing 137,000 early deaths and 7.6 million children from developing asthma on a global scale.

It subsequently became apparent that this seemed to be the legal stance of other authorities throughout Europe and in the rest of the world.

[1] The abbreviations used in this Preface (MARPOL, etc) will be spelled out and explained in the book itself.

[2] See at https://www.ft.dk/samling/20161/almdel/mof/spm/738/svar/1413565/1768008/index.htm.

This prompted me to instigate the work of which this book is the result: to completely clarify the obligations for flag States and the possibilities for non-flag States (coastal States and port States) to enforce MARPOL Annex VI within the jurisdictional framework of UNCLOS.

This was work on a private level that in many respects overlapped my employment, first at the Danish EPA, from 2012–16, where I worked on the implementation and enforcement of MARPOL Annex VI and the Sulphur Directive, then at the DMA for almost two years, where I worked, inter alia, with Port State Control (PSC), before starting work with Maersk in 2017, with a full focus on compliance with, and enforcement of, MARPOL Annex VI and future IMO regulatory measures on greenhouse gases (GHGs). My work at the DMA and Maersk has also allowed me to participate in several different IMO meetings over the last few years, primarily in the MEPC committee and PPR subcommittee, which, along with other areas of work, have applied themselves to the consistent implementation of the 2020 strengthened sulphur limit.

My private MARPOL Annex VI obsession has also overlapped with my teachings in international law, as this work has given me a unique insight into the details of UNCLOS, the case law of the ICJ, the writings of the ILC, diverse literature and led to inspiring talks with other teachers and different professors.

Most of the book relates to enforcement of the sulphur regulations in MARPOL Annex VI through UNCLOS, especially Part II of the book, which contains the essence of my theory on allowing port States to enforce violations on the high seas. (Part I introduces the legal basis of UNCLOS, MARPOL, PSC, the EU Sulphur Directive, basic principles of international law, etc.)

The rest of the book pertains to enforcement of other existing IMO regulations, as well as of the expected future IMO regulatory measures on reducing GHGs from shipping, perhaps, as a final measure, resulting in a complete prohibition on the use of fossil-based marine fuel. This is analysed in the context of UNCLOS (Part III), but also in the light of deeming a potential prohibition on the use of fossil fuel a jus cogens norm and an erga omnes obligation (Part IV).

It is important for me to make a clear distinction between those two key focal points of the book: enforcement of the IMO's sulphur regulations through UNCLOS; and enforcement of the IMO's future regulations on GHGs, including deeming a potential ban on fossil fuel a jus cogens norm.

I have presented the essence of my enforcement theories regarding the sulphur rules – found in Part II of the book – at many different conferences and venues, including at an EU level (ESSF and EMSA), at the Transatlantic Maritime Emissions Research Network (TRAMEREN) Conference at Copenhagen University and bilaterally to different authorities, professors of law, to the IMO, maritime lawyers, PSCOs, etc. All have stated that they acknowledge and accept the theory that there is an existing legal basis within UNCLOS for allowing non-flag States also to enforce the sulphur limits of MARPOL Annex VI on the high seas.

The theory regarding whether a future IMO ban on fossil fuel can be recognised as a jus cogens norm – found in Part IV – has not been presented at the above-mentioned venues or to other parties.

These two main points of the book also set themselves apart from each other in as much as the jurisdictional basis for enforcing the sulphur limits of MARPOL Annex VI through UNCLOS exists today, whereas the recognition of new jus cogens norms is a lengthy process. My structure might be somewhat 'unpolished' compared to other legal writings; it is, however, my intent and profound hope that Part IV of the book can serve as an inspiration for others to carry on with the development of environmental jus cogens norms. A future IMO regulatory prohibition on the use of fossil fuel is – given the devastating effects of GHG pollution and the special legal status of ships within international law, and the fact that these 'polluters' continually call into ports under different jurisdictions – the perfect starting point for establishing (recognising) and enforcing such environmental peremptory (jus cogens) norms.

Or to put it differently, Part IV is merely the proverbial snowball, which I have tried to start rolling down a hill, hoping that others will contribute to it and polish, elaborate and correct it.

ACKNOWLEDGEMENTS

There are so many people with whom I have worked throughout the years regarding enforcement of the sulphur (and GHG) regulations that it is impossible to mention all of them. However, some special thanks must be given to some of the individuals who inspired me and helped me enormously, including by contributing their valuable insights on topics for this book.

First of all, my mentor, friend and colleague from my EPA days – Dorte. Your patience in explaining the harmful effects of sulphur, how ships operate, combustion processes, etc was never-ending. You always had time to answer my many questions – often repeating answers you had given me before due to my poor memory. Also, your undeniably impressive knowledge of MARPOL Annex VI, including the background to the rules and the ongoing negotiations at the EU and IMO levels, has been an invaluable help to me. In addition, you found time to help me with reading through the first chapters of this book to ensure that my introduction and the reiteration of the sulphur regulations were not completely off track. You truly set the bar. Thank you.

Dea, you took me and my work and lifted it to a whole new international level when you introduced me to Maersk and the IMO. Your insight and knowledge range from being able to solve the most complex political situations on a global scale, to the most detailed understanding of how an engine or a scrubber operates. You have never ceased to amaze me with your expertise, whilst still being a kind, funny and generous person. You have taught me so much, and for that I shall always be grateful.

Simon, not only did you give me a chance at Maersk, but you have always been nothing but supportive and a true inspiration. Your knowledge of the EU is endless, as is your insight into all regulatory matters pertaining to GHGs. I look forward to learning more and sharing more laughs along the way.

Peter and Simone from the DMA, who knew that trying to get a sulphur-drone to patrol Danish waters could create so many problems – but also so many laughs. I always found your high spirits and dedication inspiring.

Anders and Astrid, despite being busy professors of international law and my employers at the University of Copenhagen, you always had time to listen to my crazy notions and give valuable feedback, for which I am grateful.

Beatriz, not only did you give me a chance to talk at the TRAMEREN Conference, but you have also spent hours trying to help me understand how academic writings should be formulated, and shared your literature with me. Thank you.

Rosa, thank you for giving me a chance to talk at different venues within the EU and EMSA. Those were such amazing experiences. Your dedication is truly motivating, as seen before and during IMO meetings, where your focus and work ethics are unparalleled.

Frederick, Dorota and Aicha from IMO Legal, thank you for taking the time to listen to me on so many occasions, and thank you all for sharing your enormous knowledge of the IMO and its regulations.

Jacob and Christian, thank you for always having the time to talk enforcement at a national level. Your insight on Danish criminal law and case law has been important in applying the universal provisions of UNCLOS and international law at a practical, national level.

Lina, thank you for contributing to my knowledge of EEDI, SEEMP and EEOI.

A big 'thank you' to all my students in international law over the last seven years. Your myriad frustrating/brilliant questions, often pertaining to obscure footnotes, references and case law of which I had never even heard, prompted me to vastly expand my knowledge, far, far, far beyond the bounds of the curriculum. Thanks to all of you for keeping me on my toes.

Thanks to my publisher, Hart Publishing, for giving me a chance, supporting me and accepting my delays. Sinead, if only you had a penny for each time I sent you an email promising delivery of the book! I am most grateful. My thanks also go to Sasha, Rose, Richard and Emma. Last, but not least, Catherine: a mere 'thank you' seems hollow considering the invaluable help you have given me regarding correcting and copy-editing this book. Your knowledge (and patience with me and my grammar) is seemingly endless.

Anne and Peter, thank you for introducing me to the practical aspects of a ship's fuel system, its operations, bunkering, fuel oil quality and the concept of 'cat fines'.

Also, thanks to all my colleagues in Public Affairs at Maersk for creating the absolute best working environment and giving me the opportunity every day to learn.

Honourable mentions must also include Palle, Clea, Peter, Pernille, Lars, Christian, Lene, Roger, Martin, Irina, Jonas, Sara, Henning and Kaare. Thank you all for always inspiring me and having the time to explain complicated matters to me.

On a personal level I must thank my parents for always believing in me. You have taught me to work hard and commit myself. Your continuing love and support mean the world to me. I owe it all to you. Thank you so much.

Anette, my beautiful fiancée (I hope wife, once this book is published), words cannot begin to express my gratitude for your never-ending support. You allowed me to sit for hours on end, with my head in the computer or in a book, without the slight sign of discontent. On the contrary, you always made sure that I had everything I needed when I worked into the night, making coffee and snacks for me before going to bed by yourself. I have nothing but love and respect for you, for helping me throughout this work. My thanks to you and Julie.

Finally, and most of all, I must thank my two daughters – Esther and Arendse. Not just for being a continual source of inspiration – and because you are the kindest, most loving, smartest, most beautiful, funniest children a father could ever be blessed with – but also because, on a completely practical level, I can unequivocally state that this book would never have seen the light of day if it had not been for your input.

Esther, I remember vividly how I had reached an impasse with this work in 2016, as, at that point, it was something I worked on only in my spare time. I had the jurisdictional basis for port States to enforce the sulphur regulations on the high seas through the interpretation of article 218(1) of UNCLOS. But it seemed futile, as the flag State principle dictates that the flag State always has primary jurisdiction in accordance with article 92. One day we were casually talking about my work and you said, in your infinite wisdom, that 'we' (lawyers, etc) should do the same with the ships and States as you do in school: *A child who always ignores what the teachers say in school is sent to a timeout and loses its privileges for playing, etc for the rest of the day. Why couldn't 'we' do the same?* In that very second the wording of article 228(1) came to mind (a provision I had merely glanced at previously) and the last piece of the enforcement puzzle fell into place. One of your drawings also inspired me to formulate one of the premises of chapter 15 regarding a potential ban on fossil fuels. Thank you, Esther. I would, without a doubt, have abandoned this work if you had not shown me the solution. I owe this to you.

Arendse, Part IV of this book was inspired by you. One evening, when I was putting you to bed, you asked me about the polar ice caps melting and how polar bears cannot survive because of 'our' (Mankind's) pollution and global warming. You had heard your big sister talking about it at dinner, because she had learned about it at school. I had at that point toyed with the idea of examining whether IMO regulations on GHGs, in the future, could be considered jus cogens norms given the adverse effects of GHGs and the special legal nature of ships within international law. I had, however, dismissed this thought as it would fall outside the scope of the book, which at that point had an exclusive focus on enforcement of the sulphur regulations of MARPOL Annex VI. Also, the subject of recognising environmental jus cogens norms had not always been well received, so the notion of writing about a topic that fell somewhat outside my normal area of expertise, as well as being controversial, had not been that appealing. However, as I talked to you that evening, and saw you hugging your stuffed polar bear, I decided to expand the book by adding Part IV. Not just because of the inspiration I got from your question about the polar bears, but also because of your own personality. You climb, jump and run every single second of the day. Sometimes you fall. Sometimes you cry. But I have never seen you give up on anything. The tree that threw you off today better be prepared to be climbed tomorrow! So why not take a stab at the environmental jus cogens issue? What is the worst that could happen? It may be bruising; perhaps even leave a scar. But we will never stop climbing those trees. Thank you for teaching me that, Arendse.

Jesper Jarl Fanø
Aunslev, May 2019

CONTENTS

PART II
SPECIAL OBLIGATIONS AND JURISDICTIONS UNDER UNCLOS PART XII TO ENFORCE AIR POLLUTION REGULATIONS

PART V
CONCLUSION

PART I

Enforcement of Emissions Legislation through UNCLOS

1

Introduction

Legislation that lowers the permissible sulphur content in marine fuel from a maximum of 3.5% to 0.5% may appear ordinary, rudimentary and unproblematic. This is, however, far from the reality. It is not ordinary or rudimentary, as such strengthened regulation can prevent 137,000 people worldwide from facing a premature death and 7.6 million children from developing asthma.[1] Nor is it unproblematic, as some shipowners can save billions of dollars each year by not adhering to this regulatory measure.

It is therefore essential that this sulphur limit, which is regulated in Annex VI to the International Convention for the Prevention of Pollution from Ships ('the MARPOL Convention'),[2] can be enforced effectively so that the global health benefits can be achieved. This obviously necessitates that shipowners who violate the regulations are not rewarded with higher profits but instead are penalised – appropriately and dissuasively.

Ships have traditionally enjoyed a special status within international law, regarding the exercising of jurisdiction over them. This distinct status has led to speculation as to whether the strengthened sulphur regulations can be enforced, and by whom, particularly when a ship violates the rules mid-ocean, far from any State.

It is the purpose of this book to establish that the United Nations Convention on the Law of the Sea (UNCLOS)[3] provides a legal basis for ensuring that the regulation of sulphur emissions by MARPOL Annex VI can be enforced effectively, regardless of where the violation occurs and which State the ship hails from.

The conclusions on enforcement through UNCLOS are also relevant and applicable for enforcing other international maritime legislation for protection of the environment, including impending legislative measures on reducing greenhouse gas (GHG) emissions from ships.

[1] See M Sofiev et al, 'Cleaner fuels for ships provide public health benefits with climate tradeoffs' (2018) 9:406 *Nature Communications* 4, available at https://www.nature.com/articles/s41467-017-02774-9#ref-link-section-d1456e583.

[2] International Convention for the Prevention of Pollution from Ships (adopted 11 February 1973, as modified by the Protocol of 17 February 1978, entered into force 2 October 1983) 1340 UNTS 61 (MARPOL), IMO Publication: IMO-520E.

[3] United Nations Convention on the Law of the Sea (adopted 10 December 1982, entered into force 16 November 1994) 1833 UNTS 3.

I. The Layout of the Book

This book is divided into five Parts consisting of 19 chapters.

Chapter 1 of Part I contains a brief introduction to the historic development of the MARPOL Convention and UNCLOS, the impact of sulphur pollution on human health and the potential savings for shipowners in not complying with the regulations. The challenges faced in detecting violations of and enforcing the regulations are also described.

The following chapters in Part I provide a basic introduction to the relevant regulatory frameworks of Annex VI of the MARPOL Convention (chapter 2) and UNCLOS (chapter 3). The practical enforcement of MARPOL, in accordance with UNCLOS, through Port State Control is analysed (chapter 4). The regional legislation under the European Union's (EU's) Sulphur Directive[4] is also examined, as the Directive has strong ties to MARPOL Annex VI (chapter 5). The final chapter of Part I describes the basic jurisdictional principles of international law and how these relate to a State's enforcement of international regulations (chapter 6).

Part II of the book contains a detailed analysis of the specific obligations and extended rights of different States to enforce rules for the protection and preservation of the marine environment, such as MARPOL Annex VI, in accordance with the special provisions of part XII of UNCLOS.

Following a general introduction to part XII of the Convention (chapter 7), the analysis will focus on the extended flag State obligations referred to in article 217 of UNCLOS (chapter 8) and the extended jurisdiction for coastal States according to article 220 (chapter 9) and for port States according to article 218 (chapter 10).

This gives way to an in-depth analysis of how UNCLOS, pursuant to article 228, governs questions of overlapping jurisdiction between States when enforcing environmental regulations in accordance with the legal basis of part XII (chapter 11).

The following chapter contains a brief overview of the procedural safeguards and general principles of dispute resolution found within UNCLOS (chapter 12). The final chapter of Part II offers a conclusion on the Part II analysis, including how the regulation of sulphur in marine fuels under Annex VI can be effectively enforced, irrespective of where a violation may occur (chapter 13).

In Part III of the book, the conclusions of the analysis in Part II are applied to enforcement of other international environmental maritime regulations (chapter 14). The conclusions of Part II are also put into perspective in relation to the enforcement of the anticipated future international regulation of GHG emissions from ships (chapter 15).

Part IV attempts to look at the enforcement of specifically international GHG regulation of ships from a completely different perspective. While Parts I–III of the book have a clear focus on analysing the enforcement of emissions regulations

[4] Directive 2016/802 of the European Parliament and of the Council of 11 May 2016 relating to a reduction in the sulphur content of certain liquid fuels [2016] OJ L132/58.

pertaining to sulphur and GHG through the 'classical' jurisdictional basis found in UNCLOS, Part IV discusses whether it is conceivable that the forthcoming GHG regulations pertaining to ships can, within the foreseeable future, attain the status of legislation protecting international recognised peremptory norms, that is jus cogens norms. The principles of jus cogens and erga omnes, and the ties to universal jurisdiction, are also explained (chapter 16).

The analysis itself is based on examined case law from the International Court of Justice (ICJ) and statements from the International Law Commission (ILC), read in conjunction with international scientific reports and predictions as to the negative climate developments due to continued release of GHG (chapter 17).

Finally, discussion is made of what the legal consequences would be if international regulations ensuring the reduction of GHG emissions from ships were to be deemed legislation protecting jus cogens norms, including the applicability of erga omnes enforcement principles (chapter 18).

In the last part of the book, Part V, an overall summary and conclusion is presented (chapter 19).

II. Man-made Pollution Emanating from Ships

Man-made pollution has resulted in the contamination of oceans from discharges of oil, chemicals, plastics, invasive species and other damaging substances, with devastating consequences for fragile undersea ecosystems and all living creatures therein.

One of the most harmful forms of man-made pollution is nevertheless *air pollution*, because it degrades the quality of the air on which living creatures and plants are so deeply dependent. For humans, inhalation of contaminated air may lead to the development of various diseases and disorders that, ultimately, can be life-threatening.

Some forms of air pollution can also lead to global warming, extreme weather, increased UV radiation, melting of the polar ice caps and rising sea levels, which can threaten the habitations and environment needed for the existence of most living things, including human survival. Furthermore, the socio-economic costs that such pollution can inflict on a society are gargantuan.

These are all examples of direct consequences resulting from anthropogenic air pollution, which includes such pollution emanating from ships.

As the destructive effects of man-made pollution became apparent, it naturally resulted in some States unilaterally attempting to reduce pollution through various measures, including national legislative measures. Yet much pollution, including air pollution, represents the total cumulative contamination emanating from various sources on land and at sea. Therefore, many international governing bodies, especially under auspices of the United Nations (UN), haven taken initiatives to seek the reduction of global air pollution, including the International

Maritime Organization (IMO), which is the UN's specialised agency responsible for ships and shipping. These initiatives include developing international rules and regulations through multilateral agreements[5] such as conventions and treaties.

The enforcement of IMO regulations on air pollution have, however, been the subject of much debate, as such enforcement must be carried out in accordance with basic principles of international law respecting the sovereignty and jurisdiction of all relevant States. Ships have a special place and status within international law, as distinct jurisdictions apply depending on a ship's flag and where it sails. This area of international law is usually referred to as the *law of the sea*.

Such jurisdictional principles pertaining to the international law of the sea are codified in UNCLOS, and consequently any State's enforcement of the IMO's environmental regulations must be exercised in accordance with the jurisdictional framework of UNCLOS, particularly the provisions of part XII of the Convention.[6]

III. The IMO

The IMO was established as a specialised UN agency in 1948, in accordance with article 57 of the UN Charter. It was founded as the Intergovernmental Maritime Consultative Organization (IMCO), but in 1982 its name was changed to the International Maritime Organization.

The IMO consists of a General Assembly and several permanent committees and sub-committees addressing different maritime topics, such as safety in the Maritime Safety Committee (MSC) and legal issues in the IMO Legal Committee (LEG).

Of particular relevance to this book is the IMO's work on regulating environmental matters in the Marine Environmental Protection Committee (MEPC) and in the sub-committee on Pollution Prevention Response (PPR). The MEPC was created in 1973 and initially focused on regulating pollution of the sea in the form of spills and discharges of oil, toxic materials and other harmful substances. Yet in recent decades the MEPC has increasingly widened its legislative scope to encompass measures on the discharge of harmful substances into the atmosphere originating from ships. This includes regulations pertaining to sulphur and nitrogen, and, most recently, taking early steps towards regulating GHG emissions, including carbon dioxide (CO_2).

[5] An example of such an international agreement was the 'Action Plan' developed at the UN Conference on the Human Environment in Stockholm in 1972 (hereinafter 'the Stockholm Conference'). The participating States undertook commitments to reduce pollution, including air pollution and pollution of the seas. This Action Plan included a list of 109 recommendations, which the States should implement to address these problems. This has since contributed to the IMO's developing different international regulations to ensure the reduction of pollution coming from ships.

[6] M Dixon, *International Law* (Oxford University Press, 2007) 235.

IV. Development of International Maritime Rules for the Protection of the Environment

The focus of international maritime laws changed significantly throughout the 1960s and 1970s. The rules of the IMO (then IMCO) had primarily dealt with ensuring the safety of ships in relation to the dangers for passengers, crews and other vessels.[7] But several shipwrecks during those years,[8] resulting in significant oil spills and extensive damage to the marine environment, prompted a change in the regulatory focus of the IMO.

Developments in the construction of ships had, after the Second World War, led to their carrying more cargo, including liquid cargoes such as oil. Consequently, when such transport ships (tankers) were lost at sea, it resulted in more damage to the marine environment.

Trying to mitigate the risk from transporting such dangerous cargoes, the IMO adopted several international conventions addressing these concerns. One of the first of these conventions for protection of the marine environment against oil pollution was the International Convention for the Prevention of Pollution of the Sea by Oil (OILPOL), which entered into force on 26 July 1958.[9] OILPOL essentially banned tankers from discharging oil within 50 nautical miles of a coast. Also, from 1969, the Convention Relating to Intervention on the High Seas in Cases of Oil Pollution Casualties ('the Intervention Convention') provided coastal States with special jurisdiction to counter pollution hazards from oil spills outside the coastal State's territory.[10]

In 1972, the Convention for the Prevention of Marine Pollution by Dumping of Wastes and Other Matter ('the London Convention') was adopted.[11] It sought to address the pollution that occurred when ships, platforms, etc were intentionally dumped in the sea as waste. A Protocol to the Convention was adopted by the IMO in 1996.[12]

In 1973, extensive IMO work culminated in the adoption of one of the most comprehensive international regulatory frameworks for the protection of the

[7] The first version of the International Convention for the Safety of Life at Sea (SOLAS), dating from 1914, was made as a direct consequence of the sinking of the *Titanic* in 1912. The latest version of SOLAS was adopted on 1 November 1974 and entered into force 25 May 1980, 1184, 1185 UNTS 2.

[8] One of those incidents was the loss of the *Torrey Canyon* off the coast of England in 1965.

[9] International Convention for the Prevention of Pollution of the Sea by Oil (adopted 12 May 1954, entered into force 26 July 1958) UNTS.

[10] The Convention Relating to Intervention on the High Seas in Cases of Oil Pollution Casualties (adopted 29 November 1969, entered into force 6 May 1975) 970 UNTS 211 ('the Intervention Convention'). In 1973, the scope of the Intervention Convention was widened to include a Protocol Relating to Intervention on the High Seas in Cases of Marine Pollution by Substances Other Than Oil (adopted 2 November 1973, entered into force 30 March 1983) 1313 UNTS 3.

[11] Convention for the Prevention of Marine Pollution by Dumping of Wastes and Other Matter (adopted 13 November 1972, entered into force 30 August 1975) 1046 UNTS 120.

[12] The Protocol to the Convention on the Prevention of Marine Pollution by Dumping of Wastes and Other Matter, 1972 (adopted on 17 November 1996, entered into force 24 March 2006) (1997) 36 ILM 1 prohibits dumping of all waste, apart from certain specified exceptions.

marine environment from ship pollution – the MARPOL Convention. At the same time, work was ongoing within the UN to create a new regulatory framework Convention for the Law of the Sea, UNCLOS. The increased attention given to pollution from ships during this period also affected this UN work, resulting in the establishment of strengthened enforcement paradigms within part XII of UNCLOS.

The Action Plan adopted in Stockholm in 1972 at the UN Conference on the Human Environment made direct references to these – at that time – ongoing works within the UN and the IMO (IMCO).[13] Recommendation no 86(e) encouraged all States to participate in the work, creating and adopting international rules and regulations that could control and reduce such pollution, including pollution of the marine environment. The recommendation directly referred to States' participation in the IMO (IMCO) Conference on Maritime Pollution, as well as the United Nations Conference on the Law of the Sea, which later led to the adoption of, respectively, the MARPOL Convention and UNCLOS. The recommendation encouraged States to:

> Participate fully in the Intergovernmental Maritime Consultative Organization 1973 (IMCO) Conference on Maritime Pollution and the Conference on the Law of the Sea scheduled to begin in 1973.

Also, recommendation no 92 of the Action Plan invited all participating States to support the work of the UN and IMCO, with the development of rules that could protect the marine environment and human health. Therefore, both the early preparatory work on UNCLOS[14] and works relating to the MARPOL Convention[15] contain direct references to recommendation no 92.

A. The MARPOL Convention

The MARPOL Convention was adopted on 17 February 1973 and entered into force on 2 October 1983. Nevertheless, it needed supplementing with an Amending Protocol as early as 1978, since several maritime disasters[16] in the period 1973–78 showed the need for clarification and strengthening of the regulations adopted in 1973. The Amending Protocol also entered into force in October 1983.

The MARPOL Convention is thus sometimes referred to as MARPOL Convention 73/78, but this book will use the shorter title 'MARPOL Convention'.

[13] See n 5.

[14] See UN General Assembly Resolution on protection of the marine environment, UNGA 78, 50.

[15] See IMCO, 'General principles for assessment and control of marine pollution recommended by the United Nations Conference on the Human Environment (Recommendation 92)', MP/CONF/INF.6, 7 May 1973, 1.

[16] This included the loss of the *Metula* off the coast of Chile in August 1974, the *Argo Merchant* off the coast of Massachusetts (US) in December 1976 and the *Amoco Cadiz* off the coast of France in March 1978.

References to Annex VI to the Convention will be cited as 'MARPOL Annex VI' or simply 'Annex VI', depending on the context.

The MARPOL Convention contains several basic provisions that apply to the Convention, the Protocol and the Annexes, including definitions of certain basic expressions and concepts (article 2), and obligations regarding the dissuasive enforcement of violations (article 4) and for the certification and inspection of all ships (article 5). The Convention also entails provisions that seek to clarify the relationship between MARPOL and international maritime law, that is UNCLOS (article 9).

When it was adopted the MARPOL Convention included five Annexes, which to this day still impose specific requirements on ships to protect the marine environment from various sources of pollution. The five original Annexes covered 'Regulations for the Prevention of Pollution by Oil' (Annex I), 'Regulations for the Control of Pollution by Noxious Liquid Substances in Bulk' (Annex II), 'Regulations for Prevention of Pollution by Harmful Substances Carried by Sea in Packaged Form' (Annex III), 'Regulations for Prevention of Pollution by Sewage from Ships' (Annex IV) and 'Regulations for Prevention of Pollution by Garbage from Ships' (Annex V). Only the first two Annexes entered into force together with the Convention and the amending Protocol in 1983, the other three being non-mandatory *operational Annexes*. The remaining three Annexes have since entered into force,[17] making their contents mandatory for the Contracting States.

B. MARPOL Annex VI – Regulations for Prevention of Air Pollution from Ships

After seeing and experiencing the tremendous, devastating impacts on the marine environment following different maritime disasters, the focus of the international community in the 1960s and early 1970s was on preventing such pollution from ships, to which focus the MARPOL Convention and its Annexes I–V bore testament. Then, in the late 1970s and especially in the early 1980s, another source of pollution attracted international attention, including from the IMO. This time, it was a form of pollution that can be difficult to see with the naked eye, and as a result of which the immediate damage from a single violation can be hard to determine – *air pollution*.

Studies revealed that the air in many areas was contaminated by man-made pollution in the form of harmful particles and substances, including substances that also could destroy the ozone layer.[18] It was believed that some of these airborne

[17] MARPOL Annex V entered into force on 31 December 1988; MARPOL Annex III entered into force on 1 July 1992; and MARPOL Annex IV finally entered into force on 27 September 2003.

[18] See at https://www.theguardian.com/environment/2015/apr/18/scientist-who-discovered-hole-in-ozone-layer-warns.

pollutants, apart from having a negative impact on the environment due to acidification and eutrophication, also had a direct adverse effect on the human health by way of causing respiratory diseases and the development of cancers.

As a consequence, a series of international conventions were adopted following the Stockholm Conference, in an attempt to reduce man-made air pollution, including The Convention on Long-range Transboundary Air Pollution 1979 ('the LRTAP Convention'),[19] The Vienna Convention for the Protection of the Ozone Layer 1985[20] and The United Nations Framework Convention on Climate Change 1992[21] (UNFCCC or 'the Rio Declaration).[22] These measures were, however, largely aimed at reducing airborne pollution originating in land-based sources. During the 1980s and 1990s, the IMO (MEPC) expanded the scope of its regulatory protection scheme to encompass the reduction of air pollution from ships.

In 1991 the work of the IMO resulted in Resolution MEPC A.719(17) on Prevention of Air Pollution from Ships, which in 1997 led to the adoption of a new annex (Annex VI) to the MARPOL Convention. MARPOL Annex VI entered into force on 19 May 2005.

The objective of this Annex was to regulate and reduce emission levels of certain airborne pollutants primarily arising from the combustion of marine fuels, including emissions of sulphur oxide (SOx), nitrogen oxide (NOx), volatile organic compounds (VOCs) and ozone-depleting substances (ODS). Some of these requirements, including the regulation of sulphur in marine fuels, were strengthened in 2008 by the MEPC's adoption of Resolution MEPC.176(58), which revised MARPOL Annex VI. In 2011, through the adoption of Resolution MEPC.203(62), Annex VI was expanded to include regulations on increasing the energy efficiency of new ships relating to the reduction of CO_2.

Annex VI therefore sets itself apart from the first five Annexes of the MARPOL Convention, by changing the regulatory focus from hindering tangible pollutants of the marine environment with visible impacts to seeking an overall reduction of air pollution from international shipping because of its cumulative effects.

At the end of 2018, 158 States, which together account for 99.01% of the total merchant fleet in the world, had ratified the MARPOL Convention. In addition, 94 States, which together constitute 96.71% of the world's total merchant fleet, have ratified MARPOL Annex VI.[23]

[19] The Convention on Long-range Transboundary Air Pollution (adopted 13 November 1979, entered into force 16 March 1983) 1302 UNTS 217.

[20] The Montreal Protocol on Substances that Deplete the Ozone Layer (adopted 16 September 1987, entered into force 1 January 1989) 1522 UNTS 3 was adopted as a protocol to the Vienna Convention for the Protection of the Ozone Layer (adopted 22 March 1985, entered into force 22 September 1988) 1513 UNTS 323.

[21] The United Nations Framework Convention on Climate Change (adopted 9 May 1992, entered into force 21 March 1994) 1771 UNTS 107.

[22] See A Henriksen, *International Law* (Oxford University Press, 2017) ch 10.

[23] The ratification status of the different IMO conventions can be seen on the IMO's homepage at http://www.imo.org/en/About/Conventions/StatusOfConventions/Pages/Default.aspx.

Although the detailed content of MARPOL Annex VI will be reviewed in chapter 2 of this book, a brief introduction to regulation 14.1 of Annex VI, containing the global limits for the maximum allowed sulphur content of marine fuels, will be given here, as the question of enforcement of this regulation forms the basis of this book.

C. Regulation 14 of MARPOL Annex VI

Ships' emissions of sulphur particles are a major source of air pollution, which has a significant negative impact on human health and on the environment. Regulation 14.1 of MARPOL Annex VI therefore aims at the continuous reduction of the sulphur content of marine fuels.

Regulation 14.1 provides as follows:

> The sulphur content of any fuel oil used on board ships shall not exceed the following limits:
>
> 1. 4.50% m/m prior to 1 January 2012;
> 2. 3.50% m/m on and after 1 January 2012; and
> 3. 0.50% m/m on and after 1 January 2020.

Regulation 14.1.1 states that from the point of entry into force of the Annex in 2005 up to 31 December 2011, a global maximum limit of 4.50% (m/m)[24] sulphur in marine fuels applied. From 1 January 2012 the regulation was strengthened, with the permissible sulphur limit being reduced to 3.50% pursuant to regulation 14.1.2.

From 1 January 2020 the sulphur limit is once more to be lowered, this time significantly, to 0.50% according to regulation 14.1.3.[25] It is this regulation that has led to speculation about how the limit can be enforced effectively after coming into force.

It should be noted that an amendment to regulation 14.1, coming into force in March 2020,[26] will delete the references to the previous limit values of 4.50% and 3.50% pursuant to regulation 14.1.1 and regulation 14.1.2. Regulation 14.1 will thereafter consist of only one subparagraph and only one *global* limit value in the form of the 0.5 % limit, which, at that time, will have entered into force three months earlier. The reasoning behind this change is that the provisions

[24] Regulation 14.1 refers to the maximum allowed sulphur content in terms of '% m/m'. The abbreviation 'm/m' means 'mass/mass', and '% m/m' refers to 'mass/mass percent'. In other words, when, for example, reg 14.1.3 refers to marine fuels not having a maximum sulphur content exceeding '0.5% m/m', it means that a specific amount of fuel (fuel-mass) may not consist of more than 0.5% sulphur (sulphur-mass). This book, for reasons of simplicity, only refers to the percentage of sulphur, ie '0.5%', when referring to the global limit in reg 14.1.3.

[25] The 0.5% sulphur limit was adopted by Resolution MEPC.176(58).

[26] This change will be effectuated in connection with a planned revision of the regulation as part of the implementation of the so-called *carriage ban*, which is described later in this chapter.

(subparagraphs) that refer to the previous limits are obsolete and should be removed from regulation 14.1. Nevertheless, this book will refer to the subparagraphs applicable at the time where the 0.5% limit enters into force, that is, regulation 14.1.3.

D. EU Sulphur Directive

As noted in section I, the EU also regulates the sulphur content of marine fuels in Directive 2016/802 relating to a reduction in the sulphur content of certain liquid fuels, often referred to as the 'Sulphur Directive'.

The Directive has, since it was first adopted in 1993,[27] undergone several changes and amendments, until a codified version was presented in 2016.[28] It is the provisions of this 2016 version that are meant in this book when reference is made to the Sulphur Directive.

The Sulphur Directive is reviewed in chapter 5 of this book, but it will not be subjected to the same detailed analysis as regulation 14.1.3 of MARPOL Annex VI. The reason for this is that MARPOL Annex VI has global applicability, meaning that ships must comply with it wherever they sail, whether in the middle of the Pacific or the Atlantic Oceans, whereas the geographical applicability of the EU Sulphur Directive is limited to areas under the jurisdiction of EU Member States, and it is therefore often referred to as *regional* regulation.

V. Sulphur Pollution – Harmful to Humans and the Environment

That sulphur emissions are among the airborne substances regulated in Annex VI is based on scientific research proving that sulphur pollution can have a harmful impact on human health and on the environment.[29] This is directly specified, inter alia, in the fourth paragraph of the preamble to the EU Sulphur Directive:

> [E]missions from shipping due to the combustion of marine fuels with a high sulphur content contribute to pollution of the air by sulphur dioxide and particles that are harmful to human health and the environment, and which contribute to the formation of acid rain.

Furthermore, it is stated:

> Without the measures provided for in this directive the emissions from shipping before long would be higher than emissions from all land-based sources.

[27] Council Directive 93/12/EEC of 23 March 1993 relating to the sulphur content of certain liquid fuels [1993] OJ L74/81.

[28] Directive (EU) 2016/802 of the European Parliament and of the Council of 11 May 2016 relating to a reduction in the sulphur content of certain liquid fuels [2016] OJ L132/58.

[29] The previously described international regulation of air pollution, eg through the LRTAP Convention, has led to the reduction of sulphur pollution from land-based sources. Sulphur pollution originating from shipping has the same negative consequences.

To better understand the harmful effects of sulphur emissions, a brief introduction to the scientific origin of sulphur is needed.

Sulphur is a natural element that has the atomic number 16 in the periodic table, with the atomic character 'S'. Sulphur is naturally found in organic material, including in fossil fuels such as coal, natural gas and oil. Oil that is processed into marine fuel therefore contains natural deposits of sulphur.

When marine fuel is burned to create energy and propulsion of a ship, the combustion process starts a chemical reaction causing the sulphur (S) bound in the fuel to react with oxygen (O_2) in the air, resulting in the formation of sulphur dioxide (SO_2).

In the atmosphere, sulphur dioxide is converted to sulphate, which can be deposited either as salts or as sulphuric acid. Sulphuric acids are one of the main reasons that precipitation polluted with sulphur becomes acidic (acid rain). Sulphate also contributes to the formation of particulate matter. As the exact state of the released sulphur oxide can vary, it is often referred to as 'SOx' where the 'x' indicates the unknown state of the sulphur.[30] This book uses 'sulphur' and 'SOx' as general and generic terms to cover different sulphur components in the exhaust fumes from ships emitted into the atmosphere via the ships' funnels (smokestacks). Similarly, 'GHG' is used to cover the diverse types of greenhouse gas, including CO_2, methane, etc.

A. How Sulphur Emissions from Ships Affect Human Health

Sulphur oxides can, as previously described, travel in an airborne state over long distances[31] until, for example, they are absorbed by humans via inhalation, at which point they can contribute to the irritation of the eyes and mucous membranes, and eventually lead to chronic respiratory disorders and diseases, including asthma, chronic bronchitis, etc. They can also cause cardiovascular diseases and result in premature death.

Certain vulnerable groups of people, including the elderly, sick, children, infants and pregnant woman, are considered as being at increased risk of suffering such adverse effects of sulphur pollution.[32] A Finnish report from 2016[33] estimated excess sulphur pollution from shipping alone to be responsible for more than 100,000 premature deaths each year, with most of these casualties occurring in highly populated coastal and harbour areas, particularly in Asia and Africa.[34]

[30] See at https://ww2.arb.ca.gov/resources/sulfur-dioxide-and-health.

[31] See at https://www.nilu.no/en/research/atmosphere-and-climate/long-range-transport-of-air-pollutants/.

[32] See Sofiev et al, n 1.

[33] JJ Corbett et al, 'Health Impacts Associated with Delay of MARPOL Global Sulphur Standards' (Finnish Meteorological Institute, 12 August 2016), available at http://shippingwatch.com/article9069067.ece/binary/Svovl_tidlig_doed.pdf, where the report was attached as an annex to a Finish submission to the IMO.

[34] The Finnish report also assesses, from a socio-economic angle, the costs for society associated with each premature death resulting from sulphur pollution from ships to be approximately $277,000.

The term 'excess sulphur pollution' is in this context used to describe the global consequences of the 0.5% sulphur limit's not coming into force or not being enforced, meaning that the shipping sector would continue to use marine fuels with a content of up to 3.5% sulphur. The reason for applying the term 'excess sulphur pollution' is that the Finnish report examined the global consequences of postponement of the entry into force of the 0.50% limit from 2020 to 2025, leaving the 3.50% limit to apply for a further five years.[35]

In 2018, two years after the Finish report, the disturbing consequences of excess sulphur pollution were confirmed and – unfortunately – uprated, as a study predicted that excess sulphur pollution every year would result in up to 137,000 premature deaths worldwide.[36] The report also noted that approximately 15,000 (14,800) of these premature deaths due to the development of lung cancer could be directly attributed to atmospheric sulphur emissions from shipping.

The same study further calculated how many cases of asthma among children could be avoided. It concluded that 7.6 million children will avoid developing asthma each year if the shipping industry uses 0.5% sulphur in marine fuels instead of 3.5%.

If ships were to continue to use much cheaper marine fuels with higher sulphur content, that is 3.5% or more, this would lead to the same harmful consequences for human health as described in the two reports occurring *every year on a permanent basis*. These calculations will be used in this book to describe the potential health benefits that will not be fully achieved if the 0.5% limit is not complied with due to lack of effective enforcement.

B. How Sulphur Emissions from Ships Affect the Environment

When sulphur emissions from ships, after undergoing the previously described chemical reactions, fall as acid rain, this can have significant negative consequences for the environment.

If acid rain falls over land it can have damaging consequences for trees, plants, crops, etc, as they absorb the acidic water, which affects their growth and eventually causes them to wither away. This is particularly so in areas where the earth is

[35] The reason for this discussion of a five-year postponement of the entry into force of the limit was the review provision (reg 14.8) in MARPOL Annex VI, which stipulated that a study of the global fuel market should be made to determine the availability of 0.5% fuel on the market in 2020. If Parties, based on the study, decided that it was not possible for ships to comply due to lack of compliant fuel, the regulation would allow for the delaying of the entry into force of the 0.5% limit until 2025. The fuel availability study ('Assessment of fuel oil availability', carried out by CE Delft and presented in submission MEPC70/INF.6 at MEPC70 in 2016, available at https://www.cedelft.eu/publicatie/assessment_of_fuel_oil_availability/1858) showed that there would be sufficient quantities of 0.5% fuel in 2020, which led the MEPC in 2016 to conclude that the entry into force of reg 14.1.3 should not be postponed. The Finnish health study was submitted to the IMO as a part of these discussions. Reg 14.8 is discussed in ch 2.

[36] See Sofiev et al, n 1, 4–5.

not very calcareous and therefore not able to efficiently neutralise the sulphur, for example in Sweden, where acid rain in the past has caused widespread deforestation of woodland areas.[37]

If acid rain falls in streams and lakes it can acidify the water by lowering its pH value, which can have severe consequences for the local ecology and the animal life of those streams and lakes.

Acid rain can also acidify seawater, especially in coastal areas, bays and ports, due to the reduced ebb and flow of the water, which can lead to a greater concentration of pollution in confined areas. This can have a destructive impact on the biodiversity, particularly in fragile ecosystems, including having adverse effects on plankton, corals and other organisms that also provide habitats and food for other species and marine life.[38]

Acidification can further have a negative impact on the biodiversity and marine life of the open oceans. This has occurred in places to such an extent that scientists refer to some parts of the oceans as being *dead zones*,[39] where ecosystems and animal life have been destroyed or are under threat of destruction. This is attributed, inter alia, to the discovery of the SUP05 bacterium[40] in deep areas of the oceans.[41] This bacterium can oxidise sulphur and tie it into organic material, including into the tissue of living organisms and animals like clams. SUP05 is a bacterium that is always present in the sea, but when introduced to certain other substances such as sulphur, it interacts with them, causing the bacterium to multiply exponentially, leading to further oxidisation resulting in deleterious effects to the ecology and marine life in these areas.

As the number of ocean dead zones are presumed to be increasing, in November 2017 this, amongst other reasons, prompted 15,364 researchers from 184 countries to sign an open letter, 'World Scientists' Warning to Humanity: A Second Notice',[42] warning the international community against such devastating environmental developments.

VI. What Shipowners Can Gain by Not Complying with the Sulphur Regulations

It should be obvious, due to the previously cited scientific research proving the harmful effects of sulphur pollution from ships, that regulations seeking to reduce

[37] See at https://www.britannica.com/explore/savingearth/acid-rain/.

[38] See at https://ec.europa.eu/jrc/en/news/ocean-acidification-exacerbated-emissions-ships-major-shipping-routes.

[39] Seeathttps://www.cbc.ca/news/technology/dead-zone-microbe-measures-ocean-health-1.814677.

[40] See at http://aem.asm.org/content/79/8/2767.full.

[41] See at https://www.ncbi.nlm.nih.gov/pmc/articles/PMC3554405/.

[42] See at https://academic.oup.com/bioscience/article/doi/10.1093/biosci/bix125/4605229.

these pollutants, such as regulation 14.1.3 of MARPOL Annex VI, ought to be complied with. The responsibility for ensuring such compliance lies with the shipowners.

In this book, the term 'shipowner' is used as a generic reference to the individual, legal person or entity (often a company) responsible for a ship, including being responsible for instructing the ship to use non-compliant fuel, and who or which stands to gain economically by not complying with the regulations.[43] UNCLOS also uses the same term.[44]

It is predicted that there will be a significant financial incentive for shipowners not to adhere to the more stringent 0.5% sulphur limit, as marine fuel with a low sulphur content will be much more expensive than high-sulphur marine fuel.[45] Different estimates and calculations have been made setting out the potential price difference between 0.5% fuels and 3.5% fuels, as the latter will continue to be available on the market after 2020.[46] These estimates place the price spread between 0.5% and 3.5% marine fuel in the range of $150–$200 per ton.[47] Based on such predictions, a large containership could save around $750,000 sailing from a port in Asia to a port in Europe[48] when using non-compliant 3.5% marine fuel instead of compliant 0.5% fuel. A shipowner with a large fleet of merchant ships could therefore save billions of dollars annually by failing to comply with the 0.5% limit.

This would create an uneven playing field for shipowners that comply with regulation 14.1.3 of MARPOL Annex VI, as some expect to spend more than $2 billion extra on fuel each year.[49] Some calculations have shown that the container

[43] Different commercial agreements, for instance the chartering of a vessel, can result in different constellations of responsibilities when operating a commercial ship, eg in accordance with the International Safety Management Code (ISM Code). It is not the intention of this book to illustrate the different persons, companies or legal entities that might be liable for a ship's violation. The relevant authorities (environmental, maritime, public prosecutors, etc) in the country where a violation is detected will be able, on a case-by-case basis, to utilise the jurisdictional basis of UNCLOS, as described in this book, to prosecute the correct person or legal entity responsible for it.

[44] See art 115 of UNCLOS.

[45] Due to the lack of demand, 0.5% marine fuel will not be a commodity available in larger quantities on the global fuel market before the coming into force of the new regulation, hence making the exact price of this fuel difficult to assess up until that point.

[46] Marine fuels with a high sulphur content, such as 3.5% products, will continue to be available after 1 January 2020. One of the reasons for this is that it will be legal for a ship to use fuel with a high sulphur content exceeding 0.5% if the vessel is employing an approved Exhaust Gas Cleaning System (EGCS), which cleans and purifies the emissions/gases from the ship by spraying water into the exhaust fumes, resulting in these having the same low levels of sulphur as those of a ship using 0.5% fuel. Many of these EGCS systems are so-called 'scrubber systems'. The rules and requirements for using such systems are described in ch 2.

[47] B Schieldrop, IMO 2020 Report #2, Skandinaviska Enskilda Banken (SEB) (25 October 2018) 9. This report even suggests a spike in the price spread between 0.5% and 3.5% fuel in 2020 of $360 per ton, before dropping down to the $200 range from 2021 onwards (ibid table 3). See at https://webcache.googleusercontent.com/search?q=cache:A-3iUin-QPIJ:https://webapp.sebgroup.com/mbs/research.nsf/alldocsbyunid/65BE99C18C39C835C12583310043C55F/%24FILE/2018_10_25_SEB_IMO_2020_Enough_MFO_05_if_the_price_is_right_Final.pdf+&cd=1&hl=da&ct=clnk&gl=jp.

[48] See at https://www.bloomberg.com/news/articles/2018-08-29/maersk-sees-fuel-bill-soaring-by-2-billion-from-2020-rules.

[49] ibid.

shipping industry alone will face an extra $10 billion fuel bill every year.[50] These costs will to a considerable extent be passed on the customers through increased freight rates or special surcharges.[51]

Non-compliant shipowners would not face these extra fuel bills, allowing them to generate greater profits for themselves and their shareholders, and at the same time to offer reduced freight rates to customers compared to compliant competitors.

Lack of effective enforcement of regulation 14.1.3 could therefore quickly create a vicious circle, where compliant shipowners are faced with the bleak outlook of either being strong-armed out of the shipping market by non-compliant competitors, or becoming non-compliant themselves in an attempt at self-preservation and survival in an uneven competitive market. As most international shipowners in such a scenario steadily become non-compliant, or disappear, the environmental and human health benefits from using maximum 0.5% instead of maximum 3.5% marine fuels are consequently also lost.

So, from environmental, human health and business viewpoints, strong enforcement of regulation 14.1.3 of MARPOL Annex VI is key. Enforcement that relies on the jurisdictional grounds embodied in UNCLOS.

VII. UNCLOS – A *Lex Superior* Framework Convention of the Sea

To obtain the full potential beneficial environmental and health-related impacts of the 0.5% limit, and to avoid commercial speculation in non-compliance, the Annex VI sulphur regulations must be enforced effectively. This means that regulation must be controlled and that any violation that is detected must be sanctioned so severely that it discourages future violations, especially considering the economic incentive for shipowners not to comply with the regulations.

This requirement for control and effective enforcement falls upon States, that is, the relevant government authorities for dealing with such violations, often the environmental or maritime (port) authorities who – sometimes in collaboration with the police and public prosecutors – bring such cases before a court, where the penalty normally consists of a fine. Many States provide the defendant with the option of paying the fine before going to court, thereby allowing the person or company to avoid additional court fees and the expense of legal representation, etc. Such fines can be issued by the police or as administrative fines directly imposed by the responsible (environmental or maritime) authority.

[50] See Report from AlixPartners, *2019 Global Container Shipping Outlook* (February 2019) 5, available at https://www.alixpartners.com/insights-impact/insights/2019-global-container-shipping-outlook/.

[51] See at https://www.maersk.com/en/news/2018/09/17/maersk-to-change-fuel-adjustment-surcharge-ahead-of-the-2020-sulphur-cap.

Irrespective of any State's national legal system, all violations of international maritime rules, such as MARPOL Annex VI, must be enforced in accordance with the jurisdictional framework provided for in UNCLOS, as the Convention specifies which rights and obligations different States have at sea. UNCLOS is therefore often referred to as a framework convention[52] or as being the Constitution of the Sea,[53] as it codifies many basic customary principles of international law (ie the law of the sea), including different enforcement paradigms for different States, depending on whether these are designated as *flag*, *coastal* or *port States*.

UNCLOS bestows different rights and imposes different obligations on each of these three types of State, including prescribing different jurisdictions for enforcing violations of international environmental regulation. One State can of course be a flag State, a coastal State and a port State if it has a ship registry, a coastline and one or more ports. It depends on the specific circumstances of a given situation whether a State is considered a flag, coastal or port State in terms of which UNCLOS regulations can be invoked.

If a State wishes, or is under duty, to implement, control or enforce regulation over ships that are registered in the State, and which are thus sailing under (flying) that State's (country's) flag, the State is deemed a *flag State*. On the other hand, if a State wants to implement, control or enforce rules governing *foreign ships*,[54] for example when they are sailing in areas of the sea under the State's jurisdiction, that State is considered a *coastal State* or a *port State*. This means that a State that is party to MARPOL Annex VI must ensure that the rules of the Annex, including regulation 14.1, are enforced in accordance with UNCLOS over ships flying its flag (as a flag State) and in areas under its jurisdiction (as a coastal or port State).

No national, regional or international regulation can therefore be applied and enforced at sea by a State, without this being in accordance with UNCLOS. Thus, the rights and responsibilities set out in UNCLOS regulate the grounds for controlling, stopping, investigating, prosecuting and sanctioning vessels that do not comply with these rules.

If there is a discrepancy between other regulations (national or international) and UNCLOS, the latter must be followed, as to a considerable extent it codifies the fundamental principles of the law of the sea. UNCLOS must therefore be considered *lex superior* to other rules and regulations when it comes to determining a State's sovereignty and jurisdiction over ships under its flag and over different maritime areas and the foreign ships therein. National authorities and courts must

[52] See Advocate General Wahls, Advisory Opinion, in Case C-15/17 *Bosphorus Queen Shipping Ltd Corp v Rajavartiolaitos*, 28 February 2018, ECLI:EU:C:2018:557, para 59. This case is reviewed in ch 5.

[53] H Ringbom, 'Enforcement of the Sulphur in Fuel Requirements: Same, Same but Different', available at https://webcache.googleusercontent.com/search?q=cache:EBk8-0aReA0J:https://www.duo.uio.no/bitstream/handle/10852/61600/SO-Artikel-Ringbom.pdf%3Fsequence%3D4%26isAllowed%3Dy+&cd=1&hl=da&ct=clnk&gl=no, 20.

[54] The term 'foreign ships' refers to ships not flying the flag of the State but flying a foreign flag.

respect the jurisdictional paradigms laid out in this Convention when enforcing legislation at sea, in this instance the 0.5% sulphur limit.

Furthermore, some of the regulations under UNCLOS could be said to have the status of *lex specialis* themselves, as the Convention contains specialised provisions that, unlike some national legislation, distinguish what sort of a violation a ship has committed, for example whether it is a *discharge* or a *dumping* violation.

A. Flag State Jurisdiction on the High Seas – A Historical Principle

As described, a *flag State* is the State where a ship is registered, and it therefore sails under the flag of that State.

The right of States to let ships register and fly their flags has, since the adoption of the Declaration recognising the Right to a Flag of States having no Sea-coast in 1921, also applied to landlocked States.[55] This was later codified in article 90 of UNCLOS, giving rights to all States, including landlocked States, to let ships sail under their flags.

A ship can, pursuant to article 92, only sail under the flag of one State, and it is thereby, apart from exceptions set out in UNCLOS, etc, also subject to the exclusive jurisdiction of that State when sailing on the high seas.[56]

Article 94 stipulates that the flag State shall effectively exercise its jurisdiction over ships flying its flag, including ensuring that these ships comply with national and international rules on safety, working environment, crewing, etc. The flag State is therefore, pursuant to article 94(6), also obligated to investigate and take necessary action when foreign States inform the flag State of a violation committed by a ship under its flag.[57]

Article 94 also requires a flag State to maintain a register of ships flying its flag, with relevant information about the ships. Shipowners often pay a fee to the flag State for the registration of a ship, and continue to make payments to the State in the form of paying taxes (tonnage tax).

A flag State's responsibility to ensure that ships flying its flag comply with international environmental regulations, including MARPOL Annex VI, is regulated in the specific (*lex specialis*) provisions covering this in part XII of UNCLOS,

[55] Declaration recognising the Right to a Flag of States having no Sea-coast (adopted 20 April 1921, entered into force 8 October 1921) 7 LNTS 73. See NK Mansell, *Flag State Responsibility – Historical development and contemporary issues* (Springer-Verlag, 2009) 21.

[56] The 'high seas' is a legal term used to describe areas of the sea that are located outside any State's territorial waters and Exclusive Economic Zone (EEZ). See UNCLOS, art 86. These areas are therefore not subject to the territorial jurisdiction of any State. Examples of such areas are large parts of the Pacific, Atlantic and Indian Ocean.

[57] NM Hosanee, '*A critical analysis of flag State duties as laid down under article 94 of the 1982 United Nations Convention on the Law of the Sea*', The United Nations-Nippon Foundation Fellowship Program (2009–2010).

articles 192–237. According to article 217, this includes special obligations for the flag State to investigate, effectively enforce violations and promptly notify the IMO and other States of their effective enforcement. Article 217 is discussed in more detail in chapter 8 of this book.

The common denominator of articles 92, 94 and 217 is that they codify a historical founding principle within the law of the sea,[58] referred to as *the flag State principle*. The flag State principle dictates that, apart from codified exceptions, for example found in part XII of UNCLOS, only a flag State has jurisdiction when a ship under its flag violates an international rule whilst sailing on the high seas.

In practice, the flag State principle not only bestow flag States with an exclusive right to enforce international rules, it also entails an obligation to do so, as explicitly stated in articles 94 and 217.

B. Open Registry States – Flag States that Fail to Enforce Effectively

Despite the clear requirements in UNCLOS for flag States to effectively enforce international rules on ships flying their flags when sailing on the high seas, some do not fully live up to these obligations. The reason for this is often related to the aforementioned fact that shipowners pay a registration fee and a tax for having a ship registered in a flag State. These fees and taxes represent a significant revenue base for some flag States. Hence, some States offer (implicitly) particular advantageous terms for shipowners in an attempt to gain as large a merchant fleet as possible under their flag. These advantageous terms can, for example, involve the flag State's not exercising its jurisdiction over ships flying its flag when they violate international laws, or at least not exercising such jurisdiction effectively. Such flag States are often referred to as being 'open registry States'.[59] Another term used to describe such flag States is that they offer a 'flag of convenience'.[60] This book uses the term 'open registry flag States'.

It could be a potential win–win scenario for shipowners and open registry flag States, if the latter do not effectively penalise violations of the 0.5% sulphur limit. Shipowners could save billions of dollars on fuel expenses each year, and delinquent flag States could attract new shipowners, giving those flag States a larger income from the extra registered ships.

Open registry flag States could potentially also attract new non-compliant shipowners even if they enforce the regulations under MARPOL Annex VI, if this

[58] There are references to the principle in the verdict of the Permanent Court of International Justice in the *Lotus case* from 1927 (*SS Lotus (France v Turkey)* PCIJ Rep Ser A No 10 (7 September 1927) – see ch 6) and it was codified in art 11(1) of the Convention on the High Seas 1958 (adopted 29 April 1958, entered into force 30 September 1962) 450 UNTS 11, which today is implemented in art 97 of UNCLOS (see ch 3).

[59] See Mansell, n 55, 18, fn 14.

[60] See Henriksen, n 22, 165.

enforcement is ineffective. As savings for a shipowner could be in the range of $750,000 for a single non-compliant trip from Asia to Europe, a potential fine from a flag State to a shipowner of, for example, $50,000[61] for such an infringement would – although constituting a high fine compared to fines usually issued for violations of international regulations in other regulatory areas, such as safety, crewing, etc – nonetheless amount to *ineffective enforcement*, as the shipowner could pocket a $700,000 saving for the violation.

This could quickly lead to the previously envisioned vicious circle, forcing otherwise compliant shipowners to re-register their fleets with an open registry flag State to survive in an uneven competitive market, completely negating the potential environmental and health benefits expected from the 0.5% sulphur limit in MARPOL Annex VI.

It is therefore important to determine whether other States than flag States, that is coastal or port States, can exercise jurisdiction over such violations of regulation 14.1.3 of Annex VI committed by foreign ships, to counter any lack of enforcement by open registry States and prevent the vicious-circle effect. This author believes that part XII of UNCLOS provides such a jurisdictional basis, which will be analysed in Part II of this book.

VIII. Challenges with Detecting and Proving Violations on the High Seas

The purpose of this book is to clarify whether there is a legal basis for non-flag States to effectively enforce MAROL Annex VI,[62] ensuring that violations are met with effective sanctions (fines). There is, however, another aspect of ensuring the effective enforcement of the 0.5% sulphur limit that this book does not address in detail, regarding practical enforcement, that is, *detecting violations*.

The initial test of whether the fuel on board is non-compliant is carried out by examining the Bunker Delivery Note (BDN), which is issued by the fuel supplier to the ship each time marine fuel is 'bunkered' (taken on board). This follows from regulation 18.5 of MARPOL Annex VI.

The BDN specifies the sulphur content of the delivered fuel, and a (port) State's authority will begin a control (inspection) of a ship by examining this document. The inspection will also focus on whether the BDN itself meets the formal requirements of appendix 5[63] of MARPOL Annex VI.

[61] Ringbom, n 53, 21, fn 74.

[62] The conclusions of Part II regarding enforcement of the sulphur limit of reg 14.1.3 can be applied to the enforcement of other emission regulations, including future GHG measures, which are described in Part III of this book.

[63] The BDN is described further in ch 2.

If a BDN shows that fuel being used has a non-compliant sulphur content, or if the BDN is missing, flawed or can be suspected of being falsified, it will give the authorities grounds for examining the fuel in greater detail by drawing a fuel sample and analysing it. Fuel samples can also be drawn and analysed on grounds that do not directly relate to the BDN. In fact, the drawing of fuel samples is an integral part of Port State Control (PSC) that many port States perform. This PSC is described in chapter 4.

Fuel samples are drawn from the ship's fuel system, usually as close as possible to the engine where the fuel is combusted.[64] The fuel sample is afterwards taken to a laboratory for an analysis that determines the precise sulphur content of the fuel.

If the analysis shows that non-compliant fuel is being used, and the ship has not yet departed, the port State can demand that the ship debunker (remove) the non-compliant fuel and bunker compliant fuel instead. The port State can further demand that the ship (shipowner) post economic security (bail) to cover any potential fine before the ship is permitted to sail. The posting of economic securities by non-compliant ships is a key factor in achieving effective legal enforcement of MARPOL Annex VI.[65]

Yet a fuel sample taken from the main engine fuel oil system merely provides evidence as to whether the ship was using compliant fuel coming into port. It can be difficult to detect and prove any prior violation of the sulphur regulations, especially infringements that may have occurred days or weeks earlier on the high seas.

It should be noted in this regard that many ships have several fuel tanks on board that can contain different types of marine fuel, such as fuels with 0.1% and 0.5% (and 3.5%) sulphur content. The ship can switch the fuel tank that supplies the last (day/settling) tank before the fuel reaches the engine. It is therefore possible that a ship might sail for weeks on end, for instance when crossing the Pacific, on non-compliant 3.5% fuel and then, perhaps 24 hours prior to arrival in port, change to compliant 0.5% fuel stored in another tank on board.

Such a switch, carried out shortly before entering port, could nevertheless potentially lead to a *positive* (ie non-compliant) test result, as 3.5% residues in the combustion system could contaminate the 0.5% fuel, resulting in a higher sulphur value. In spite of that, given the chance of potentially saving billions of dollars a year, some non-compliant shipowners may approach such switchovers with scientific accuracy to determine when a fuel changeover from a non-compliant to a compliant fuel could be made without resulting in a positive test result. This leads to challenges with *practical enforcement* of regulation 14.1.3, regarding the detection and proof of non-compliance during a voyage on the high seas, for example in the middle of the Pacific.

[64] Fuel samples are therefore normally drawn from the day (/settling) tank, as these tanks often feed fuel directly into the engine.

[65] PSC and the possibilities for detaining ships and requiring economic security (bail) are discussed in ch 4.

A. Developing Advanced Methods and Technologies for Detection

Because of the abovementioned challenges in proving non-compliance, research and development of alternative technical solutions to aid such detection have evolved into a fast-growing area of business. Some of the most promising and scientifically recognised solutions will briefly be touched upon in the following to underline that one or more of these solutions, perhaps used in conjunction with fuel sampling, will be able to detect and prove non-compliance during any part of a voyage, including on the high seas.

One of the most developed and promising solutions for external detection of infringements of the sulphur limit is the use of a *remote sensing technology*, which measures the SO_2/CO_2 ratio in the exhaust fumes (emissions) coming out of a ships funnel. Too high a SO_2/CO_2 ratio in the exhaust fumes is a clear indication that non-compliant fuel is being used.

The devices utilising this remote sensing technology are often referred to as *sniffers*, and they can be fitted to bridges,[66] aircraft, helicopters, other ships and drones. Research has shown that sniffers attached to long-range solar-operated drones might be an effective way of detecting non-compliant ships in the middle of an ocean, as these drones can operate for extended periods of time, flying over ships and shipping lanes, measuring funnel output and combining the data with AIS data.[67] These combined data could immediately identify a ship not complying with the sulphur regulations and inform the authorities in the ship's next port of call. Further development of this technology and positive empirical data on the level of confidence resulting from the measures could lead to remote sensing (sniffers) being used as the sole evidence for proving non-compliance.

Another possibility is to install continuous monitoring of the emissions output. Noting this internal (on-board) detection technology is often abbreviated to CEMS ('Continuous Emission Monitoring System'), and it is already applied on most ships using an Exhaust Gas Cleaning System (EGCS),[68] which cleans the emissions from ships' exhaust gases, as described in chapter 2.

[66] In Denmark, the Danish Environmental Protection Agency has used a fixed sniffer attached to the Great Belt Bridge to detect non-compliant ships sailing through the Great Belt. See at https://eng.mst.dk/about-us/news/news-archives/2017/okt/two-shipping-companies-reported-to-the-police-for-sulphur-pollution/. It should be noted that some in some areas, including the waters off Denmark, a lower 0.1% sulphur limit already applies. This is described in ch 2.

[67] AIS refers to the Automatic Identification System that all merchant ships must use. The AIS data are transmitted continuously from a transponder on board a ship, sending GPS updates on the ship's position, along with information about the ship, eg the ship's unique identification number (IMO number), its name, its next port of call, etc.

[68] Regulation 4.1 of MARPOL Annex VI allows for the use of alternative means of compliance, eg approved technologies that reduce the sulphur emission levels to levels equivalent to using compliant fuel in accordance with the conversion table (table 1) implemented in MEPC Resolution MEPC.259(68).

Other promising means of detecting violations anywhere on the open seas is through *satellite monitoring*. Satellites have already been launched from different companies, and have shown themselves to be able to monitor and measure large emissions of SOx, NOx, CO_2, etc on a busy sailing route. Still, this technology must be developed further to be able to identify the emissions from specific ships. The satellite data could be used in conjunction with the aforementioned sniffers attached on a helicopter, plane or drone, using the sniffer to carry out exact measurements of a particular ship as a follow-up to a satellite's indication of general non-compliance in a particular area.

Development of a *fuel calculator* has also been mentioned as a possible means of detecting non-compliance. It is a method of calculation that is meant to determine the amount of fuel that a ship *should* have used, and which then compares the result to the ship's information on the amount of fuel *actually* used to find any discrepancies. This to prevent ships from filling a fuel tank with unregistered non-compliant fuel and burning that fuel before coming into the next port.[69] The method uses various data for the calculation, including the ship's size and engine, its cargo, speed, weather conditions, etc.

As mentioned, the fuel sampling that is drawn in port requires that the samples are sent to a laboratory for testing according to a specific reference method.[70] It is a procedure that can take several days to achieve a result, depending on the distance to the laboratory, etc. During this time the ship from which the sample was drawn has usually continued its journey. This can complicate the ensuing investigation and court proceedings. *Hand-held fuel-sampling devices* could resolve some of these challenges, as they allow authorities to conduct an on-board analysis of the fuel and instantaneously receive measurement of the sulphur content. Even though hand-held devices might not enable better detection of non-compliance on the high seas, they will offer port State authorities an easier and faster way of determining non-compliance, thereby giving them (clear) grounds for detaining a ship for further investigation during PSC. Hand-held fuel-sampling devices are already used by some port State authorities, but could be developed further and applied by other (or all) port States.

B. Ban on Having Non-compliant Fuel on Board – The 'Carriage Ban'

Finally, one of the most effective and important means of detecting and proving non-compliance had already been achieved in 2018 through a procedural change

[69] Fuel calculating was initially developed for ensuring compliance in SECA zones (described in ch 2), where a 0.10% sulphur limit applies.

[70] For reference methods see ISO 3685:1998 or ISO 12185:1996 and ISO 8754:2003; see app 5 to MARPOL Annex VI.

of the wording of regulation 14.1 of MARPOL Annex VI. This procedural change is often referred to as the acceptance of the 'carriage ban'.

To understand the reasoning behind the carriage ban, the precise wording of regulation 14.1 must be recalled: 'The sulphur content of any fuel oil used on board ships shall not exceed the following limits …' The phrase 'any fuel oil used on board' indicated that it was not illegal to have non-compliant fuel in some fuel tanks on board if the fuel *used*, that is the fuel fed directly into the engine, was compliant. As a result, fuel samples taken during a PSC were often, as described, taken from the day (/settling) tank or from the part of the fuel system that feeds directly into the engine. There was no reason for examining the sulphur content in other fuel tanks on board, as it was not illegal to *carry* non-compliant fuel in those tanks due to the wording of regulation 14.1.

This could potentially have caused severe problems in proving non-compliance, as it would be possible for shipowners, with intent to avoid compliance, to perfect the aforementioned changeover procedures, allowing the ship to undertake an entire voyage, perhaps lasting several weeks, on non-compliant fuel, finally changing over to compliant fuel just before sailing into port.

Resolution of this problem was sought by the IMO and its Member States, by amending the wording of regulation 14.1 to extend the prohibition to encompass fuel used 'or carried for use', thus rendering it illegal to carry fuel with more than 0.50 % sulphur content in any fuel tank on board.[71] Tankers, etc are still allowed to sail with such fuel carried as cargo in their cargo tanks.[72]

The introduction of the carriage ban will facilitate the detection of non-compliance and ease the burden of proof for many port States, as it will enable fuel sampling of several or all fuel tanks on board, making it impossible to engage in changing over from non-compliant to compliant fuel just before arriving in port.

C. IMO Work on Consistent Implementation of Regulation 14.1.3 of MARPOL Annex VI

The PPR committee has, under the auspices of the MEPC, also discussed several different means of effective detection as part of the Committee's ongoing work on ensuring *consistent implementation of regulation 14.1.3 of MARPOL Annex VI*. One of these means is developing a standard FONAR template and reporting system.

[71] The changing of the wording (implementing the carriage ban) was done in accordance with an expedited IMO procedure where it was first presented and agreed upon in the PPR committee at PPR5 in February 2018. The MEPC committee approved the carriage ban at MEPC72 in April 2018 before finally adopting it at MEPC 73 in October 2018. Following a mandatory 16-month implementation period, the carriage ban enters into force on 1 March 2020. As previously mentioned, at that time the other procedural changes to reg 14.1, ie the deletion of references to the former sulphur limits of 4.5% and 3.5% pursuant to regs 14.1.1 and 14.1.2, will also come into force.

[72] Ships using an approved EGCS (see n 46) will, needless to say, also be allowed to carry non-compliant fuel, as they are using an alternative means of compliance. This is described in ch 2.

A FONAR is a 'Fuel Oil Non-availability Report', which a ship can use to inform its flag State and the next port State of any non-availability of compliant fuel in the last port of call, requiring the ship to bunker and use non-compliant fuel.[73]

A FONAR issued on a truthful basis should allow the ship to use the non-compliant fuel without being penalised, provided the FONAR requirements are met. This means that the ship took every conceivable action to try to obtain compliant fuel elsewhere, including from other fuel suppliers, and that the ship bunkered only just enough non-compliant fuel to get it to the next port, meaning that it would be illegal to fill all tanks with the non-compliant fuel.

The FONAR system ensures that ships cannot justify any non-compliance by pleading the lack of compliant fuel at their last port of call; this must be reported using a FONAR. This system should, when used in conjunction with the ship's sailing plan, indicate whether a ship or shipowner deliberately, perhaps on a recurring basis, sought out ports and regions where compliant fuel is usually unavailable. This renders the use of a FONAR invalid.

The FONAR is not an exception as such from the sulphur regulations, as port States are merely required to take it into consideration. The authorities are conversely always within their rights to reject a FONAR.

The MEPC (PPR) work on consistent implementation of regulation 14.1.3 of MARPOL Annex VI has, in addition to FONARs, also included work on developing and amending guidelines on: designating sampling points for fuel sampling; PSC; confidence levels for fuel samples; safety implications; EGCSs, fuel oil suppliers; and a non-mandatory Ship Implementation Plan.[74]

D. Most Plausible Means of Detecting Violations of the Sulphur Limit

The abovementioned solutions show that detecting and proving violations of the 0.5% limit are achievable, including on the high seas.

The carriage ban will, upon entering into force in March 2020, undoubtedly have an immense impact on ensuring the practical enforcement of regulation 14.1, as it will complement the other possible detection methods described. For instance, if a violation on the high seas is detected by using a sniffer attached to a long-distance drone, perhaps initially patrolling the area due to data provided by satellite, this information on non-compliance can be used to target the ship for a PSC in the next port, where all fuel tanks on board, by way of the carriage

[73] The legal basis for the FONAR reporting system is found in reg 18.2 of MARPOL Annex VI, which addresses claims of non-availability from ships not able to bunker compliant fuel. This regulation is discussed in ch 2.

[74] See at http://www.imo.org/en/OurWork/Environment/PollutionPrevention/AirPollution/Pages/Index-of-MEPC-Resolutions-and-Guidelines-related-to-MARPOL-Annex-VI.aspx.

ban, will be subject to fuel sampling, for example by using a hand-held detection device, perhaps backed up by a fuel calculation. The possible application of these detection methods as part of PSC, pertaining to their constituting *clear grounds* for detailed inspections, is discussed in chapter 4.

It should be noted that the sniffer technology could eventually be considered so well-advanced, and the sniffer results so reliable, that the information provided by the sniffer would by itself be sufficient evidence for proving non-compliance on the high seas and issuing a fine on that basis. So if the next port State the ship visits does not react to the information and does not target the ship for a PSC, other States can enforce the detected violation at a later stage when the ship calls into their ports, as the evidence provided by the sniffer(-drone) is sufficient to prove that a violation occurred on the high seas.

This author therefore believes that the challenges with the *practical enforcement* of regulation 14.1.3, that is in detecting and proving non-compliance by a ship at any point during its voyage, can be overcome.

This leaves one challenge unresolved before effective enforcement of the 0.5% limit is completely ensured: establishing that there is a *legal basis* in UNCLOS for non-flag States, that is coastal and port States, to ensure the sanctioning of violations, irrespective of where they are committed.

IX. Conclusion

The development of the MARPOL Convention and its Annexes reflects in many ways the general legislative advances within maritime law for protection of the marine environment and human health.

First, the original focus on the visible, tangible pollution of the sea by oil, etc has changed to include contamination of the atmosphere by airborne pollutants such as sulphur emissions from ships, as such emissions are harmful to both the environment and human health. Sulphur emissions from ships are regulated in Annex VI to the MARPOL Convention. Regulation 14.1 of MARPOL Annex VI sets the global limit values for the maximum sulphur content allowed in marine fuels. From earlier limit values of 4.5% and 3.5%, the limit has dropped significantly to 0.5%, which, if enforced effectively, will prevent 137,000 premature deaths and millions of children from developing asthma every year, particularly in highly populated coastal and port areas.

Still, some shipowners will have a huge economic incentive for not complying with the regulations, especially if they fly flags from an open registry flag State, as such a State might not meet its obligations to penalise violations of the 0.5% limit effectively, thereby neglecting to adhere to the flag State principle.

The detection and proving of violations that take place on the high seas are likely to be achieved by several different means used in conjunction, such as CEMS or sniffers and fuel sampling.

Part XII of UNCLOS provides a jurisdictional basis for coastal and port States to enforce compliance with regulation 14.1 of Annex VI if the flag State fails to do so.

The jurisdictional basis of part XII of UNCLOS is also of relevance when looking at enforcement of other international (IMO) rules for protection of the environment, including future legislative measures for reducing GHGs.

2

The Regulation of Sulphur in MARPOL Annex VI

Chapter 1 presented a brief introduction to the MARPOL Convention[1] and its Annex VI, especially in connection with regulation 14.1, concerning the challenges faced in detecting and enforcing the 0.5% sulphur limit. However, Annex VI contains several other complex regulations relating to the overall purpose of reducing the sulphur emissions from ships. Some of these will be touched upon in this chapter, but references will be made to many of these provisions throughout the book, notably in chapter 4 on Port State Control (PSC) enforcement.

I. Regulation 1 – All Ships are Covered by Regulation 14

Regulation 1 of MARPOL stipulates that the sulphur limits in Annex VI apply to *all ships*, as regulation 14 is not listed as one of the exceptions mentioned in the provision.

Furthermore, States cannot, at their own discretion, exempt a ship or type of ship from complying with the regulation, as MARPOL Annex VI does not allow for this. Exemptions can only be made in accordance with the explicit possibilities of the Annex relating to situations of force majeure and ships on which approved exhaust gas cleaning systems (EGCSs) have been installed. These exemptions are discussed in sections II and III of this chapter.

II. Regulation 3 – Exemptions in Cases of Force Majeure or Damage to the Ship

Regulation 3.1.1 of Annex VI specifies that a ship is not required to follow the requirements of the Annex, including regulation 14, if it poses a danger to the

[1] International Convention for the Prevention of Pollution from Ships (adopted 11 February 1973, as modified by the Protocol of 17 February 1978, entered into force 2 October 1983) 1340 UNTS 61 (MARPOL), IMO Publication: IMO-520E.

safety of the ship and crew, or if the ship is involved in saving lives at sea. In other words, the ship is exempt from following the rules in cases of force majeure.

Regulation 3.1.2 also exempts ships from adhering to the regulations in the Annex if an unlawful emission is caused by damage to the ship or its equipment. This necessitates all reasonable measures having been taken to avoid the damage, and that the damage was not an intentional or reckless result of the owner's or master's[2] actions or knowledge.

III. Regulation 4.1 – Exhaust Gas Cleaning Systems: Equivalent Compliance Solutions

As previously mentioned, ships have the option of complying with the sulphur regulations by installing and using an *equivalent* solution. The most widely used equivalent solution is the EGCS.

An EGCS is a system that cleans the exhaust gas of the ship, thereby reducing the sulphur emissions to a level equivalent to using fuel compliant with regulation 14 of Annex VI. This allows ships with an EGCS to use cheaper, high-sulphur marine fuel that does not comply with the sulphur limits.

Resolution MEPC.259(68) of the Marine Environmental Protection Committee (MEPC) includes an conversion table (table 1), which shows the emission levels (limits) that a ship using a EGCS must meet in order for them to be the equivalent of those produced by ships using fuel with a sulphur content of 3.5%, 0.5%, 0.1%, etc.

Even though the use of an EGCS may represent a formal exemption from complying with regulation 14, it does not set a lower regulatory bar for combatting air pollution as the EGCS technology reduces the sulphur emissions from the ship's exhaust fumes to a level equivalent to those released by ships using fuels compliant with Annex VI.

Such EGCSs must be approved by the competent administration (authorities) of the ship's flag State in accordance with to regulation 4.1. The flag State administration may also approve the installation and use of new emission reduction systems still under development. However, these must be tested over a specified trial period laid down by regulation 3.2. Equivalent emission levels must be achieved during that period. Regulation 4.3 stipulates that the administration, when evaluating and approving such systems, should do so in accordance with the guidelines issued by the International Maritime Organization (IMO).

[2] This book uses the term 'master' to describe the person in charge on board a ship, as this is the term used in the United Nations Convention on the Law of the Sea (adopted 10 December 1982, entered into force 16 November 1994) 1833 UNTS 3 (UNCLOS), eg in arts 27, 94 and 211. Another, more common term for this person is 'captain'.

Such 'Guidelines for Exhaust Gas Cleaning Systems' have been issued by the MEPC in the form of Resolution MEPC.184(59) (2009), as amended by the aforementioned Resolution MEPC.259(68) (2015), which, inter alia, contains detailed requirements for such emission reduction systems and how these should be approved. In short, the guidelines refer to their being approved through either Scheme A or Scheme B. Scheme A takes a type approval approach. The focus is on the system's fulfilling several specific requirements, including documentation that the system is compliant when using a fuel with the maximum sulphur content (eg 3.5 %) specified by the EGCS manufacturer. The fuel used by the ship should not exceed this recommended maximum as otherwise it is not guaranteed that the emissions can meet the specified limits. Scheme B is based on continuous documentation to show that the emissions comply with the specified limits. This is done by installing a Continuous Emission Monitoring System (CEMS), which ensures that emissions comply with the limit values set out in table 1 of Resolution MEPC. 259(68). All ships with an EGCS should monitor the wash water discharge to demonstrate compliance with the limits in the EGCS guidelines.[3]

It might seem like the obvious solution for all shipowners to comply with the sulphur limits of Annex VI by using an EGCS, as their ships can then use cheaper high-sulphur fuel, but several other factors come into play. An EGCS is often very costly to acquire, install and operate, as it may require extra capacity, including manpower for continued maintenance and surveying of the system. The installation of an EGCS may also require the ship to dock while the system is being installed, which can come at a high price for shipowners if the ship is out of commission during that time. The constructions of some ships may also present special practical challenges in relation to installation of such a system. Furthermore, the risk of a malfunctioning EGCS can be a deciding factor in comparison with the stability ensured by using compliant fuel. The business case for using an EGCS therefore varies immensely from shipowner to shipowner, depending on the type of ship, its age and its operational pattern.

A. Scrubber Systems – A Widely Used EGCS

One of the most widely used and applied EGCSs are the so-called 'scrubber systems'.

A 'scrubber' is normally a type of EGCS technology that sprays water/liquids onto the uprising exhaust gas, which leads to the 'scrubbing out' (washing out) of a substantial amount of sulphur (and NOx and particulate matter) from the exhaust gas before it leaves the funnel and interacts with the external atmosphere.

[3] A circular was drafted at Pollution Prevention Response (PPR) 6 in February of 2019, amending the EGCS Guidelines to clarify what recourse should be taken in the event of EGCS breakdown. For instance, it was determined that a ship with a malfunctioning EGCS should not be allowed, by a port State, to leave a port either before the EGCS is repaired or before the ship uses compliant fuel.

Scrubber systems usually come as one of three types:[4] (i) an open loop system; (ii) a closed loop system; or (iii) a hybrid system.[5] Open loop scrubber systems often use seawater to scrub the exhaust gas. The seawater is afterwards discharged back into the sea. This 'discharge wash water' must meet the criteria set out in the IMO guidelines. Open loop systems usually have the lowest operating costs, but some States regulate (limit) such open loop discharges in their internal and territorial waters, as made possible by UNCLOS.[6]

Closed loop scrubber systems recirculate the wash (scrubber) water in a closed system, where additives (often alkaline) are added to the re-used water. The operating costs for such systems are often higher, but on the other hand the system can operate in waters where national or international regulations apply to restrict the use of open loop scrubbers. The closed loop wash water must, when it is necessary to change it, be discharged (delivered) in port to an appropriate reception facility. States that are parties to Annex VI undertake an obligation to ensure that there are port facilities available for receiving such wash water, laid down by regulation 17.1.2 of Annex VI.

Hybrid scrubber systems are systems that can function in both closed and open loop modes. These systems often come at a higher price, but they allow the ship to operate in open (ie the cheapest) mode whenever possible, while still allowing the ship to call at all ports and waters since it can switch to the closed loop mode when needed.

Both open and closed loop scrubber systems must, if approved pursuant to the previously described scheme B, use a CEMS installed in the funnel, ensuring continued emission compliance compared to ships using compliant fuel in accordance with the conversion factors implemented in table 1 of MEPC Resolution MEPC.259(68).[7]

IV. Regulation 14.3 and 14.4 – Special Sulphur Limits in SECA Zones

Annexes I, II, IV and V of the MARPOL Convention enable the designation of specific areas of the sea as 'Special Areas', where distinct and strict regulations apply to the discharging of oil, harmful substances, sewage and waste. The foundation

[4] Other scrubber modes include 'dry scrubbers', which do not use water or other liquids but clean the exhaust gases with hydrated lime-treated granulates.

[5] See https://www.idcconsorzio.com/all-you-need-know-about-scrubbers/.

[6] See ch 7.

[7] The EU Sulphur Directive (Directive 2016/802 of the European Parliament and of the Council of 11 May 2016 relating to a reduction in the sulphur content of certain liquid fuels [2016] OJ L132/58) contains special regulations for the maximum allowed sulphur content in marine fuel used when operating an open loop system. This is described in ch 5.

for such a designation relates to special concerns regarding the protection of the marine environment within such a particular area of the sea.[8]

When MARPOL Annex VI was adopted in 1997, a new protection concept was introduced in the form of designating certain areas of the sea as Emission Control Areas (ECAs) in which special emission limits apply. Regulation 14.3 allows for the creation of Sulphur Emission Control Areas (SECAs) and regulation 13.6 of Annex VI for the creation of Nitrogen Oxide Emission Control Areas (NECAs). Regulation 14.4.3 stipulates a strict limit of maximum 0.1% sulphur content in marine fuels when sailing in a SECA. This limit came into force on 1 January 2015.

Present SECAs include the Baltic, North Sea[9] and North American waters,[10] including Canadian and Caribbean waters pursuant to regulations 14.3.1–14.3.3.[11]

A new SECA can be designated in accordance with the criteria and procedures specified in regulation 14.3.4, appendix 3 to MARPOL Annex VI and regulation 14.7. The Regional Marine Pollution Emergency Response Centre (REMPEC)[12] for the Mediterranean initiated a study in 2018 to assess the benefits, costs and feasibility of a new SECA in the Mediterranean.[13]

It should be noted that the enforcement problems addressed in this book primarily concern the enforcement of the global 0.5% sulphur limit in regulation 14.1.3 on the high seas. SECAs are located within waters under the jurisdiction of coastal States, that is in internal and territorial waters and in Exclusive Economic Zones (EEZs).[14] The challenges for enforcing MARPOL Annex VI in these areas therefore differ from those pertaining to high-sea enforcement. That being said, Part II of this book will also analyse the jurisdiction of coastal States, primarily pursuant to article 220 of UNCLOS.[15] This will encompass enforcement of a 0.1% limit in a SECA, as well as of a 0.5% limit outside a SECA.[16]

[8] An evaluation of coastal State jurisdiction in such Special Areas is to be found in ch 3.

[9] A change to the wording of reg 14.3.1 will come into force on 1 March 2020 as part of implementation of the carriage ban (see ch 8, section VIII.B). The reference to reg 1.11.2 in MARPOL Annex I and reg 1.14.6 of MARPOL Annex V defining the Baltic and North Sea as 'emissions' control areas is changed, as MARPOL Annexes I and V define these areas as Special Areas. MARPOL Annex VI merely re-uses the geographical delimitations of the areas as defined in these Annexes.

[10] See Resolution MEPC.202(62).

[11] The 0.1% limit is also implemented in the EU Sulphur Directive regarding the aforementioned SECAs in European waters; see art 6(2)(b) of the Directive.

[12] See http://www.rempec.org/rempec.asp?pgeVisit=New&theID=6.

[13] Available at https://www.assarmatori.eu/rempec-presented-the-study-of-cost-and-benefit-analysis-for-the-introduction-of-a-seca-area-in-the-mediterranean/?lang=en.

[14] The general jurisdiction of coastal States in internal and territorial waters and in the EEZ is described in ch 3. The special jurisdiction for coastal States to enforce the sulphur regulations in MARPOL Annex VI in these areas is discussed in ch 9.

[15] It was the enforcement of the 0.1% limit, after it came into force in 2015, that gave rise to the challenges associated with issuing fines that, at a minimum, strip all the shipowner's savings from the entire trip.

[16] Although this book focuses on enforcement of the sulphur regulations under reg 14, including coastal State jurisdiction and enforcement in SECAs, many of the same principles for implementation and enforcement apply to coastal State jurisdiction in NECAs; see reg 13 of MARPOL Annex VI.

V. Regulation 14.8 – The 2020 Sulphur Limit was Not Delayed Until 2025

Regulation 14.8 of MARPOL Annex VI presented the possibility for the IMO (MEPC) to postpone the entry into force of the 0.5% limit from 2020 to 2025, if a global analysis (fuel availability study) showed that there would be insufficient amounts of compliant 0.5% fuel available to meet the needs of the international shipping sector in 2020. The reason for this was that 0.5% fuel was a product not widely available before 2020 as there was no demand for it.

The fuel availability study showed that there *would* be sufficient amounts of fuel on the global market in 2020 for the global merchant fleet to meet the 0.5% limit.[17] The MEPC therefore decided, at MEPC70 in 2016, by adoption of Resolution MEPC.280(70), that the entry into force of regulation 14.1.3 was *not* to be postponed until 2025.

It should be noted that even though the study showed that there would be adequate amounts of compliant fuel in 2020 seen from an *overall global perspective*, it was not unlikely that some regions, countries or specific ports might have difficulties suppling enough compliant 0.5% fuel over a limited period after the entry into force until the fuel supply market had completely transitioned to the new production demand. Such transitional issues are alleviated by regulation 18.1, which requires all parties to Annex VI to take all reasonable steps to promote and ensure that compliant fuel is available in their ports and off-shore terminals, and to inform the IMO of this availability.

The possibility of the non-availability of compliant fuel in a port or region is also addressed in regulation 18.2 of Annex VI by allowing ships to present a *claim of non-availability* to flag and port States. Regulation 18.2 has led to the development of a Fuel Oil Non-availability Report (FONAR) template and reporting system, as described in chapter 1.

VI. Regulation 18.2 – Claims of Non-availability

As mentioned previously, regulation 18.2 addresses the possibility of ships not being able to bunker compliant fuel in certain ports. The regulation encourages authorities in the next port State at which the ship calls to take this into consideration before penalising the ship for its use of non-compliant fuel, provided the ship has met the conditions of the regulation for claiming non-compliance. These conditions in regulation 18.2 will be integrated in the FONAR template, as highlighted in section VIII.C of chapter 1.

When a ship wishes to claim non-availability, it must forward the required information, that is the FONAR, to its 'Administration' (flag State) and the next port

[17] CE Delft, 'Assessment of fuel oil availability' in submission MEPC 70/INF.6 at MEPC70 in 2016, available at https://www.cedelft.eu/publicatie/assessment_of_fuel_oil_availability/1858.

of destination (next port State), as laid down in regulation 18.2.4. Regulation 18.2.5 require any party to Annex VI to inform the IMO of a claim of non-availability (FONAR) forwarded to them.

Regulation 18.2.1.1 stipulates that the ship must present a record of its actions trying to obtain to compliant fuel, which, pursuant to regulation 18.2.1.2, shall include information on the ship's attempts to purchase compliant fuel from alternative sources, that is from other fuel suppliers.

Regulation 18.2.2 specifies that the ship is *not* required to deviate from its planned route to purchase compliant fuel. This delimits the scope of regulation 18.2.1.2, regarding finding alternative fuel sources, to attempts made within the same port.

Even though it is not directly specified in the regulation, it is assumed, following discussions on this at the MEPC and PPR, that a ship cannot claim non-availability if any kind of compliant fuel is available in a port. This means that if a ship cannot purchase 0.5% marine fuel in a port but a 0.1% distillate fuel is available, the ship must purchase the 0.1% fuel even though that fuel is more expensive than 0.5% fuel. A ship can thus not claim non-availability (and issue a FONAR) if any kind of compliant fuel was available in port.

The competent authorities of the port State may, in accordance with regulation 18.2.3, take a claim of non-availability into consideration as part of determining whether a ship should be prosecuted for its violation. The port State may also take other circumstances into consideration, for instance if the ship or shipowner previously has claimed non-availability after calling into the same port or region where compliant fuel is often unavailable, as it shows a pattern of the shipowner's trying to circumvent the sulphur regulations of Annex VI. The FONAR reporting system plays a crucial role in determining such premeditated attempts from shipowners to continuously plan bunkering in ports or regions known for their lack of compliant fuel, as each FONAR will be entered into a database accessible to all IMO Member States.[18]

It should be noted that regulation 18.2.3 is not to be considered an exemption, as the port State is *not* required to accept a ship's claim of non-availability as a circumstance precluding wrongfulness for not complying with regulation 14.1. The port State is merely required to take the claim (FONAR) 'into account'.

VII. Documentation for the Fulfilment of the Requirements in Annex VI

MARPOL Annex VI contains several different provisions that require ships to have specific documents on board at all times, to show compliance with requirements of the Annex. This also applies to proving compliance with the sulphur limits under regulation 14.

[18] This is the GISIS (Global Integrated Shipping Information System) database described in ch 4.

This documentation is required to include an up-to-date fuel changeover log, an International Air Pollution Prevention Certificate (IAPP Certificate) and an International Energy Efficiency Certificate (IEE Certificate), as well as Bunker Delivery Notes (BDNs) for all purchased fuel.[19]

A. Regulation 14.6 – Fuel Changeover Procedure and Log

Regulation 14.6 of MARPOL Annex VI requires ships to a carry a written procedure for fuel changeover, which describes the operation of switching between different fuels located in different tanks. This procedure should indicate whether a ship is able to comply with the different sulphur limits when sailing in and out of SECAs.

When such changeovers are preformed, the ship must log information about the changeover in a fuel changeover log, including information on when and where the switch was made and the specifications of the sulphur content in the switched fuels.

The fuel changeover log and procedure must be presented during a PSC, which can be of particular relevance when calling into a port in a SECA zone. If the log is kept correctly, it can document whether the ship has switched between fuel complying with the global 0.5% sulphur limit and that complying with the 0.1% SECA limit.

The changeover log also gives an indication of whether a changeover was carried out in a timely manner, ensuring that the fuel system on board has been 'flushed' with 0.1% fuel before entering the SECA. A late changeover may cause fuel residues from using fuel above the SECA limit (eg 0.5%) to contaminate the 0.1% fuel entering the day tank, causing the sulphur content of the fuel used to rise above 0.1%.

If the changeover log indicates that a correct changeover was not performed, or that this was not done in a timely manner, or if the ship cannot present a written procedure for conducting such changeovers, then this could give port State authorities grounds for conducting a detailed PSC inspection[20] of the ship, including taking fuel samples.

B. Regulation 18.5 – Bunker Delivery Notes

Regulation 18.5 of Annex VI stipulates that all ships must carry a document (the BDN) showing what type of marine fuel (bunker) the ship has acquired. The BDN

[19] Ensuring that ships comply with these requirements for documentation will often be part of the initial inspection during a PSC, as described in ch 4.

[20] Grounds for PSCs conducting detailed inspections are described in ch 4.

contains information and specifications on the fuel purchased, including information on the sulphur content therein ensuring compliance with regulation 14.1 or 14.4.

The BDN is issued by the fuel supplier, and it is signed by fuel supplier's representative and the person on board ship responsible for receiving the fuel, for example the chief engineer, upon delivery. The fuel supplier and the ship respectively receive the original and a copy of the BDN.

A BDN must contain the information set out in regulation 18.5 and appendix 5 to MARPOL Annex VI, which is:

- The name and IMO number of the ship that bought the fuel.
- The name of the port where the fuel was delivered, together with the date of delivery.
- The name, address and telephone number of the fuel supplier.
- Full name and specifications of the product delivered.
- The amount of fuel delivered, specified in metric tons.
- Information on fuel density at 15°C, kg/m3.[21]

And perhaps most importantly in this context:

- Information on the sulphur content in the fuel purchased, specified in % m/m.[22]

In 2017, MEPC 71 decided to amend appendix 5 to Annex VI to include a box relating to fuel delivered to ships with an EGCS in the BDN information.[23] After the amendment came into force on 1 January 2019, ships using an EGCS must notify the fuel supplier of that fact. The fuel supplier will then 'check off' the aforementioned box on the BDN, stating that marine fuel, with a sulphur content exceeding the then allowed limit under regulation 14.1, has been delivered to the ship with an EGCS.

The ship must keep a BDN on board for a minimum of three years in accordance with regulation 18.6, as the ship must present the BDN during a PSC under regulation 18.7.1.

If a ship cannot present a BDN, or the BDN does not meet the requirements of appendix 5, it should lead to an immediate detailed inspection and the drawing of a fuel sample for testing. A missing or faulty BDN could by itself, regardless of the fuel sample test result, constitute a violation of Annex VI, which can be penalised by a port State by imposition of a fine.

[21] App 5 refers to the fuel density being tested in accordance with ISO Standard 3675:1998 or ISO Standard 12185:1996.

[22] App 5 refers to sulphur content's being tested and measured in accordance with ISO Standard 8754:2003.

[23] This amendment to app 5 to MARPOL Annex VI was made by Resolution MEPC.286(71) entitled 'Amendments to MARPOL Annex VI (Designation of the Baltic Sea and the North Sea Emission Control Areas for NOX Tier III control and the information to be included in the bunker delivery note)'.

i. Regulation 18.8 – The MARPOL Sample

Regulation 18.8.1 of Annex VI requires that a fuel sample be drawn from the bunker fuel supplied.[24] A sample is often referred to as 'the MARPOL sample'. It is also stipulated that the MARPOL sample must be sealed and kept on board for a minimum of 12 months.

The sample can, under regulation 18.8.2, be used during a PSC to verify the sulphur content as declared on a BDN.[25] Nevertheless, port authorities often choose to focus on the fuel sample they draw directly from the fuel oil system, as it presents reliable evidence to prove that non-compliant fuel is being used.[26]

Some shipowners may draw a second representative sample intended for commercial purposes. This sample is analysed by the shipowner to ensure that the delivered fuel is 'on spec'. If the supplied fuel does not meet the sulphur content stated in the BDN, or if the fuel does not meet the commercial standard,[27] it can lead to a civil case against the fuel supplier. This commercial analysis is important to shipowners, as 'off spec' fuels can have a negative impact on the ship's performance, and perhaps even damage the engine due to catalytic ('cat') fines, etc.[28]

C. Regulations 5 to 9 – IAPP and IEE Certificates

Regulation 5 of MARPOL Annex VI states that all ships of 400 Gross Tonnage (GT)[29] or above must undergo a periodic survey[30] to ensure that they, including their installations, equipment, etc, adhere to the different requirements of the Annex as listed in regulations 5.1–5.5.

Regulation 6 contains provisions ensuring that ships that have undergone and passed a survey in accordance with regulation 5, and are issued an IAPP Certificate in accordance with regulation 6.1 and an IEE Certificate in accordance with

[24] Regulation 18.8.1 refers to the MARPOL sample's being drawn in accordance with Resolution MEPC.182(59) '2009 Guidelines for the sampling of fuel oil for determination of compliance with the Revised MARPOL Annex VI'.

[25] A MARPOL sample must analysed in in accordance with the procedure described in reg 18.8.2. See app 6 to Annex VI, stating, inter alia, that the testing laboratory must be accredited to perform the analysis in accordance with ISO Standard 17025.

[26] After the carriage ban enters into force, port authorities will have a legal ground for drawing fuel samples from several or all fuel tanks on board.

[27] The commercial standard is set in accordance with ISO 8217.

[28] See https://www.exxonmobil.com/en/marine/technicalresource/marine-resources/cat-fines-removal.

[29] 'Gross tonnage' is a measure of a ship's overall internal volume. (Not be confused with 'deadweight tonnage' (DT).)

[30] Such regulations requiring ships to undergo periodical surveys and certifications are often integrated as a part of the IMO's specialized Harmonized System of Survey and Certification (HSSC). See Resolution A.1104(29) from 2015. This system ensures that the different IMO conventions requiring ships to undergo different periodical surveys are coordinated according to ship types, often following a five-year interval. This allows a ship to meet several different international regulations for survival and certification at the same time, including regs 5–9 of MARPOL Annex VI.

regulation 6.2, which certify that the ship complies with the requirements set out in Annex VI.

Point 2.3.1 of the IAPP Certificate requires the ship to confirm whether it complies with the sulphur regulations of regulation 14.1 by using compliant fuel (point 2.3.1.1 of the IAPP) or if it uses an approved EGCS for compliance (point 2.3.1.2).

Point 2.3.2 of the IAPP Certificate requires the ship to confirm whether it complies with the 0.1% SECA limits by using compliant fuel in accordance with regulation 14.4 (point 2.3.2.1), or if it uses an approved EGCS (point 2.3.2.2).

The IAPP Certificate must be drawn up in accordance with regulation 8.1, in a form corresponding to that set out in appendix 1 to MARPOL Annex VI, including point 2.3 addressing compliance with regulation 14. The certificate is issued for a period not exceeding five years, as laid down in regulation 9.1 of Annex VI.

The IEE Certificate is drawn up in accordance with regulation 8.2, based on the form set out in appendix 8 to MARPOL Annex VI. The certificate is, as a general rule, valid throughout the lifetime of the ship, according to regulation 9.10.

The IAPP and IEE Certificates must be presented upon request during a PSC, as laid down in regulations 10.1 and 10.5. If an IAPP Certificate is not presented or the certificate does not comply with the regulations, it can lead to a detailed PSC inspection and possibly a fine.

D. Regulation 17 of MARPOL Annex I – Oil Record Book, Part 1

Although this chapter focuses on the regulations under MARPOL Annex VI, brief reference should be made to MARPOL Annex I, as it contains provisions that implicitly enable effective enforcement of the sulphur regulations under Annex VI.

Regulation 17.1 of MARPOL Annex I requires all oil tankers of 150 GT and above, and *all* other ships of 400 GT and above, to carry – and continually update – an Oil Record Book (ORB), Part I.[31] This must contain information on all fuel bunkerings in accordance with regulation 17.2.5 of Annex I, including information on when and where the fuel is taken on board, and the quantities and specifications of the fuel purchased (point H/26 of ORB, Part I).

The ship is also required to log additional information in the ORB, Part I relating to the bunker fuel on board, such as cleaning of fuel tanks, ballasting and discharging ballast from fuel tanks, collected oil residues, etc, under regulation 17.2 of MARPOL Annex I.

The ORB, Part I must be presented upon request during a PSC in accordance with regulations 17.6 and 17.7 of MARPOL Annex I, and can, due the content and information mentioned above, also contribute towards establishing whether a

[31] The Oil Record Book, Part II, is reserved for tankships.

ship is complying or not complying with the sulphur regulations under MARPOL Annex VI, including carrying out a detailed PSC.

VIII. Regulation 10 and Regulation 11 – PSC, Detection and Enforcement of Violations

Regulation 10.1 specifies that ships can be subjected to an inspection (PSC) in a foreign port to determine if the requirements of Annex VI have been fulfilled. If this is not the case, the ship can be detained until the non-compliance has been rectified, according to regulation 10.2, including ensuring that compliant fuel is used.

Regulation 11 stipulates several obligations for States that are party to Annex VI to work together to detect violations and inform each other of detected violations, in their capacities either as port (PSC) States or as flag States.

Regulations 10 and 11 are discussed in more detail in chapter 4.

Furthermore, regulation 11.4 is studied in chapter 8, regarding the obligation for flag States to enforce Annex VI and inform relevant port States and the IMO of this enforcement; while regulation 11.6 is examined in chapter 10 with regard to its reference to enforcement of MARPOL Annex VI always being in accordance with contemporary interpretations of international law *mutatis mutandis*, that is, with the necessary changes being applied.

IX. Conclusion

MAPROL Annex VI (and Annex I) contains several regulations that are relevant when looking into effective enforcement of the sulphur limits under regulation 14, mainly pertaining to documentation that it is mandatory for ships to have on board. This includes BDNs dating back three years, showing the sulphur content of all fuel purchased. Fuel changeover logs and procedures indicate a ship's ability to perform timely changeovers between 0.1% SECA fuel and 0.5% fuel. A FONAR should also be presented to port authorities, when applicable.

Compliance with the sulphur regulations of MARPOL Annex VI can also be achieved by using EGCSs such as scrubbers.

Irrespective of whether a ship uses an EGCS or compliant (0.1% or 0.5%) fuel, the ship must be able to document that it complies with the applicable regulations of MARPOL Annex VI, and relevant guidelines, during a PSC.

3

UNCLOS

The United Nations Convention on the Law of the Sea (UNCLOS)[1] was adopted on 10 December 1982 but did not enter into force until 16 November 1994. Prior to its entry into force, several States had expressed dissatisfaction with the contents of part XI of the Convention concerning the 'Area', which, under article 1(1)(1) of UNCLOS, refers to 'the seabed and ocean floor and subsoil thereof, beyond the limits of national jurisdiction'. Or in other words, the deep-seabed beneath the high seas. The dispute, which revolved around exploration of the Area, resulted in 1994 in the adoption of a supplementary agreement to part XI, which entered into force on 28 July 1996.[2] A second supplementary agreement to UNCLOS entered into force on 11 December 2001, relating to the conservation and management of migratory species.[3]

One hundred and sixty-seven States,[4] as well as the European Union, have ratified UNCLOS, and 14 States are signatories to the Convention but have not yet ratified it.

UNCLOS consists of 17 parts (I–XVII), with 320 articles and nine Annexes. In general, parts I–VII (except part IV on the special regimes relating to archipelagic States[5]) will be studied in this chapter as they codify basic principles of international law pertaining to the law of the sea and different jurisdictional principles for coastal, port and flag States.

The provisions of part XII of UNCLOS, relating to protection of the marine environment, are analysed in detail in Part II of this book in chapters 7–13, although some provisions of part XII regarding Port State Control (PSC) will

[1] 1833 UNTS 3.

[2] Agreement relating to the implementation of Part XI of the Convention (adopted 28 July 1994, entered into force 28 July 1996) 1836 UNTS 3.

[3] Agreement for the implementation of the provisions of the Convention relating to the conservation and management of straddling fish stocks and highly migratory fish stocks (adopted 4 August 1995, entered into force 11 December 2001) 2167 UNTS 3.

[4] See at http://www.un.org/depts/los/reference_files/chronological_lists_of_ratifications.htm.

[5] As this chapter merely seeks to establish the general basic legal principles for enforcement of environmental regulations and not an in-depth description of all UNCLOS provisions, pt IV on archipelagic States is considered too specific to be covered here. The same applies to pt VIII on the 'Regime of islands', pt IX on enclosed or semi-enclosed seas, pt X on the right of access of landlocked States to and from the sea and freedom of transit, pt XI on the Area, pt XIII on marine scientific research, pt XIV on development and transfer of marine technology, pt XV on general provisions and pt XVI on final provisions. Pt XV on dispute settlement is discussed in ch 12 of this book.

also be addressed in chapter 4. This chapter will also, where applicable, make reference to relevant provisions of part XII, mainly article 217 relating to flag States, article 218 relating to port States, article 220 relating to coastal States and article 228 for determining how overlapping jurisdiction between flag States and other (port or coastal) States is resolved.

I. The 1958 Conventions

In 1958 the United Nations (UN) adopted four Conventions and an Optional Protocol to regulate (codify) various parts of the law of the sea. The four Conventions and the Optional Protocol were:

- the Convention on the Territorial Sea and the Contiguous Zone (CTS);[6]
- the Convention on the High Seas (CHS);[7]
- the Convention on Fishing and Conservation of the Living Resources of the High Seas (CFCLR);[8]
- the Convention on the Continental Shelf (CCS);[9]
- the Optional Protocol of Signature concerning the Compulsory Settlement of Disputes (OPSD).[10]

The regulatory basis set out in the four 1958 Conventions and the Optional Protocol has since, to a large extent, been absorbed by UNCLOS.[11] For instance, the CHS embodied the flag State principle in its article 11(1) concerning collisions by ships on the high seas, which today is directly codified in article 92 of UNCLOS.

The 1958 Conventions are, in principle, still in force, but due to its widespread acceptance, UNCLOS is considered as being *lex superior*,[12] this also being directly stated in UNCLOS article 311(1), which provides:

> This Convention shall prevail, as between States Parties, over the Geneva Conventions on the Law of the Sea of 29 April 1958.

[6] Convention on the Territorial Sea and the Contiguous Zone (adopted 29 April 1958, entered into force 10 September 1964) 516 UNTS 205.

[7] Convention on the High Seas (adopted 29 April 1958, entered into force 30 September 1962) 450 UNTS 11.

[8] Convention on Fishing and Conservation of the Living Resources of the High Seas (adopted 29 April 1958, entered into force 20 March 1966) 559 UNTS 285.

[9] Convention on the Continental Shelf (adopted 29 April 1958, entered into force 10 June 1964) 499 UNTS 311.

[10] Optional Protocol of Signature concerning the Compulsory Settlement of Disputes (adopted 29 April 1958, entered into force 30 September 1962) 450 UNTS 169.

[11] See T Treves, *1958 Geneva Conventions on the Law of the Sea* (United Nations Audiovisual Library of International Law, 2008) 4, available at http://legal.un.org/avl/ha/gclos/gclos.html (English version).

[12] UNCLOS could also be considered *lex specialis* to the 1958 Conventions, as the UNCLOS regulations on these matters are more detailed, accurate and up-to-date, as well as being *lex posterior* because of the more recent entry into force of UNCLOS.

This is further recognised by the International Law Commission (ILC) in its 2006 Report to the UN General Assembly, in which the Commission reflects on article 311:

> Clauses in the subsequent treaty that expressly abrogate the earlier treaty. An example would be article 311(1) UNCLOS according to which between parties to it and to the 1958 law of the sea conventions the former shall prevail.[13]

II. The Law of the Sea is Dynamic and Constantly Evolving

UNCLOS[14] represents the further development of the international law of the sea since 1958, not only in respect of amending and clarifying the legal matters addressed in the 1958 Conventions, but also in respect of the development of new principles within the law of the sea.

An example is the right of coastal States to invoke a 200 nautical mile (nm) Exclusive Economic Zone (EEZ), where the coastal State has certain specific and limited rights and obligations, for instance relating to exploiting and preserving the natural living and non-living resources in the zone, that is, fishing and exploration of the deep-seabed. The rights of the coastal State in the EEZ also extend to enforcing rules and regulations for the protection of the marine environment, which means exercising jurisdiction over foreign ships in the EEZ.

The right for coastal States to claim an EEZ, unlike the right to exploit the continental shelf contained in the CCS, was not yet a recognised right within international law in 1958. This right was recognised in 1982 in part V of UNCLOS.[15] This was in great part attributable to the International Court of Justice's (ICJ's) acknowledgement of the principle in the *Fisheries Jurisdiction case* in 1974,[16] where the ICJ accepted that Iceland had a right to the establish a 50 nm fishing zone.[17]

This example illustrates that international law is dynamic and under constant development, which is also explicitly affirmed in article 235(3) of UNCLOS,

[13] ILC, *Fragmentation of International Law: Difficulties Arising from the Diversification and Expansion of International Law* (finalised by M Koskenniemi), A/CN.4/L682, 13 April 2006, report presented at meeting no 58 in Geneva (2006) 136, ex no 5.

[14] UNCLOS is occasionally referred to as 'UNCLOS III', which refers to its having been adopted at the Third United Nations Conference on the Law of the Sea that took place between 1973 and 1982. The First Conference was held in 1956 and the Second in 1960, without resulting in any adoption. This book simply uses 'UNCLOS', as it refers to the *Convention* and not to the *Conference* on the Law of the Sea.

[15] GG Schram, 'Havretten og de nordiske lande' (1984) 1 *De nordiske juristmøter* 401, 407.

[16] *Fisheries Jurisdiction (United Kingdom v Iceland)*, Jurisdiction of the Court, Judgment [1973] ICJ Rep 3.

[17] L Bernard, 'The Effect of Historic Fishing Rights in Maritime Boundaries Delimitation', LOSI Conference Papers from the Law of the Sea Institute, UC Berkeley–Korea Institute of Ocean Science and Technology Conference held in Seoul, Korea, May 2012, 8.

which provides that 'States shall cooperate in the implementation of existing international law and the further development of international law'. Such legal developments are often made in line with the developments made by the international community relating to new technological advances, for example of more and bigger ships, and new acquired needs, such as exploiting natural resources in new areas.

New advances and needs also lead to new challenges and dangers, especially to the environment in view of the pollution emanating and generated from ships and from the exploitation of natural resources. This prompted the development of part XII of UNCLOS, which represented a monumental legal advancement within international law as it laid down a new set of framework obligations and rights regarding the enforcement of legislation for the protection of the marine environment. This included introducing a new legal entity in article 218 in the form of *port States*, which were granted a new and broader jurisdictional scope in comparison to the already well-established coastal and flag State entities.

This was also emphasised in the closing speech at the UN Conference in December 1982, following the adoption of UNCLOS, where conference President Tommy Koh Thong Bee stated that 'the Convention contains important new rules for the protection and preservation of the marine environment from pollution'.[18]

The special jurisdictions developed in part XII of UNCLOS were unlike anything seen in the 1958 Conventions, and can certainly be attributed to the increased focus on protecting the marine environment in the 1960s and 1970s, as described in chapter 1. In fact, many of the codified exceptions to the flag State principle, to which article 92 refers, are found in part XII of UNCLOS, of which article 218(1),[19] article 220[20] and article 228(1)[21] are particular examples, article 217 – entailing the flag State principle for enforcing environmental legislation – directly referring to flag States having to respect these exceptions.[22]

International law, including the law of the sea, is thus constantly evolving to meet the more newly developed needs of the international community, which is in line with the premise set out in Part II of this book as it, through analysis of articles 217, 218, 220 and 228, seeks to clarify whether there is a legal basis for ensuring that the more 'recent'[23] need of the international community to enforce legislation preventing air pollution from ships, including regulation 14 of MARPOL Annex VI, can be met – especially bearing in mind that effective

[18] Schram, n 15, 405.

[19] Art 218 is analysed in ch 10.

[20] Art 220 is analysed in ch 9.

[21] Art 228 is analysed in ch 11.

[22] See art 217(4). Art 217 is analysed in ch 8.

[23] See ch 1 for a description of the shift in environmental focus in the 1980s and 1990s towards addressing air pollution.

enforcement of regulation 14 will prevent 137,000 early deaths and 7.6 million children from developing asthma each year.[24]

III. UNCLOS – Protecting against Pollution of the 'Marine Environment'

It should be noted that the references in UNCLOS to protecting against 'pollution of the marine environment', particularly in part XII of the Convention, include pollution that causes a danger to human health. This is specified in the definition of the term in article 1(1)(4), which reads:

> 'pollution of the marine environment' means the introduction by man, directly or indirectly, of substances or energy into the marine environment, including estuaries, which results or is likely to result in such deleterious effects as harm to living resources and marine life, hazards to human health, hindrance to marine activities, including fishing and other legitimate uses of the sea, impairment of quality for use of sea water and reduction of amenities …

This means that any references in UNCLOS to international rules and regulations for the *protection and preservation of the marine environment* include MARPOL Annex VI, because of the reference to the introduction 'by man, directly or indirectly, of substances or energy into the marine environment' which results in 'hazards to human health'. This covers polluting substances such as sulphur released during the combustion of marine fuels, which can cause an early death and numerous different respiratory diseases. It also covers the release of greenhouse gases (GHGs).[25]

Enforcement of Annex VI, including the regulation 14 sulphur limits, is therefore covered by the special provisions of part XII of UNCLOS, including the wide jurisdictional basis for port States under article 218.[26] This conclusion is confirmed by the IMO Legal Committee:

> The power to impose sanctions conferred by IMO regulations on the port State (notably in the MARPOL Convention) should be related to the rights and obligations provided in part XII of UNCLOS.[27]

[24] International law in general could also develop to meet the same need of the international community for ensuring that legislation preventing destructive air pollution is enforced. For instance, ensuring enforcement of future IMO measures for reducing GHG emissions from ships through the further development of the jus cogens and erga omnes principles, as described in Part IV of this book.

[25] Y Tanaka, 'Regulation of Greenhouse Gas Emissions from International Shipping and Jurisdiction of States' (2016) 25 *Review of European, Comparative and International Environmental Law* 337.

[26] The application of art 218(1) on violations of emission regulations such as reg 14 is described in detail in ch 10.

[27] IMO, 'Implications of the United Nations Convention on the Law of the Sea for the International Maritime Organization' (LEG/MISC.8, 30 January 2014) 13.

All references in this book to UNCLOS provisions mentioning international regulations for the protection of the marine environment therefore automatically include MARPOL Annex VI, pursuant to the definition in article 1(1)(4) of the Convention.

IV. The Connection between UNCLOS and MARPOL Annex VI

It should be noted that besides the conclusions previously drawn pertaining to Annex VI's being subject to the provisions of, inter alia, part XII of UNCLOS, both UNCLOS and the MARPOL Convention contain indirect references to each another.

Although the MARPOL Convention[28] was adopted prior to UNCLOS, article 9(2) of the MARPOL Convention makes direct reference to UN Resolution 2750 C(XXV), which formed the basis leading to the development of UNCLOS.[29] Article 9(2) reads:

> Nothing in the present Convention shall prejudice the codification and development of the law of the sea by the United Nations Conference on the Law of the Sea convened pursuant to resolution 2750 C(XXV) of the General Assembly of the United Nations nor the present or future claims and legal views of any State concerning the law of the sea and the nature and extent of coastal and flag State jurisdiction.

The provision affirms that nothing in MARPOL Convention 'shall prejudice the codification and development of the law of the sea', nor 'future claims and legal views'. Thus, the MARPOL Convention and its Annexes are subject to, and bound by, the international principles on the law of the sea later set out in UNCLOS, and how these are interpreted in the future.

Article 9(2) of the MARPOL Convention refers to coastal and flag States, but there is no reference to port States, because the concept of port States did not exist as such until UNCLOS introduced it in 1982. But even though article 9(2) does not refer to port States directly, the reference to 'future claims and legal views of any State concerning the law of the sea' includes being subject to the jurisdiction of the port State entity that was developed later.

It should be noted that article 1(2) of the MARPOL Convention lays down that 'unless expressly provided otherwise, a reference to the present Convention constitutes at the same time a reference to its Protocols and to the Annexes'. This means

[28] International Convention for the Prevention of Pollution from Ships (adopted on 11 February 1973, as modified by the Protocol of 17 February 1978, entered into force 2 October 1983) 1340 UNTS 61 (MARPOL), IMO Publication: IMO-520E.

[29] LM Paul, 'Using the protective principle to unilaterally enforce transnational marine pollution standards', Report of the Proceedings of the Second International Conference on Marine Debris, Honolulu, Hawaii, 2–7 April 1989, 1055.

that since the MARPOL Convention is subject to the principles and provisions of UNCLOS, the same applies to its Annexes, including MARPOL Annex VI.

An implicit reference to the MARPOL Convention is found in the last article (article 237) of part XII of UNCLOS. Article 237(1) states that the provisions of part XII of UNCLOS are without prejudice to the obligations conferred by other conventions for the protection of the marine environment accepted by a State before becoming party to UNCLOS. However, article 237(2) emphasises that obligations assumed according to previous accepted conventions, such as the MARPOL Convention of 1973, must be enforced in a manner consistent with the general principles and objectives of UNCLOS, including the provisions of part XII. The MARPOL Convention and all of its Annexes, including the later adopted VI, must therefore be enforced in accordance with part XII of UNCLOS.

This corresponds with regulation 11.6 of MARPOL Annex VI, which refers to the regulations of the Annex being enforced and safeguarded in accordance with the international law for the protection of the marine environment in force at a given time. This ties directly to part XII of UNCLOS, especially section 6 on *enforcement* (which includes articles 217, 218 and 220) and section 7 on *safeguards* (which includes article 228).

Regulation 11.6 of Annex VI states:

> The international law concerning the prevention, reduction, and control of pollution of the marine environment from ships, including that law relating to enforcement and safeguards, in force at the time of application or interpretation of this Annex, applies, mutatis mutandis, to the rules and standards set forth in this Annex.

MARPOL Annex VI, including regulation 14, should therefore always be enforced in accordance with part XII of UNCLOS, with the necessary changes (*mutatis mutandis*) being applied.[30] This means that the conclusion and interpretations contained in Part II of this book pertaining to the scope and applicability of articles 217, 218, 220 and 228 also encompass enforcement of the sulphur limits of Annex VI.

Before analysing these special jurisdictions of part XII, the basic jurisdictions pursuant to parts II–VII (except part IV) of UNCLOS, and their relevance for enforcing the sulphur limits in Annex VI, should be examined.

V. Coastal State Jurisdiction

A coastal State has the right, within certain limits, to claim different maritime areas and zones in which different jurisdictions apply. The right for a coastal State to exercise jurisdiction over a foreign ship and the limitations on doing so are

[30] Case C-308/06 *Intertanko*, 20 November 2007, Advocate General Juliane Kokoot, Advisory Opinion, ECLI:EU:C:2007:689, paras 86–87.

therefore highly dependent on where the ship commits a violation and whether it afterwards continues to a port in the coastal State.

The relevant parts of UNCLOS specifying a coastal State's jurisdiction in internal waters, the territorial sea and the EEZ are described in the following subsections, as these are key areas when concentrating on enforcement of environmental legislation such as the sulphur regulations in MARPOL Annex VI. A brief introduction to the contiguous zone and the continental shelf will also be provided.

The portrayal of coastal States' jurisdiction in this chapter primarily relates to the general jurisdictions in these areas pursuant to parts II–VII of UNCLOS. The specific jurisdiction under part XII for coastal States to adopt national laws in their internal and territorial waters and in the EEZ pursuant to articles 211 and 212 is examined in chapter 7 and the jurisdiction for enforcing regulations pursuant to article 222 and, primarily, article 220 is studied in chapter 9.

A. Delimitation of Maritime Areas and Zones – Measured from the Baseline

The different areas and zones are measured in nautical miles, with 1 nm being equivalent to 1,852 kilometres (km). This means that 12 nm equal 22,224 km and 200 nm equals 370,400 km.

The nautical-mile delimitations are measured from the *baseline*, which in article 5 of UNCLOS is defined as being the low-water line along the coast, that is the coastline. But as many coastal States have a very uneven (indented) coastlines, UNCLOS provides the option of drawing a *straight baseline* pursuant to article 7, as it would be difficult to make exact measurements by using the normal baseline in article 5.

Article 7 allows for the drawing of a straight baselines between joining appropriate points from where the measurements are made. This is a codification of the practice accepted by the ICJ in the *Anglo-Norwegian Fisheries case*,[31] where the Court recognised that Norway had the right to delimit its coastline by applying this principle.[32]

Articles 8–13 of UNCLOS set out other measures for determining baselines pertaining to internal waters, the mouths of rivers, bays, ports, roadsteads and low-tide elevations.

Article 14 allows a coastal State to determine its baselines through a combination of any of these methods to suit different conditions.

[31] *Fisheries case (United Kingdom v Norway)* [1951] ICJ Rep 116.

[32] B Geirr Harsson and G Preiss 'Norwegian Baselines, Maritime Boundaries and the UN Convention on the Law of the Sea' (2012) 3(1) *Arctic Review* 117, available at https://arcticreview.no/index.php/arctic/article/view/28.

B. Internal Waters

A coastal State's *internal waters* are those on landward side of the baseline under article 8,[33] for instance on the landward side of a straight baseline drawn in accordance with article 7. This includes landward ports and harbours.

Article 2(1) of UNCLOS stipulates that a coastal State has full sovereignty over these waters, as the wording equates sovereignty over internal waters with the sovereignty a State has over its land. Article 2 reads:

1. The sovereignty of a coastal State extends, beyond its land territory and internal waters and, in the case of an archipelagic State, its archipelagic waters, to an adjacent belt of sea, described as the territorial sea.
2. This sovereignty extends to the air space over the territorial sea as well as to its bed and subsoil.
3. The sovereignty over the territorial sea is exercised subject to this Convention and to other rules of international law.

The jurisdiction of the coastal State over internal waters encompasses the air above and the subsoil beneath these waters, as stated in article 2(2). A coastal State therefore has full *territorial jurisdiction* in internal waters under article 2(1), allowing it to *prescribe*, *enforce* and *adjucate on* laws and regulations within this area. (See chapter 6 for discussion of these principles.) This includes full rights to adopt (prescribe) and enforce (including adjudicating on) national sulphur rules, although these rights are covered by the *lex specialis* provisions of article 212 and article 222.

C. The Territorial Sea (Territorial Waters)

The concept of special regulations applying to the *territorial sea* (territorial waters) was, prior to the adoption of UNCLOS – and alongside regulations on the contiguous zone – codified in the 1958 CTS.[34]

Article 3 of UNCLOS allows all coastal States to claim a territorial sea with a breadth of up to 12 nm, whereas previously the maximum allowed breadth was 3 nm.

It is optional for a coastal State to claim a territorial sea, but unless otherwise expressed, it is tacitly assumed that a coastal State has such a sea.

If two coastal States have waters adjacent to each and their 12 nm territorial seas overlap, article 15 stipulates that, unless otherwise provided for by an agreement or by historical rights, the sea between the two States should be divided equally.

[33] Art 8(1) refers to pt IV, on archipelagic States, as having exceptions (eg art 47) to this.
[34] See n 6 above.

The coastal State furthermore enjoys full sovereignty in that area under article 2(1), which also covers the air above and subsoil beneath in accordance with article 2(2) (see section V.B).

This also gives the coastal State *territorial jurisdiction* in these waters. Yet, unlike in its internal waters, the coastal State must respect certain exceptional limitations on its jurisdiction in the territorial sea under article 2(3). These exceptions include the right for foreign ships to make an *innocent passage* pursuant to articles 17–19 and the right to make a *transit passage* where a strait used for international navigation courses through a coastal State's territorial waters pursuant to part III (article 38).

These exceptions, now codified in UNCLOS, represent customary principles of international law and must therefore also be respected by States that are not party to the Convention.

i. The Right of Innocent Passage

The principle of *the right of innocent passage* can be traced back to the eighteenth century[35] and is now embodied in articles 17–19 of UNCLOS. This right allows a foreign ship to pass directly through the territorial sea without any hindrance by the coastal State pursuant to article 17, thereby limiting that State's territorial jurisdiction.

Foreign ships must fulfil certain requirements according to articles 18–19 for a passage to be deemed innocent. Under article 18, the passage must be *continuous and expeditious*, without entering internal waters.

Article 8(2) stipulates that if the drawing of straight baselines in accordance with article 7 results in the enclosing of internal waters that had not previously been considered as internal waters, the right of innocent passage shall exist in those waters.

Article 18(2) specifies that a foreign ship is not allowed to stop or anchor during its passage, unless it is necessary due to ordinary navigation, force majeure or danger to the ship, or to render assistance to persons, ships, etc in distress.

Article 19(1) defines the term 'innocent' by demanding that such passage must not infringe on the *peace, good order or security of the coastal State*. Article 19(2)(a)–(h) set out a non-exhaustive list of examples of conduct and actions rendering a passage harmful, including any threat or use of force, propaganda, fishing, wilful and serious pollution, research activities, and acts interfering with communication systems or the loading or unloading of any commodity or person contrary to the customs, fiscal, sanitary or immigration laws of the coastal State.

[35] A 2007 report from the Danish Department of Justice refers to the principle of innocent passage as dating back to 1894 when it was referenced in *Institut de droit international*: Danish Justice Department, Report no 1488 on jurisdiction (2007), section 13.2.4.

Articles 17–19 should be read in conjunction with article 21, which provides coastal States with jurisdiction to adopt laws and regulations in conformity with the provisions of UNCLOS pertaining to passage through the territorial sea.

Article 21(3) requires the coastal State to make such laws and regulations public. Foreign ships must comply with such laws and regulations despite performing an innocent passage, in accordance with article 21(4).

Article 21(1) has clear ties to article 19, as it lists numerous regulatory matters that can be legislated by the coastal State and which are applicable to innocent passage. These are regulatory matters that coincide and overlap with many of the prohibited conducts and actions listed in article 19. Article 21(1)(a)–(h), inter alia, refer to ensuring the safety of navigation, preventing infringement of fisheries laws, illegal scientific research and infringements of customs, fiscal, immigration or sanitary laws.

When examining the legal basis for coastal States to adopt or implement sulphur rules in the territorial sea, such as the 0.5% or 0.1% limits in Annex VI, article 21(1)(f) allows for the adoption of national laws, or implementation of international regulation, for 'the preservation of the environment of the coastal State and the prevention, reduction and control of pollution thereof'.[36]

It should be noted that article 212 and article 222, which are discussed in chapter 7, are considered as being *lex specialis* provisions for coastal States to adopt and implement the regulation of sulphur in the territorial sea. Under article 222, this regulation is enforced in accordance with article 220.

Nevertheless, if a foreign ship undertakes an innocent passage as stipulated in articles 17–19, the coastal State cannot take enforcement measures in response to violations of the sulphur regulations unless these are implemented in accordance with article 21(1)(f), explicitly obligating foreign ships as laid down in article 21(4).

This conclusion is supported by article 220(2) – allowing coastal States to undertake physical inspection of a vessel sailing in the territorial sea if the State has *clear grounds* for believing that a violation of an environmental rule has occurred[37] – which directly provides that such inspections must respect 'the relevant provisions of Part II, section 3'.

Coastal States can thus only exercise jurisdiction to stop and prosecute a foreign ship making innocent passage for a violation of environmental legislation if the legislation has been adopted and publicised in accordance with the provisions of article 21.

This conclusion applies to merchant ships passing through the territorial sea from a foreign port without entering internal waters, in accordance with article 27(5), which, as a rule, bars criminal enforcement over such ships. However,

[36] Advocate General Juliane Kokott, Advisory Opinion, n 30, para 121, discussed further in ch 5.

[37] Y Tanaka, *The International Law of the Sea*, 2nd edn (Cambridge University Press, 2015) 95.

this provision also exempts such enforcement (imposing fines, etc) of environmental regulations following the direct reference of the article to the provisions of part XII, encompassing article 220(2).[38]

Although nationally adopted environmental regulations, for example national sulphur limits in territorial waters, could be enforced over foreign ships making an innocent passage, the same does not apply to coastal State laws requiring or banning the use of exhaust gas cleaning systems. Such nationally adopted laws would, according to article 21(2), not be enforceable by the coastal States, as this provision prohibits national regulations pertaining to the *design, construction, manning or equipment* of foreign ships, unless these rules are 'giving effect to' (ie implementing) generally accepted international (IMO) standards. Nationally adopted rules banning the use of open loop scrubbers would, for example, not have any effect on ships making an innocent passage, unless as a result of an implemented IMO regulation, as this would otherwise constitute a national requirement relating to the *equipment* of the ship.

ii. The Right of Transit Passage through a Strait in the Territorial Sea

If part of a coastal State's territorial sea encompasses a strait used for international navigation (an 'international strait'), part III of UNCLOS sets limitations on the coastal State's right to exercise its territorial jurisdiction to investigate and prosecute foreign ships in that strait. These limitations are in most respects more restrictive than the those applying to the coastal State with regard to innocent passage.

An international strait is a strait in the territorial sea used for international navigation from one part of the high seas or EEZ to another part of the high seas or EEZ (see article 37).[39] Article 35(a) underlines, like article 8(2), that straits in internal waters are not subject to the same limitations as straits in the territorial sea, unless the use of straight baselines, in accordance with article 7, resulted in territorial waters becoming enclosed and deemed internal waters.

Article 38(1) requires a coastal State to respect a foreign ship's (or airplane's) right to a direct and unhindered *transit passage* through an international strait.[40] In return, under article 38(2), the ship must make a *continuous and expeditious transit of the strait.*

[38] Art 27 is part of pt II, section 3, subsection B (arts 27–28) relating to merchant ships and government ships operated for commercial use, whereas the regulations in section 3, subsection A (arts 17–26) apply to all ships.

[39] Art 35(c) provides that any longstanding international conventions still in force relating to a strait will be considered *lex specialis* in connection to the regulations in pt III of UNCLOS. The Copenhagen Convention of 1857 regulating tolls in the Sound of Oresund is an example of such an agreement: *Traktat ang. Afløsningen af Sund- og Belttolden* TRA nr 21000 af 14/03/1857 [The Copenhagen Convention (adopted 14 March 1857, entered into force 1 April 1857)], available at https://www.retsinformation.dk/eli/retsinfo/1857/21000.

[40] Art 38(1) excepts straits running between an island and the mainland of a State if a route of similar convenience exists on the seaward side of the island.

The right of transit passage is *lex specialis* compared to the right of innocent passage, which is stated explicitly in article 45(1)(a) and implicitly in article 38(3).

Article 39 sets out numerous criteria and demands a foreign ship must meet to be able to invoke the right of transit passage. For example, Article 39 (1) stipulates the transit must be continuous and expeditious, without delay, unless rendered necessary by force majeure and provided it does not pose any threat to the coastal State. Article 39(2)(b) also directly requires transiting ships to comply with international (IMO) regulations for the *prevention, reduction and control of pollution from ships*, which covers MARPOL Annex VI.

Article 42(1) allows the coastal State to adopt national laws to which foreign ships must adhere whilst making a transit passage, as laid down in article 42(4).[41]

Article 42(1)(b) specifically refers – as does article 21(1)(f) – to laws pertaining to *the prevention, reduction and control of pollution*. But while article 21(1)(f) allows coastal States to adopt national laws on all environmental matters applicable to innocent passage, for example establishing national sulphur limits, article 42(1)(b) only refers to the coastal State's *giving effect to* (implementing) *applicable international regulations regarding the discharge of oil, oily wastes and other noxious substances in the strait*. This means that these 'national laws' of the coastal State must implement international (IMO) regulations. This limits the coastal State's jurisdiction for adopting and enforcing national environmental regulations over foreign ships making a transit passage, compared to its jurisdiction over those making an innocent passage.

The reference to the 'applicable international regulations regarding the discharge of oil, oily wastes and other noxious substances' in article 42(1)(b) must be seen as a direct reference to MARPOL Annexes I and II. Conversely, coastal States cannot enforce any national or international (MARPOL Annex VI) sulphur regulations for ships making a transit passage.

One might ask how the limitations in article 42(1)(b) relate to article 39(2), obligating transiting ships to comply with all international regulations for the prevention, reduction and control of pollution from ships covering MARPOL Annex VI. This seemingly leaves a legal void regarding whether violations of regulation 14 of Annex VI by foreign ships can be enforced in an international strait. In the view of this author, the answer is to be found in article 233 in section 7 of UNCLOS part XII, as it relates to *safeguards* for enforcing such environmental legislation in an international strait.

Article 233 states as a main rule that nothing in sections 5, 6 and 7 affects the legal regime of straits used for international navigation, thereby including article 220(2) in section 6. Nevertheless, article 233 subsequently makes a direct reference to article 42(1)(a) and (b), stating that if a foreign ship violates these rules, that is discharges oil, oily wastes or other noxious substances into the sea,

[41] Art 43(b) encourages neighbouring coastal States bordering a strait to work together to prevent, reduce and control pollution from ships.

and this causes or threatens to cause major damage to the marine environment of the straits, then the States bordering the straits may take appropriate enforcement measures and, if so, shall respect *mutatis mutandis* the provisions of section 7 of part XII of UNCLOS.

The reference to *appropriate enforcement* must be a reference to the coastal State's stopping, investigating and, where the evidence so warrants, instigating legal proceedings against a ship violating these rules. The reference to this respecting *mutatis mutandis* the provisions of the section means that the procedural safeguards of section 7 must – with the necessary changes applied – be complied with by the coastal State when proceeding against violations of article 42(1)(b), that is, violations of MARPOL Annexes I or II.

Circling back to the previously asked question in which article 39(2) was compared with article 42(1)(b), the answer seems to be that article 233 allows a coastal State to take action in response to a foreign ship's violation of MARPOL Annex I or II in the strait, provided this violation also causes, or threatens to cause, major damage to the marine environment of the strait. The coastal State therefore cannot proceed against violations of MARPOL Annex VI, including regulation 14, committed by a foreign ship during its transit passage, as this regulation is not encompassed by article 42(1). The main rule of article 233 therefore dictates that a coastal State cannot exercise jurisdiction over such violations.

A foreign ship's infringement of article 39(2), for instance by violating regulation 14 of MARPOL Annex VI, can therefore only be enforced by a flag State under article 217. Article 217 is examined in chapter 8.

D. The Contiguous Zone

Article 33(1) of UNCLOS allows a coastal State to invoke a *contiguous zone*, where it has limited jurisdiction to prevent, control and bring enforcement proceedings against any infringement of its national laws relating to *customs, fiscal, immigration or sanitary matters.*

Article 33(2) stresses that the contiguous zone may not extend beyond 24 nm from the baseline. As the coastal State can claim a territorial sea extending up to 12 nm from the baseline, the contiguous zone merely adds an extra 12 nm in which these regulations can be enforced, as the coastal State has full territorial jurisdiction – with the aforementioned exceptions regarding innocent and transit passage – in its territorial sea and therefore also over these listed regulatory matters.

As the coastal State does not have full sovereignty over all regulatory matters in the contiguous zone, the added 12 nm cannot be viewed as part of the coastal State's territory as such, but merely as an area where the coastal State enjoys certain extended rights to regulate and enforce in the specific legislative areas listed in article 33(1). This list must be presumed to be exhaustive because of its clear and closed wording referring to these four regulatory areas.

As the list makes no reference to enforcement of any rules and regulations regarding the protection and preservation of the marine environment, the coastal State's jurisdiction in the contiguous zone does not confer any widened jurisdiction regarding the taking of measures against violations of MARPOL Annex VI occurring in that area.

E. The Continental Shelf

Part VI of UNCLOS enables a coastal State to claim certain sovereign rights over its *continental shelf* to exploit the natural resources therein, such as oil and minerals, and to protect these rights by regulatory measures.

The continental shelf is, in short, the natural underwater prolongation of a coastal State's mainland into the sea. The continental shelf can, in terms of UNCLOS, range up to 200 nm from the baseline under article 76(1).

If scientific geological data can confirm that a continental shelf extends beyond 200 nm, the continental shelf rights can equally be extended up to 350 nm from the baseline, provided the coastal State can meet the conditions for asserting such extended jurisdiction, pursuant to article 76(5)–(8).

Article 77(1) stipulates that the coastal State has the right to explore and exploit the natural resources found in the subsoil of the continental shelf.

The concept of a (coastal) State claiming certain sovereign rights over its continental shelf can be dated back to the period after World War II, when technological advancements allowed for the better extraction of oil and other substances located in the seabed. This led to the American President Harry S Truman's proclaiming, in 1945, that the United States invoked such sovereign rights over its continental shelf.[42] This was afterwards embodied in the previously mentioned 1958 CCS.[43]

The water above the continental shelf is considered *high seas* (unless it overlaps with a coastal State's 200 nm EEZ[44]), as emphasised by Harry S Truman's proclamation that 'The character as high seas of the water above the continental shelf and the right to their free and unimpeded navigation are in no way thus affected.'[45]

As the coastal State's continental shelf jurisdiction pursuant to part VI focuses on the exploration of the subsoil and does not confer any rights for enforcing violations of environmental legislation in waters above the shelf, these provisions of part VI will not be examined further.[46]

[42] See the White House News Release, 'The Policy of the United States with Respect to the Natural Resources of the Subsoil and Sea Bed of the Continental Shelf' (28 September 1945).

[43] See n 9 above.

[44] The water above a prolonged continental shelf from 200 nm to 350 nm will therefore always have high seas status.

[45] HW Roberts, 'International Law – The High Seas, The Continental Shelf, and Free Navigation' (1957) 35 *North Carolina Law Review* 524, 532.

[46] Pt XII of UNCLOS contains few references to the continental shelf, eg art 210(5)and art 216(1) refer to dumping on the continental shelf.

F. The Exclusive Economic Zone

A coastal State is entitled to claim an EEZ, which is regulated in part V of UNCLOS.

An EEZ can, according to article 57, extend up to 200 nm from the baseline. It thereby overlaps the first 12 nm of territorial sea where the coastal State has, with certain exceptions, full territorial jurisdiction given its sovereignty in that area under article 2.

The additional 188 nm, measured from the 12 nm to the 200 nm from the baseline, can therefore – like the contiguous zone and the continental shelf – not be regarded as an extension of the coastal State's territory as it does enjoy full sovereignty in the zone.[47]

The EEZ merely establishes a *special legal regime*, as stated in article 55. This bestows coastal States with limited sovereign rights for exploiting and managing living and non-living resources in the zone, creating artificial islands, and conducting scientific research and protecting the marine environment, pursuant to article 56(1).

Article 58(1) confers different rights and duties on other States in a coastal State's EEZ, such as allowing all States, and the ships flying their flags, to enjoy the *freedom of navigation* (and overflight), by referring directly to these rights codified in article 87 pertaining to sailing on the high seas. Article 58(2) also refers to application of the rights, obligations and exceptions regarding the high seas set out in articles 88–115, when compatible, to sailing in the EEZ.

Article 58(3) refers to compliance by all States, and therefore by the ships flying their flags, with nationally adopted laws and implemented international rules whilst sailing in an EEZ. Article 211[48] of part XII refers to a coastal State's right to implement international regulations and, in accordance with article 211(6), adopt national laws for the protection of the marine environment in the EEZ. If the EEZ is covered with ice then article 234 applies.

Article 73 presents coastal States with a jurisdictional basis for punishing violations of a coastal State's laws and regulations protecting its sovereign rights in the EEZ, regarding the exploitation, conservation and management of the living resources therein. This includes boarding, inspecting, instigating legal proceedings against and arresting a vessel until reasonable economic security is posted. That being said, article 73 makes no reference to the coastal State's enforcement of its laws and regulations for *protection of the marine environment*, although a coastal State has jurisdiction over such matters in the EEZ under article 56(1)(b)(iii).

Enforcement must therefore be carried out in accordance with the principles for coastal State enforcement laid out in part XII of UNCLOS, as article 56(1)(b)

[47] *Troubled Waters: Inquiry Into the Events Surrounding Crimes Committed at Sea*, Report of the Australian Parliament, The House of Representatives, Standing Committee on Social Policy and Legal Affairs (2013) 27.

[48] Tanaka, n 37, 133.

refers to the coastal State's having jurisdiction over these matters, *as provided for in the relevant provisions* of UNCLOS. This refers to the special coastal State jurisdiction found in article 220 for bringing proceedings for violations of environmental legislation, including violations of MARPOL Annex VI in the EEZ, as described in chapter 9. It is noted that the geographical scope of article 222, relating to enforcement of air pollution regulations, is limited to areas and to the air space under the sovereignty of the coastal State, which, in accordance with the principles of article 2 and article 58, does not cover the EEZ. This is discussed further in chapter 7.

A coastal State's enforcement of national and international laws and regulations for the protection of the marine environment in the EEZ, for example enforcing the sulphur limits under regulation 14 of MARPOL Annex VI, is to be carried out in accordance with article 220.[49]

VI. The High Seas

Part VII of UNCLOS (articles 86–120) regulates the *high seas*, which, according to article 86, are defined as those areas of the sea that are not deemed to be internal or territorial waters or part of an EEZ. The high seas are also referred to as 'international waters' or the 'open sea'.

Article 87 refers to the *freedom of the high seas*, which means that that the high seas are open to all States. Article 87(1)(a)–(f) present a non-exhaustive list of examples of what the freedom of the high seas comprises, including the rights to freedom of: (i) navigation; (ii) overflight; (iii) laying submarine cables and pipelines;[50] (iv) constructing artificial islands and other installations; (v) fishing; and (vi) conducting scientific research.

The right of all States to enjoy the freedom of the high seas is reinforced by the principles under articles 88 and 89, which stipulate (respectively) that the high seas can only be used for peaceful purposes[51] and that no State can claim sovereignty over any part of the high seas. The concept of the high seas being free and open to all States, and free from any claims to sovereignty, was first declared by

[49] This corresponds with the *travaux préparatoires* of the MARPOL Convention, which refer to coastal States' being required to protect the areas adjacent to their territorial waters against pollution, ie the EEZ embodied in the later adopted UNCLOS. See 'General principles for assessment and control of marine pollution recommended by the United Nations Conference on the Human Environment (Recommendation 92)' MP/CONF/INF.6, IMCO (1973) 6, item 17; http://www.imo.org/en/KnowledgeCentre/ReferencesAndArchives/IMO_Conferences_and_Meetings/MARPOL/MP_CONF_1973/Pages/default.aspx.

[50] Arts 112–115 of pt VII lay down the requirements for the laying of such cables and pipelines.

[51] Art 301, found in pt XVI of UNCLOS containing the 'General Provisions' of the Convention, also refers to the 'Peaceful uses of the seas', establishing that all States must exercise their rights and perform their duties without any use or threat of force.

Queen Elizabeth I in 1580, when Spain protested against English maritime activities in an area of the western Atlantic over which Spain had made claims. Queen Elizabeth declared:

> The use of the sea and air is common to all; neither can any title to the ocean belong to any people or private persons forasmuch as neither nature nor regard of the public use and custom permitteth any possession thereof.[52]

The Dutch lawyer Hugo Grotius cemented the concept of the freedom of the high seas in his study *Mare Liberum* ('freedom of the seas') in 1609,[53] when he stated that '[t]he open sea cannot be subject to the sovereignty of any state, access to all nations is open to all, not merely by the permission but by the command of the Law of Nations'.[54] That no State can claim sovereignty over the high seas is, according to Grotius, attributed to the fact that the high seas are considered '*Res Communis*', that is, the common heritage of mankind.[55]

A. Flag State Jurisdiction on the High Seas – The Flag State Principle

Under article 90 of UNCLOS, every State has the right to sail ships flying its flag on the high seas. This is conditional on there being a genuine link between the ship and the State, and article 91 provides that States must have fixed conditions for granting their nationality to a ship and for registering them.

As mentioned in chapter 1, article 92(1) of UNCLOS codifies the so-called *flag State principle*, which stipulates that ships can only sail under the flag of one State and that that State has, except for explicit accepted exceptions codified in UNCLOS or other treaties, exclusive jurisdiction over the ship on the high seas.

It should be noted that warships and ships used solely for governmental, non-commercial service are always immune from the jurisdiction of any State other than the flag State under articles 95–96. Such ships are also covered by the exceptions referred to in article 92(1).

The principle of a State's having jurisdiction over ships flying its flag can be traced back to the early 1800s, especially in the United Kingdom, where an Act for the registration of British ships was adopted in 1823 stating that these ships were subject to the jurisdiction of the State, including whilst sailing on the high seas.[56]

[52] W Vitzhum, 'From Rhodian Sea Law to UNCLOS III' in *Marine Issues: From a Scientific, Political and Legal Perspective* (Kluwer Law International, 2002) 7–8.

[53] *Mare Liberum* was published anonymously in 1609, but Hugo Grotius wrote it in 1604.

[54] Roberts, n 45, 524.

[55] See Advocate General Wahl, Advisory Opinion, in Case C-15/17 *Bosphorus Queen Shipping Ltd Corp v Rajavartiolaitos*, 28 February 2018, ECLI:EU:C:2018:557, para 52, note 13, where reference is made to Grotius' formulating this in his *Mare Liberum*.

[56] JNK Mansell, *Flag State Responsibility – Historical development and contemporary issues* (Springer-Verlag, 2009) 18.

The flag State principle was also recognised by the Permanent Court of International Justice (PCIJ) in the *Lotus case* in 1927,[57] and later embodied in the 1958 CHS[58] and subsequently in UNCLOS.

The flag State principle therefore permits the flag State to exercise *extraterritorial jurisdiction* over ships flying its flag. The *Lotus case* and the extraterritorial nature of flag State jurisdiction are discussed further in chapter 6.

Article 94(1) of UNCLOS describes how a flag State must effectively exercise its jurisdiction over ships flying its flag in *administrative*, *technical* and *social* matters. Article 94(2)–(3) set out examples of what these matters can cover, such as: registering the names and details of ships flying the State's flag (ie administrative matters); ensuring safety at sea by surveying the construction, equipment and seaworthiness of ships (ie technical matters); and, with regard to the manning of ships, labour conditions and the training of crews (ie social matters). Also, article 98 specifies that a flag State must ensure that all ships flying its flag are under a duty to render assistance to any ship or person in need or distress at sea.

Flag State obligations for enforcing environmental legislation are to be found in part XII of the Convention, particularly in article 217, as discussed in chapter 8.

B. Exceptions to Exclusive Flag State Enforcement on the High Seas

As noted in section VI.A, article 92 of UNCLOS also refers to exceptions to the flag State principle, inter alia, codified in UNCLOS. Some of the most relevant exceptions for this book are found in part XII, predominantly in article 218(1) and, to some extent, article 228(1). These articles are analysed in chapters 10 and 11. Yet part VII of the Convention also codifies certain exceptions to the flag State's unimpeded high seas jurisdiction (see article 97 and articles 99–111), which in brief provide as follows:

- In cases of collisions or other incidents at sea, article 97(1) and (2) allow the State of which the master or crew member deemed responsible for the collision or incident is a national to assert jurisdiction over this person, irrespective of the ship's flying another flag.
- Every State is required to take measures to prevent the transport of slaves on the high seas (see article 99); and States are required to cooperate in combatting piracy (articles 100–107) and illegal broadcasting (article 109).[59]

[57] *SS Lotus (France v Turkey)* PCIJ Rep Ser A No 10.

[58] See n 7 above.

[59] Art 108 refers to the cooperation between States to suppress *illicit drug trafficking*. This does not, in the view of this author, actually bestow any non-flag States with any extraterritorial jurisdiction on the high seas. The results of this cooperation can of course result in international treaties or conventions conferring extraterritorial jurisdiction regarding illicit drug trafficking in regulations, but art 108 in itself does not.

- Article 110(1)(a)–(c) allow a warship of any State to investigate foreign ships on the high seas if it has grounds for suspecting that such a ship has violated the aforementioned prohibitions.
- Article 110(1)(d) and (e) also allow all States to exercise jurisdiction over ships unwilling or unable to show a flag. This indirectly refers to article 92(2), which stipulates that if a ship sails under more than one flag, then it must *be assimilated to a ship without nationality*, which gives all States jurisdiction over this vessel in accordance with article 110(1)(d).

i. The Exception in Article 111 – The Right to Hot Pursuit

Article 111(1) of UNCLOS gives coastal States the right to undertake *hot pursuit* of a foreign ship on the high seas, provided the coastal State has good reason to believe that the ship has violated the State's laws whilst sailing in internal or territorial waters or in the contiguous zone,[60] and provided that the coastal State has been able to pursue the vessel uninterrupted after the violation.

This means that if a coastal State can pursue the ship uninterrupted (*in continenti*) from it own waters out onto the high seas, that State may exercise jurisdiction on the high seas by stopping, investigating and, if the evidence so warrants, detaining[61] the ship, or by escorting it back to the coastal State to instigate legal proceedings.

Article 111(2) also allows coastal States to initiate a hot pursuit for violations occurring in the EEZ or on the continental shelf, provided it is a violation of the rights over which a coastal State has jurisdiction in these areas, which includes protection of the environment in the EEZ pursuant to article 56(1)(b)(ii).

Certain conditions must be met by the coastal State when making a hot pursuit. First, the coastal State must ensure that the violation took place within the waters and zones mentioned in article 111(1) and (2), and that the ship has received a visual or auditory signal to stop as laid down in article 111(4). Secondly, under article 111(5), the pursuit must be exercised by a clearly identifiable warship or government ship or aircraft. If the pursuit is carried out by planes, the specific requirements of article 111(6) must be met.

The coastal State is, pursuant to article 111(3), required to abandon the pursuit when the foreign ship enters the territorial sea of another coastal State or its flag State.

If a coastal State stops a ship outside its own territorial waters on false grounds, it can result in the State's becoming liable for any economic loss or damage to the ship, as laid down by article 111(8).

[60] A hot pursuit relating to a violation in the *contiguous zone* must of course relate to a violation of the laws over which the coastal State has jurisdiction in that zone, ie customs, fiscal, immigration or sanitary matters (see section V.D).

[61] Such a detention could, if disputed by the flag State, be subject to the expedited judicial procedure under art 292; see art 297(1)(c) of UNCLOS.

It should be noted that article 111, unlike article 110, does not limit the right of hot pursuit to the investigation or prosecution of the crimes listed in articles 99–107 and 109 of UNCLOS (ie regarding the transport of slaves, piracy and illegal broadcasting). Article 111 refers, as previously mentioned, to *all violations* of coastal State laws and regulations within its jurisdiction, which thereby includes violations of environmental regulations in internal and territorial waters and in the EEZ.[62] This is relevant when examining the possibilities for coastal States to prosecute violations of the sulphur regulations under MARPOL Annex VI.

Chapter 9 of this book includes a section describing the possibilities for coastal States, in accordance with article 220 and article 111, to institute proceedings for violations of the 0.5% (or 0.1%) limit in internal and territorial waters and in the EEZ, by commencing a hot pursuit using a drone or plane with a sniffer attached.

VII. Special Protection of the Marine Environments and Biodiversity through UNCLOS

UNCLOS makes numerous recommendations for States to cooperate to preserve and protect the marine environment, for example in articles 118 and 119, which refer to cooperation between States to ensure the conservation and management of living resources. Article 197, in part XII of UNCLOS, also requires States to cooperate to protect the marine environment.[63]

There are several references throughout UNCLOS, including in articles 119 and 197, to this international cooperation between States being accomplished through the 'competent international organizations', which includes the IMO as the 'IMO is the competent international organization to adopt rules and standards relating to pollution from vessels and pollution by dumping'.[64]

This cooperation between States can result in the designation of specific maritime areas where different forms of protectionary measures apply that can confer obligations and rights on the participating (coastal) States. Such States are required to accept any strengthened regulation applying in their territorial waters or EEZ if the designated area covers parts of these waters and zones. This imposes a restriction on their territorial jurisdiction to legislate freely and carry out enforcement in these areas as they normally would in connection with sovereign environmental rights. They are nonetheless also provided with

[62] A Pozdnakova, *Criminal Jurisdiction over Perpetrators of Ship-Source Pollution* (Martinus Nijhoff, 2012) 113.

[63] Art 197 is found in pt XII of UNCLOS, but given its use as a legal foundation for establishing areas where special environmental regulations apply, often covering different maritime areas and zones, it is noted here as well as in ch 7.

[64] IMO, n 27, 56.

extraterritorial rights to participate in international cooperation that results in legislation that applies outside their territory, that is, on the high seas and in maritime areas and zones of other (coastal) States.

Articles 118 and 119 of part VII and article 197 of part XII emphasise the obligation for all States to participate in such international cooperation. Cooperation can be carried out under the auspices of the IMO system (Special Areas, Emissions Control Areas (ECAs) and Particularly Sensitive Sea Areas (PSSAs)) or of the UN system (Biodiversity Beyond National Jurisdiction (BBNJ)), or through ad hoc multilateral agreements (MPAs), all of which are described further in sections VII.A–C following.

A. Special Areas, ECAs and PSSAs – Cooperation and Protection through the IMO

The IMO has several legislative tools that allow it to protect the marine environment in designated areas. One example of such an IMO 'tool' is the possibility of deeming a delimited maritime area a *Special Area* in accordance with Annexes I, II, IV and V of the MARPOL Convention.

Particularly strict requirements apply in such a Special Area regarding discharges of oil, other noxious liquid substances, sewage and garbage, due to its specific oceanographic and ecological conditions. A Special Area may overlap the different areas and zones as defined in UNCLOS, for example by covering parts of a territorial sea, an EEZ and parts of the high seas. The detailed conditions for the designation of Special Areas are specified in accordance with the 2001 IMO Resolution A.927(22), 'Guidelines for the designation of Special Areas under MARPOL 73/78 and guidelines for the identification and designation of Particularly Sensitive Sea Areas'.

Its may be recalled from chapter 2 that MARPOL Annex VI uses a different term for establishing areas in need of distinct protection from certain airborne pollutants found in the emissions of ships. Annex VI refers to ECAs, in which control can cover NOx emissions under regulation 13 of Annex VI (NECAs) and SOx emissions under regulation 14.3 (SECAs), where a 0.1% sulphur limit applies in accordance with regulation 14.4.

The IMO can furthermore designate areas as being PSSAs, in which several more stringent rules and conditions apply, overlapping the regulatory scope of different IMO regulations pertaining to discharge, navigation, etc. A PSSA can be designated where there is a need for a higher level of protection of rare and unique ecosystems, or of marine animals and their habitats against *maritime activities*. In addition, considerations pertaining to an area's historical value or its value for marine biological research may also affect the assessment.

When the IMO has identified a PSSA, several Associated Protective Measures can be designed and implemented to reduce, and if possible eliminate, the threat to this Area's ecosystem and biodiversity. For example, by rerouting shipping lanes, imposing reporting obligations and issuing IMO recommendations on the use of pilots.[65] By the end 2018, the IMO had designated 17 areas as PSSAs,[66] including the Great Barrier Reef and the Baltic Sea.[67]

The specific conditions for establishing a PSSA were, until 2006, found in Annex 2 of the previously mentioned IMO Resolution A.927(22). However, those regulations have been separated out from those guidelines and are now found in the independent IMO Resolution A.982(24). It is clear from point 4.5 in Resolution A.982(24) that it is possible to create a PSSA 'inside' an existing 'Special Area' and vice versa.[68]

The designation of Special Areas and PSSAs must always be done in accordance with the principles set out in UNCLOS.[69]

B. Marine Protected Areas – Cooperation and Protection through Ad Hoc Multilateral Agreements

The term 'Marine Protected Area' (MPA) is often used as a general umbrella term[70] for describing an area of the sea where special protective measures apply.[71] Special Areas and PSSAs[72] are therefore sometimes also described as MPAs.[73]

Bilateral and multilateral agreements, following international cooperation carried out in accordance with articles 118 and 119 or article 197 of UNCLOS, that have resulted in the designation of ad hoc areas where special protective measure apply can also be referred to as MPAs. Various conventions, including the

[65] IMO Recommendation SN 1/Circ 263, 23 October 2007, recommends that certain ships, depending on their draught and cargo, use a pilot when sailing through the Great Belt and Oresund.

[66] The 17 PSSAs can be seen at http://www.imo.org/en/OurWork/Environment/PSSAs/Pages/Default.aspx.

[67] The Baltic Sea was, on the basis of a request from several of the coastal States bordering the Baltic Sea, including Denmark, Estonia, Latvia, Lithuania, Germany, Poland, Finland and Sweden, designated as a PSSA on 22 July 2005 by IMO Resolution MEPC.136(53), which entered into force on 1 July 2006.

[68] The precise wording of point 4.5 is 'In some cases a PSSA may be identified within a Special Area and vice versa. It should be noted that the criteria with respect to the identification of PSSAs and the criteria for the designation of Special Areas are not mutually exclusive.'

[69] IMO, n 27, 71.

[70] The WWF defines an MPA as 'An area designated and effectively managed to protect marine ecosystems, processes, habitats, and species, which can contribute to the restoration and replenishment of resources for social, economic, and cultural enrichment.' See at https://wwf.panda.org/our_work/oceans/solutions/protection/protected_areas/.

[71] Tanaka, n 37, 346.

[72] PSSAs are designated in accordance with IMO Assembly Resolution A.927(22), 'Guidelines for the designation of Special Areas under MARPOL 73/78 and guidelines for the identification and designation of Particularly Sensitive Sea Areas' (2001).

[73] Tanaka, n 37, 347.

Convention on the Conservation of Antarctic Marine Living Resources (CAMLR Convention)[74] and the Convention on Biological Diversity (CBD Convention),[75] designate such protected areas as MPAs in which special regulations apply.[76]

C. Biodiversity Beyond National Jurisdiction – Cooperation and Protection

Articles 116–120 of UNCLOS confer rights and obligations for all States to cooperate and adopt measures for the conservation of the living resources, including fish stocks, of the high seas.[77] For example, article 116 grants a right for all States to fish on the high seas, but explicitly refers in article 116(b) to article 63(2) and articles 64–67, regarding the rights and duties of coastal States to protect and manage stocks of certain species that live in EEZs (such as highly migratory species, marine mammals and anadromous stocks), also applying on the high seas.[78]

These principles, concerning the rights and obligations of all States in respect of protecting marine life in the high seas, were reiterated in UN Resolution 69/292 of 19 June 2015, in which the UN General Assembly agreed to develop an international legally binding instrument under UNCLOS to ensure the conservation and sustainable use of marine biological diversity of *areas beyond national jurisdiction* (ABNJ), that is on the high seas. This UN programme of work is entitled *Biodiversity Beyond National Jurisdiction* (BBNJ).

The ongoing work in the UN makes continued reference to these UNCLOS principles, for example in a statement made by the chair of the Preparatory Committee established by Resolution 69/292, referring to the BBNJ work, inter alia, having '[r]espect for the rights and obligations of coastal States over the resources under their jurisdiction, as provided for in UNCLOS' and to '[c]onservation and sustainable use of marine biological diversity of areas beyond national jurisdiction, in line with applicable UNCLOS provisions'.[79]

The aim of the Resolution is, inter alia, to ensure the conservation and sustainable use of marine biodiversity for ocean health, productivity and resilience, food

[74] Convention on the Conservation of Antarctic Marine Living Resources (adopted 20 May 1980, entered into force 7 April 1982) 1329 UNTS 47.

[75] Convention on Biological Diversity (adopted 5 June 1992, entered into force 29 December 1993) 1760 UNTS 79.

[76] KM Gjerde and A Rulska-Domino, 'Marine Protected Areas beyond National Jurisdiction: Some Practical Perspectives for Moving Ahead' (2012) 27 *International Journal of Marine and Coastal Law* 357.

[77] ibid 356.

[78] Tanaka, n 37, 238–41.

[79] 'Chair's streamlined non-paper on elements of a draft text of an international legally-binding instrument under the United Nations Convention on the Law of the Sea on the conservation and sustainable use of marine biological diversity of areas beyond national jurisdiction' from the fourth session of the BBNJ PREP COM (10–21 July 2017), available at http://www.un.org/depts/los/biodiversity/prepcom.htm, 16.

security, ecosystem services and sustainable development for present and future generations.

The UN work on this furthermore refers to article 197 of UNCLOS relating to international cooperation through competent international organizations for protection of the marine environment, which means the IMO in respect of adopting rules relating to pollution from vessels. This is sought to be achieved by designating certain areas of the high seas (ie ABNJ) as MPAs, which, as described, ties the BBNJ work to the prerogative powers of the IMO to protect certain marine areas by designating them as Special Areas or as PSSAs to which the above-mentioned statement made by the chair of the BBNJ Preparatory Committee also refers.[80]

The BBNJ work may, through the application of the legal regimes of Special Areas and PSSAs, result in coastal States' having certain extraterritorial rights in ABNJ, for example on the high seas, regarding protection of the biodiversity in these areas. However, as that work does not focus, nor have any impact, on the enforcement of air pollution regulations such as MARPOL Annex VI, this book will not address the BBNJ work further.[81]

VIII. Conclusion

In conclusion, the implementation and enforcement of the MARPOL Convention and its Annex VI must, pursuant to article 237 of UNCLOS and article 9(2) of the MARPOL Convention, be carried out in accordance with the principles and jurisdictions provided for in UNCLOS parts II–VII as described in this chapter, but also in accordance with the special provisions of part XII of UNCLOS as described in chapters 7–13 (ie in Part II of this book). This is supported by the definition of 'pollution of the marine environment' in article 1(1)(4) of UNCLOS, which refers to manmade pollution by substances or energy that presents a hazard to human health, which completely matches the characteristics of sulphur pollution.

The provisions of parts II–VI relate to a coastal State's jurisdiction in its internal and territorial waters and EEZ, with certain exceptions applying, inter alia, relating to *innocent* and *transit* passage. These set the general parameters for how coastal States can enforce legislation, in principle, also covering enforcement of environmental legislation.

Still, the *lex specialis* regulation of part XII, for example article 222 enforced in accordance with article 220, will be the relevant jurisdictional framework for

[80] 'Chair's streamlined non-paper on elements of a draft text of an international legally-binding instrument under the United Nations Convention on the Law of the Sea on the conservation and sustainable use of marine biological diversity of areas beyond national jurisdiction' from the third session of the BBNJ PREP COM (27 March–7 April 2017), available at http://www.un.org/depts/los/biodiversity/prepcom.htm, 44 and 52.

[81] See, eg, Gjerde and Rulska-Domino, n 76, 356, who state that 'The LOSC also recognizes a duty to cooperate in the conservation and management of high seas living resources; although its more specific provisions focus mainly on fish.'

determining how a coastal State can prescribe (including implementing) and enforce legislation for protection of the marine environment, such as MARPOL Annex VI, within areas under its jurisdiction, perhaps using the principles for hot pursuit found in article 111 to support this enforcement.

Conversely, unlike parts II–VI, part VII of UNCLOS provides a coastal State, with few exceptions,[82] with no extraterritorial jurisdiction for penalising violations committed outside its areas, for instance on the high seas. The provisions of part VII do reiterate the flag State principle, which establishes that flag States have jurisdiction on the high seas. But article 92 stipulates that exceptions exist to this principle, which again leads to the *lex specialis* provisions of part XII, including article 218.

The IMO's designation of Special Areas and PSSAs, the UN's work on BBNJ and ad hoc multilateral environmental agreements can, in principle, protect maritime areas that overlap territorial waters, EEZs and the high seas.

Finally, it should be noted that international law, including the law of the sea, is subject to constant development and aligns itself with the technical progress and needs of the international community, as seen in the development of the EEZ and the special environmental protection principles of part XII of UNCLOS compared to the previous 1958 Conventions. The establishment of a new legal basis for enforcing regulation of a relatively new type of pollution, the release of long-range air pollutants from ships on the high seas, would therefore be a normal evolution of the law of the sea.

If such a basis were to be found within the provisions of part XII of UNCLOS, following the analyses in chapters 7–13 (Part II) of this book, it would be immediately applicable for enforcing the sulphur limits set out in regulation 14 of MARPOL Annex VI, including the 0.5% limit on the high seas. This is based on article 9(2) of the MARPOL Convention, referring to the Convention's (and its annexes[83]) being enforced in accordance with future claims and legal views developed within international law.

This is supported by regulation 11.6 of MARPOL Annex VI, which directly states that the enforcement of the Annex, including regulation 14, should be done in accordance with international law in force at the time and how this is interpreted, with the necessary changes applied, ie *mutatis mutandis*.

This means that if it is determined, following the analysis in chapter 10 of this book, that the jurisdictional scope of article 218(1) of UNCLOS is applicable to emission violations on the high seas, it could immediately be applied to the enforcement of regulation 14.1.3 of MARPOL Annex VI, pursuant to regulation 11.6 (*mutatis mutandis*) and article 9(2) of the MARPOL Convention.

[82] None of these above-mentioned exceptions relates to the enforcement of regulations on air pollution on the high seas; rather, they focus on the transport of slaves, piracy and illegal broadcasting. The work on BBNJ does aim at protecting the biodiversity in the high seas, but does not aim at protecting the marine environment itself, nor the air above, from air pollution.

[83] See art 1(2) of the MARPOL Convention.

4

Port State Control

Many port States perform a control (inspection) of foreign merchant ships while they are at berth in a port. The purpose of these Port State Controls (PSCs) is to ensure that the foreign ships, which come into port voluntarily, comply with certain internationally accepted International Maritime Organization (IMO) and International Labour Organization (ILO[1]) rules regarding safety on board, navigational safety, manning, working conditions and the protection of the marine environment, which includes the regulations under MARPOL Annex VI.

Since the beginning of the 1980s, PSC has increasingly become more internationally coordinated between port States, to ensure that the rules which are internationally agreed upon are effectively enforced. The different IMO Conventions often have regulations referring to how compliance with the rules of the particular convention can be ensured during a PSC inspection. The IMO also issues guidelines for how PSCs in general should be carried out, thereby establishing common criteria to guarantee a unified PSC approach in all ports.

This chapter seeks to depict how these PSCs are internationally coordinated and what legal measures can be applied by PSC authorities in the event of a ship's non-compliance, such as the imposition of fines, detaining the ship, requiring economic security (bail), etc. These legal measures are also examined in light of the general framework principles set out in UNCLOS,[2] particularly in articles 219 and 224–226[3] of part XII, and how these regulations support the PSC regime. So, even though Part II of this book contains in-depth analysis of most of the part XII provisions, articles 219 and 224–226 are singled out in this chapter, given their relevance for PSC.

It should be noted that PSC has a focus on port States' inspection of *foreign* ships. This might seem to conflict with article 227 of UNCLOS, which prohibits all States from discriminating, in *any form or in fact*, against foreign ships. But a port State must always ensure that any ship flying its flag – thereby rendering that

[1] The ILO is a specialised United Nations (UN) agency that, inter alia, governs the Maritime Labour Convention (adopted 23 February 2006, entered into force 20 August 2013) 2952 UNTS 3 (MLC).

[2] United Nations Convention on the Law of the Sea (adopted 10 December 1982, entered into force 16 November 1994) 1833 UNTS 3.

[3] Art 218 contains the only direct reference to *port States* in UNCLOS, but art 226 refers directly to art 218 and describes under which conditions a (port) State may investigate a foreign vessel while in port.

(port) State a flag State – complies with the same international regulations as those ships inspected during a PSC. (See, for example, article 94[4] pertaining to safety, navigation, manning and labour conditions, and see article 217(2)–(3) relating to international environmental regulations.) These obligations apply especially while ships under its flag are at berth in a port located in the flag State, as the flag State has complete jurisdiction over such ships, by virtue of their flying its flag and being located in an area (port) under its jurisdiction. Article 217 is examined in detail in chapter 8.

It should also be noted that article 236 of UNCLOS stipulates that the provisions of part XII, including articles 219 and 224–226, do not apply to any State-operated warship, naval vessels or aircraft in government (ie non-commercial) service.

This chapter will also look into the reporting systems that many port States use to inform each other of deficiencies discovered on board a particular ship, including the IMO's *GISIS* database[5] and the EU's *THETIS* database. The principle of 'no more favourable treatment' will be examined, in part with regard to whether this principle has obtained the status of international customary law.

Finally, this chapter attempts to establish a link between these general principles of PSC and the enforcement of MARPOL Annex VI, especially regarding how effective enforcement of the 0.5% sulphur limit could be envisioned. The intention is to establish that the current PSC structure, including the application of the GISIS module, will be able to offer the 'practical enforcement' that fully supports the 'legal basis' analysed in Part II of this book.

I. Port State Control Coordinated on an International Scale

Port State Control was originally preformed sporadically from port State to port State on an inconsistent basis. In 1978, a multilateral administrative arrangement was drawn up in The Hague between a number of European port States. This agreement (The Hague Memorandum) aimed at establishing coordinated regional PSC between the participating States based on unified criteria. The goal was to ensure that all ships voluntarily calling into a port in those countries complied with

[4] M Reuß and J Pichon, 'The European Union Exercise of Jurisdiction Over Classification Societies – An International Law Perspective on the Amendment of the EC Directive on Common Rules and Standards for Ship Inspection and Survey Organisation and for the Relevant Activities of Maritime Administrations' (2007) 67 *Zeitschrift für ausländisches öffentliches Recht und Völkerrecht* 119, 122, available at https://www.zaoerv.de/67_2007/67_2007_1_a_119_144.pdf.

[5] The Global Integrated Shipping Information System (GISIS) is an IMO-administered online reporting database, where the Member States can report different findings and information relating to different IMO Conventions, including MARPOL Annex VI.

the requirements of ILO Convention no 147 on minimum standards in merchant ships,[6] pertaining to working conditions on board merchant ships.

The agreement was scheduled to enter into force in July 1978, but this was delayed due to the running aground of the Liberian oil tanker *Amoco Cadiz* off the west coast of Brittany on 16 March that year. The incident resulted in a massive oil spill, and consequently the States participating in The Hague Memorandum agreed to expand the regulatory scope of the PSC agreement to include ensuring compliance with applicable IMO (IMCO) regulations on safety and environmental protection. This extension was adopted in Paris in 1982 as a Memorandum of Understanding (MoU).[7] This agreement is therefore often referred to as 'the Paris MoU'.

This form of regional coordination of PSC through MoUs spread, following the adoption of the Paris MoU, to other parts of world. The EU has also tried to coordinate mandatory PSC measures between EU Member States by adopting regional legislation, the Port State Control Directive (PSC Directive).[8]

The international regulations with which PSCs aim to ensure compliance include, but are not limited to, the following IMO conventions:[9]

- the MARPOL Convention and its Annexes;[10]
- the SOLAS Convention;[11]
- the BWM Convention;[12]
- the STCW Convention.[13]

The Hong Kong Convention[14] on environmentally sound recycling of ships will also be encompassed when it eventually enters into force. The MLC Convention,[15] which as noted is an ILO convention, is also part of the general PSC inspection scheme.

[6] Convention (No 147) concerning minimum standards in merchant ships (adopted 29 October 1976, entered into force 28 November 1981) 1259 UNTS 335.

[7] Paris Memorandum of Understanding on Port State Control (adopted 26 January 1982, entered into force 1 July 1982) (Paris MoU). The Paris Memorandum is available at https://www.parismou.org/inspections-risk/library-faq/memorandum.

[8] Directive 2009/16/EC of the European Parliament and of the Council of 23 April 2009 on Port State Control [2009] OJ L131/57.

[9] Y Tanaka, *The International Law of the Sea*, 2nd edn (Cambridge University Press, 2015) 297–98.

[10] International Convention for the Prevention of Pollution from Ships (adopted on 11 February 1973, as modified by the Protocol of 17 February 1978, entered into force 2 October 1983) 1340 UNTS 61 (MARPOL), IMO Publication: IMO-520E.

[11] International Convention for the Safety of Life at Sea (SOLAS) 1974 (adopted 1 November 1974, entered into force 25 May 1980), 1184, 1185 UNTS 2.

[12] The International Convention for the control and management of ship's ballast water and sediments 2004 (adopted 13 February 2004, entered into force 8 September 2017) (the BWM Convention).

[13] The International Convention on Standards of Training, Certification and Watchkeeping for Seafarers 1978 (the STCW Convention) (adopted 7 July 1978, entered into force 28 April 1984) 1361, 1362 UNTS 2.

[14] The Hong Kong International Convention for the Safe and Environmentally Sound Recycling of Ships, 2009 (the Hong Kong Convention) (adopted 15 May 2009, not yet entered into force).

[15] The MLC Convention, n 1.

A. Port State Control through Memorandums of Understanding

Several other European States, and Canada and Russia, have, since its adoption in 1982, joined the Paris MoU, leading to a total 27 participating States[16] bordering the Atlantic region, thereby ensuring that most (if not all) merchant ships crossing the Atlantic on a regular basis can be subjected to a PSC.

In December 1993 a Tokyo MoU was adopted, entering into force in April 1994, which focuses on PSC in the Pacific region.[17]

Seven other regions have since established such MoUs, covering most of the world's oceans. These are the Caribbean MoU, the Mediterranean MoU, the Indian Ocean MoU,[18] the Abuja MoU,[19] the Black Sea MoU,[20] the Riyadh MoU[21] and the Acuerdo de Vina del Mar MoU.[22] It should be noted that the United States (US) adopts a singularly unilateral approach to PSC (see section I.C).

Many countries are parties to different MoUs as a result of their natural geographic locations. For example Canada is party to the Paris and Tokyo MoUs, Australia to the Tokyo and Indian Ocean MoUs, and Russia to the Paris, Tokyo and Black Sea MoUs.

Each MoU has it own underlying Memorandum (of Understanding), which sets out the conditions and principles for PSC in that region,[23] although many of these MoUs are modelled on the original *Paris Memorandum*.[24] This analysis will therefore focus on the provisions of the Paris Memorandum,[25] as the principles laid down in it will also apply to the other MoUs, including the Tokyo MoU[26] and the Indian Ocean MoU.[27] Those three MoUs (Paris, Tokyo and Indian Ocean) cover, in cooperation with the US, PSC of the majority of all merchant vessels crossing the high seas, so that they are of special relevance when looking into

[16] See at https://www.parismou.org/about-us/organisation.

[17] See at http://www.tokyo-mou.org/organization/.

[18] See at http://www.iomou.org/.

[19] The Abuja MoU covers West and Central Africa. See at http://www.abujamou.org/index.php.

[20] See at http://www.bsmou.org/.

[21] The Riyadh MoU covers the Persian Gulf. See at https://www.riyadhmou.org/.

[22] The Acuerdo de Vina del Mar MoU covers Central and South America. See at https://alvm.prefecturanaval.gob.ar/cs/ciala/home?Lang=1374097586979.

[23] For a thorough analysis of these MoUs, see A Kulchytskyy, *Legal Aspects of Port State Control* (Faculty of Law, Lund University, 2012) 22–33.

[24] It should be noted that this chapter uses both the term 'Paris MoU' and the term 'Paris Memorandum'. Although these describe the same Memorandum of Understanding, they are used/applied in different contexts for reasons of clarity. The term 'Paris MoU' (as well as 'Tokyo MoU', 'Indian Ocean MoU', etc) refers to the intergovernmental collaboration/association of (port) States for coordinating PSC. The term 'Paris Memorandum' refers to the tangible Memorandum (document) itself, ie the legislative text and the provisions and annexes therein.

[25] See n 7.

[26] The Tokyo Memorandum is available at http://www.tokyo-mou.org/organization/memorandum_of_understanding.php.

[27] The Indian Ocean Memorandum is available at http://197.230.62.214/IndianMoU.aspx?AspxAutoDetectCookieSupport=1.

effective enforcement of the sulphur limits (and future regulations on greenhouse gas (GHG) emissions).

B. Port State Control in the EU

Many EU Member States, which are required to comply with the PSC Directive, are furthermore voluntary parties to the Paris MoU. It therefore comes as little or no surprise that the PSC Directive contains several references to the Paris Memorandum, linking the two European PSC schemes. For instance, in paragraph (13) of the Preamble to the PSC Directive it is asserted that:

> The inspection system set up by this Directive takes into account the work carried under the Paris MoU. Since any developments arising from the Paris MoU should be agreed at Community level before being made applicable within the EU, close coordination should be established and maintained between the Community and the Paris MoU in order to facilitate as much convergence as possible.

Also, article 1(b) and (c) of the Directive include very clear references to the Paris MoU:

> The purpose of this Directive is to help to drastically reduce substandard shipping in the waters under the jurisdiction of Member States by:
>
> …
>
> (b) establishing common criteria for control of ships by the port State and harmonising procedures on inspection and detention, building upon the expertise and experience under the Paris MoU;
>
> (c) implementing within the Community a port State control system based on the inspections performed within the Community and the Paris MoU region, aiming at the inspection of all ships with a frequency depending on their risk profile, with ships posing a higher risk being subject to a more detailed inspection carried out at more frequent intervals.

Paragraph (10) of the Preamble refers directly to the European Maritime Safety Agency (EMSA), ensuring that EU PSC is implemented uniformly and effectively in all Member States and in accordance with the Paris Memorandum.

Article 3(1) of the PSC Directive stipulates that all ships calling at EU ports must comply with applicable international standards and conventions. Under article 2(1)(c), 'Conventions' includes the MARPOL Convention.

Article 11 requires that such ships be subject to PSC with a certain frequency. This decision on the necessary frequency, inter alia, is taken in light of the ship's risk profile, as specified in article 10. This allows for ships posing a higher risk to undergo more frequent and more thorough inspections, in accordance with article 14.

All initial PSCs must, in accordance with article 13(1)(b), verify whether outstanding deficiencies found during the previous PSCs carried out by an

EU Member State or by a State signatory to the Paris MoU have been rectified. Article 13(3) provides that a more detailed inspection must be carried out if 'the condition of a ship or of its equipment or crew does not substantially meet the relevant requirements of a Convention'.

C. Port State Control in Relation to the United States

The US has not joined any MoUs, but the US Coast Guard (USCG) conducts PSC on an unilateral basis and enforces all the same IMO (and ILO) Conventions that are enforced through the MoU regimes, including MARPOL Annex VI.[28] It should be recalled that the US and Canada also have a 200 nm Sulphur Emission Control Area (SECA) in which a 0.1% sulphur limit applies (see chapter 2).

The US PSC does have some link to the Paris Memorandum as, inter alia, it refers to the *banning list* issued under the Paris MoU. The ships banned from the Paris MoU region and ports are thereby also banned from the US region and its ports.[29]

That the US conducts unilateral PSC of the same IMO regulations, including MARPOL Annex VI,[30] is of significant importance, as US ports on the east coast comprise important destinations and hubs for many trading routes across the Atlantic basin; and this also applies to ports on the US west coast regarding trade routes across the Pacific basin.

It should be noted that US is not a party to UNCLOS, but as many of the relevant provisions of that Convention are deemed to represent customary principles of international law, the US can – and does – invoke them.[31]

D. White-, Grey- and Black-listed Flag States

The different MoUs enter their PSC findings into joint databases,[32] where the deficiencies found on each ship are recorded, including information on which flag it flies. The Paris MoU, as expressed in paragraph 3 of Annex 3 to the PSC Memorandum, uses these data to calculate the *white, grey* and *black lists* of flag States.[33] These white, grey and black lists, inter alia, provide information on

[28] See at https://www.dco.uscg.mil/Our-Organization/Assistant-Commandant-for-Prevention-Policy-CG-5P/Inspections-Compliance-CG-5PC-/Commercial-Vessel-Compliance/Domestic-Compliance-Division/MARPOL/.

[29] See at https://www.dco.uscg.mil/Our-Organization/Assistant-Commandant-for-Prevention-Policy-CG-5P/Inspections-Compliance-CG-5PC-/Commercial-Vessel-Compliance/Foreign-Offshore-Compliance-Division/PSC1/.

[30] MH Nordquist, TTB Kohn and J Norton Moore (eds), *Freedom of Seas, Passage Rights and the 1982 Law of the Sea Convention* (Martinus Nijhoff Publishers, 2009) 472.

[31] This includes art 218(1). See ch 10.

[32] The different databases used for reporting PSC findings are described in section XI of this chapter.

[33] See at https://www.parismou.org/detentions-banning/white-grey-and-black-list.

whether a flag State has many ships under its flag that do not comply with the required international conventions. These data are provided to illustrate whether there is a pattern showing that most of the ships found to be substandard during PSCs hail from the same flag States. The Tokyo MoU uses the same white, grey and black list approach.[34]

These lists are updated on an annual basis and follow a commonsensical approach. Flag States on the *white list* are considered effective at ensuring that they meet their flag State obligations pursuant to articles 94 and 217 of UNCLOS. Flag States on the *grey list* represent flag States that are less effective at ensuring such compliance. Finally, the *black-listed* flag States are, for want of a better word, 'inefficient' – to say the least – when it comes to fulfilling their obligations to ensure compliance by vessels sailing under their flags.

Whether a ship flies the flag of a white-, grey- or black-listed flag State also has an impact on the vessel's overall risk assessment profile, which, inter alia, determines how often it should be inspected.[35]

II. The Principle of 'No More Favourable Treatment'

Before going into the details of and regulations relating to PSC, the principle of 'no more favourable treatment' (NMFT) must be analysed, as this confers a right for a port State to ensure that all ships that voluntarily call into a port in the State comply with the relevant international regulations, regardless of whether the ship's flag State is party to the regulations or not. This is conditional on the IMO or ILO conventions' having incorporated the NMFT principle, and on the port State's having ratified and implemented this regulation.

The principle of NMFT is to some extent contrary to the general international law principle of *pacta tertiis nec nocent nec prosunt*, which means that a treaty only binds the parties to the treaty and does not create any obligations for a third State. This principle is embodied in articles 34 and 35 of the Vienna Convention on the Law of Treaties (VCLT).[36]

The principle of NMFT also conflicts with the flag State principle, as the latter dictates that the flag State is considered the primary legal entity that, in principle, can always exercise extraterritorial jurisdiction over ships flying its flag. It therefore seems anomalous that these ships can be forced to adhere to regulations the flag State has not accepted (ratified), as the regulations thereby implicitly bind the flag State and clash with articles 34 and 35 VCLT and the flag State principle.

[34] See at http://www.tokyo-mou.org/inspections_detentions/NIR.php.

[35] See art 12 of and Annex II to the PSC Directive, n 8.

[36] Vienna Convention on the Law of Treaties 1969 (adopted 22 May 1969, entered into force 27 January 1980) 1155 UNTS 331 (VCLT).

Then again, the NMFT principle rests on several reasonings that counter any concerns raised regarding the principle *pacta tertiis nec nocent nec prosunt*. First, NMFT is an expression of a State's right to legislate and enforce (and adjudicate) within its own territory – an expression of *territorial jurisdiction* (see chapter 6) – especially in a port in internal waters, where a State has full sovereignty according to article 2(1) of UNCLOS (see chapter 3). Second, the NMFT principle is limited to ships *voluntarily* entering a port, which means that if a ship cannot comply with the State regulations then it should not enter that State's territory – again, an expression of territorial jurisdiction. Lastly, the NMFT principle also ensures that that foreign ships that comply with the aforementioned international regulations are not put at a disadvantage compared with those ships that do not, due to their flag States' not having ratified these rules. This would introduce a distortive competitive element and could lead to a downward spiral, where shipowners seek open registry States, resulting in the worsening of safety and labour conditions and the lowering of standards for the protection of the environment, including safeguarding human health from SOx, NOx and CO_2. The reasons underlying the NMFT principle are therefore sound and in full alignment with article 227 of UNCLOS, which does not allow discrimination, as the principle ensures that all ships from all flag States are treated equally.[37]

The principle is also in complete alignment with international law, under which the territorial jurisdiction of a (port) State prevails over the *pacta tertiis nec nocent nec prosunt* principle given the State's sovereignty in its ports. Simply put, it can be compared to tourists having to abide the laws of the country they visit despite the fact that their homeland has different laws, for instance respecting the side of the road on which one drives. Or, to quote St Ambrose, 'When in Rome do as the Romans do.'[38]

A. IMO Conventions Codifying the 'No More Favourable Treatment' Principle

The application of NMFT requires, apart from the port State's having ratified the regulation, that the regulation (convention) itself codifies the NMFT principle. The MARPOL Convention refers explicitly to NMFT in article 5(4):

> With respect to the ship [*sic*] of non-Parties to the Convention, Parties shall apply the requirements of the present Convention as may be necessary to ensure that no more favourable treatment is given to such ships.

[37] As the NMFT principle relies on the port State's having ratified the international legislation that is enforced, it can ipso facto be concluded that the same State – when acting as a flag State – also enforces the same regulations over ships flying its flag; see art 94 and art 217(2)–(3).

[38] St Ambrose to St Augustine in the 4th century. See https://theculturetrip.com/europe/italy/articles/the-origin-of-the-saying-when-in-rome-do-as-the-romans-do/.

This principle also applies to the Annexes to the MARPOL Convention, in accordance with article 1(2). Further, MARPOL Annex VI contains a direct cross-reference in regulation 10.3 to article 5(4) of the Convention, applying to enforcement of the Annex, which includes the sulphur limits in regulation 14.

This means that all ships from all States must comply with the requirements laid down in MARPOL Annex VI when they call at a port in a State that is party to the Annex. This includes complying with the global 0.5% sulphur limit and the 0.1% SECA limit, where applicable.

Other IMO conventions, besides MARPOL, also refer to the NMFT principle, including article 1(3) of the SOLAS Convention, article 10(5) of the STCW Convention, article 3(3) of the BWM Convention and article 3(4) of the Hong Kong Convention. The ILO's MLC Convention also refers to the principle in article 5(7).

In addition, there are references to the NMFT principle in sections 1.2.2. and 1.5.1 of IMO Resolution 1052(27) on procedures for port State control (2011) and in article 3 of the EU PSC Directive and article 6 of the EU Sulphur Directive,[39] stating that all regulations of the Directives must be met by all ships when calling upon an EU port, irrespective of which flag they fly.

B. 'No More Favourable Treatment' – A Principle of Customary Law

In the view of this author, the NMFT principle must have obtained status as a recognised principle of international customary law as it meets the criteria for this. It fulfils the objective criterion as it constitutes *consistent repetition of a particular behaviour*,[40] the NMFT principle having been consistently applied by the IMO in its conventions since the 1970s. The subjective criterion of *opinio juris sive necessitatis* ('an opinion of law or necessity')[41] is also met, as NMFT constitutes a principle that is internationally recognised and used as applicable law. For example, many of the IMO conventions that codify the NMFT principle have over 100 States parties, which, in light of their participation in the IMO negotiations leading to the drafting of the regulations, and by their subsequent ratification of the convention, acknowledge the principle. This majority of nations apply the NMFT principle on a daily basis when carrying out PSCs through the nine MoUs (10 if we include the US unilateral PSC), as the NMFT principle allows those port States to enforce regulations and penalise detected violations by the imposition of fines and/or detention, irrespective of what flag a foreign ship flies.

[39] Directive 2016/802 of the European Parliament and of the Council of 11 May 2016 relating to a reduction in the sulphur content of certain liquid fuels [2016] OJ L132/58.

[40] A Henriksen, *International Law* (Oxford University Press, 2017) 25.

[41] ibid 27.

The NMFT principle therefore represents 'evidence of a general practice accepted as law', in light of which it should be considered an *international custom* pursuant to article 38(1)(b) of the Statute of the International Court of Justice ('ICJ Statute').[42] The list of judicial sources listed in article 38(1)(a)–(d) of the ICJ Statute is believed to represent a comprehensive, but not exhaustive, list of the sources within international law,[43] thereby also applying to the law of the sea.

C. UNCLOS in Relation to 'No More Favourable Treatment'

UNCLOS contains no reference to the NMFT principle, but article 237(1) states, as noted in the previous chapter, that nothing in part XII of UNCLOS affects the obligations assumed by a State under a previous convention for the protection of the marine environment. This includes the MARPOL Convention and article 5(4) of that Convention.

In addition, article 237(2) requires States to comply with the specific obligations for protection of the marine environment in conventions they ratify after becoming party to UNCLOS, so that the references to NMFT in the BWM Convention and the Hong Kong Convention, and in MARPOL Annex VI (regulation 10.3), are also in alignment with UNCLOS.

Finally, it is to be recalled that the 'no discrimination' principle in article 227 of the Convention does not conflict with the NMFT principle.

This author therefore believes that NMFT is a principle of international customary law.

III. The A–Z of Port State Control

A PSC is an inspection of a foreign ship that has voluntarily called into a port and is berthed there. The maritime authority in the port State that carries out PSCs will often have targeted in advance which ships at berth should be subjected to a PSC.

Representatives from the maritime authority, often referred to as Port State Control Officers (PSCOs – see further section V), identify themselves to the ship's crew and should subsequently be allowed to come aboard.

Almost all PSCs start with an initial inspection, during which the PSCOs look around the ship to see whether there are any obvious deficiencies, followed by a document inspection of the relevant certificates, records and logbooks, including checking Bunker Delivery Notes (BDNs). During the initial inspection the PSCOs

[42] The ICJ Statute is available at https://www.icj-cij.org/en/statute.

[43] See M Dixon, *International Law* (Oxford University Press, 2007) 24, who eloquently states 'Article 38 of the Statute does not provide a complete and unambiguous statement of sources of international law and it leaves several questions unanswered. Still, it would be a mistake to underestimate the importance of Article 38, not least because it is vital that a reasonably clear and precise statement of the sources be available.'

might ask the crew, and often the master and chief engineer, supplementary questions to determine whether the crew is familiar with relevant regulations, safety procedures, etc.

If the initial inspection leaves the PSCOs with clear grounds for assuming that the ship is not complying with the applicable international regulations, it can lead to a *detailed inspection* in which more thorough investigations are carried out, for example drawing fuel samples from the fuel tanks. Nevertheless, detailed inspections can also be carried out irrespective of what the findings may be on an initial inspection. For instance, if a ship, during a previous PSC in another port State, was found to have certain deficiencies, a detailed inspection may be carried out during the next PSC, to check whether these have been rectified. Also, ships that generally are found to present with several violations and/or that are flying flags from flag States that notoriously do not enforce, that is *black-listed* flag States, can automatically be subject to detailed inspections.

Sometimes detailed inspections are also carried out as part of a *concentrated inspection campaign*, during which all or several ships – within a certain period – are subjected to the same in-depth inspection of certain regulations. A detailed inspection can also be instigated automatically if the port State has received information indicating that a particular ship has violated relevant regulations. For example, if another State (the ship's flag State or a coastal or port State) provides such information, or if a pilot or another ship has detected anomalies that need further investigation.

Finally, if an initial or a detailed inspection shows that a ship is not complying with certain regulations, it can lead to sanctions in the form of *fines*. These can be supplemented by *detention* until the violation has been rectified and/or economic security (bail) has been posted by the shipowner to cover the fine, which is often imposed at a later stage.

A. The Legal Basis for Port State Control

IMO conventions often include regulations relating to how PSC is to ensure compliance with a particular convention. For example, regulations 10 and 11 of MARPOL Annex VI provide information to PSCOs, as well as to shipowners, pertaining to what measures must be complied with to ensure compliance, including how PSCOs should ensure that the ship has valid BDNs in accordance with regulation 18 of and appendix 5 to MARPOL Annex VI. These general procedures for performing PSC are described in IMO Resolution A.052(27) on the procedures for Port State Control.

The subcommittees of the IMO also issue supplementary guidelines and procedures pertaining to PSC in connection with their respective conventions. See, for instance: Marine Environmental Protection Committee (MEPC) Resolution 181(59) on guidelines for port State control under the revised MARPOL Annex VI; MEPC.1/Circ 864 (9 December 2016) on guidelines for on-board sampling for the verification of the sulphur content of the fuel oil used on board ships;[44] or MEPC Resolution 252(67) on guidelines for port State control under the Ballast Water Management Convention. The different memorandums also have regulations with which a State's PSC must comply when it becomes party to such a MoU.

Further, EU Member (port) States must ensure compliance with the PSC Directive, which, as described (see section I.B), to some extent overlaps with the regulations of the Paris Memorandum.

The PSC inspections and the subsequent means of enforcement must correspond with the legal basis of UNCLOS, especially according to articles 219 and 224–227.

B. Port State Control Officers

Paragraph 1.9.1 of IMO Resolution A.1052(27) and section 3.3 of the Paris Memorandum stipulate that PSCOs must be qualified and trained to perform these inspections. The detailed minimum criteria and qualifications for becoming a PSCO are listed in Annex 6 to the Paris Memorandum. Section 3.3 of the Paris Memorandum also stipulates that PSCOs must be authorised by the port State authorities to carry out inspections on behalf of the State.

Article 22 of Annex XI to the PSC Directive sets the minimum criteria European PSCOs must fulfil, and lays down how the Member States are required to ensure their continued training. Paragraph (10) of the Preamble to the PSC Directive proclaims that EMSA should contribute to the development and implementation of a harmonised EU scheme for the training and assessment of competence of PSCOs.

That PSCOs are required to possess these special qualifications for carrying out PSC (of international environmental regulations) and that they must be official representatives of the port State authorities aligns with article 224 of UNCLOS, which provides:

> The powers of enforcement against foreign vessels under this Part may only be exercised by officials or by warships, military aircraft, or other ships or aircraft clearly marked and identifiable as being on government service and authorized to that effect.

C. Port State Control Must Not Present a Danger to the Ship, Crew or Environment

The reasoning behind requiring that PSCOs fulfil certain criteria and that the port State must educate and train them, is not only to ensure that they can identify

[44] The MEPC guidelines relating to PSC of MARPOL Annex VI and on-board fuel sampling are expected to be revised following the MEPC's and Pollution Prevention Response's (PPR's) ongoing work on ensuring 'The consistent implementation of regulation 14.1.3 of MARPOL Annex VI' to enforce the 0.5% limit on a global scale.

any deficiency on board a ship, but also to ensure that they can carry out PSCs in a manner that does not constitute a danger to themselves, the crew, the ship or the environment. This latter requirement is also covered by the different IMO guidelines pertaining to PSC.

This is supported by article 225 of UNCLOS, which requires port States, when exercising enforcement against foreign ships, not to endanger the safety of navigation or otherwise create any hazard to a vessel, or bring it to an unsafe port or anchorage, or expose the marine environment to an unreasonable risk.

The practical scope of this article is somewhat limited in terms of PSC, as the ship is normally voluntarily safely berthed in port during the inspection, but it does underline that PSCs must not present a danger to ship, crew or environment.

D. Ships Must Report their Arrival and Departure

To ensure that port States can prepare and plan PSC inspections of foreign ships that intend to enter their ports, paragraph 3.1 of Annex 12 to the Paris Memorandum requires ships to notify their next port of call of their estimated time of arrival (ETA) and estimated time of departure (ETD) no later than 24 hours prior to ETA.

Ships that, due to their risk profile, will be subject to a detailed PSC inspection upon arrival, are required to report their ETA and ETD no later than 72 hours prior to their ETA.[45]

Paragraph 3.1 of Annex 12 also requires the ship afterwards to inform the port State of its actual time of arrival (ATA) and actual time of departure (ATD).

IV. Initial Port State Control Inspections

The *initial* part of a PSC primarily relates to checking that the ship has all mandatory documents and that these are in order. Annex 10 to the Paris Memorandum stipulates which documents are included in this initial PSC inspection.

The document control under MARPOL Annex VI covers, for example, the checking of BDNs (regulation 18 and appendix 5), the International Air Pollution Prevention Certificate (IAPP Certificate – regulations 5–9), the fuel changeover log (regulation 14.6), the International Energy Efficiency Certificate (IEE Certificate – regulations 5–9), the Ozone-Depleting Substance (ODS) Record Book (regulation 12.6), the Volatile Organic Compound (VOC) Management Plan (regulation 15.6), the manufacturer's operating manual for incinerators (regulation 16.7), the Energy Efficiency Design Index (EEDI) Technical File (regulation 20), the Ship Energy Efficiency Management Plan (SEEMP – regulation 22) and

[45] The concept of risk profiling is described in section V of this chapter, but it is noted that a ship that was detained or sanctioned during a previous PSC can be cautioned to use a 72-hour warning before calling into the next port, making the ship cognisant of this extended reporting obligation.

the Record Book of Engine Parameters (regulation 6.2 of the NOx Technical Code 2008 and regulation 13 of MARPOL Annex VI).

The documents that are of primary importance for ensuring compliance with the sulphur limits under regulation 14 are the IAPP Certificate, fuel changeover log and, obviously, BDNs, especially since the BDN template in appendix 5 to MARPOL Annex VI was amended in January 2019, thus requiring fuel suppliers to 'check off' on the BDN whether a ship has an Exhaust Gas Cleaning System (EGCS) on board if non-compliant fuel has been delivered to the ship. The ship must also attest to the fact, on the BDN, that such an approved and functional EGCS is on board.

The initial PSC will, apart from the document control, often also consist of the crew's being questioned regarding on-board procedures and international regulations, with the PSCOs looking for any observable deficiencies on board. The questioning aims to establish whether the crew is familiar with relevant regulations and procedures. This includes the crew's being able to answer any questions pertaining to an invoked Fuel Oil Non-availability Report (FONAR), such as providing information on which alternative fuel suppliers were contacted in the last port of call before bunkering non-compliant fuel.[46]

This initial PSC inspection (of compliance with international environmental regulations) follows the principle set out in article 226(1)(a) of UNCLOS, which provides, inter alia:

> States shall not delay a foreign vessel longer than is essential for purposes of the investigations provided for in articles 216, 218 and 220. Any physical inspection of a foreign vessel shall be limited to an examination of such certificates, records or other documents as the vessel is required to carry by generally accepted international rules and standards or of any similar documents which it is carrying ...

The references to article 218 and article 220 establish that the principles of inspecting a vessel in accordance with article 226 apply irrespective of whether it is a port State investigating a discharge violation taking place outside its own territory, for example on the high seas (article 218), or a coastal State investigating a violation taking place within its internal or territorial waters or in the EEZ (article 220).[47]

The flag State is also compelled to help ensure that any ship flying its flag complies with such international regulations for the protection of the marine environment, including having all the required documents on board (see article 217(2)–(3)).[48]

[46] A visible indication, which PSCOs might note during the initial part of a PSC, of the ship's compliance with the sulphur regulations of Annex VI could be the lubricants that are present on board, often seen in the engine room. The use of 0.5% fuel will often require special lubrication to be used, to ensure that the engines are functioning in the optimal and most reliable way.

[47] The reference to art 216 in art 226(1)(a) is to the obligation for all States to penalise *dumping violations*.

[48] The flag State often devolves these obligations to Classification Societies, but such an inter partes agreement does not relieve the flag State of its international ex parte obligations; see art 217(2)–(3). For more on this, see ch 8.

The last sentence of article 217(3) implicitly refers to article 226(1)(a) and the initial PSC, as it emphasises that the certificates issued by the flag State, such as the IAPP Certificate (but not BDNs), should be accepted by other States as evidence of the condition of the ship, 'unless there are clear grounds for believing that the condition of the vessel does not correspond substantially with the particulars of the certificates'.

V. Detailed Port State Control Inspection

A PSC can result in a *detailed inspection* if a PSCO, after the initial inspection, finds that there are *clear grounds* for believing that the vessel has violated one or more regulations. This means that a more detailed inspection is carried out on board the ship.

A detailed inspection can include specific surveys of equipment, for example life-saving equipment, or of the machinery or technical systems relating to navigation, etc. It can also include checking that a ship's EGCS is fully functional, or drawing fuel samples from a ship's fuel tanks in accordance with MEPC.1/Circ 864 (9 December 2016) on guidelines for on-board sampling for the verification of the sulphur content of the fuel oil used on board ships.

That *clear grounds* can lead to a detailed inspection corresponds with article 226(1)(a)(i)–(iii) of UNCLOS, which make provision for a detailed PSC inspection if:

(i) there are clear grounds for believing that the condition of the vessel or its equipment does not correspond substantially with the particulars of those documents;
(ii) the contents of such documents are not sufficient to confirm or verify a suspected violation; or
(iii) the vessel is not carrying valid certificates and records.

The last two grounds ((ii)–(iii)) are self-explanatory – missing documents, or documents omitting vital information, can lead to a detailed inspection to determine whether a violation has occurred or not, as these documents are formal testaments to the ship's actual compliance.

Article 226(1)(a)(i) refers to there being clear grounds for believing that the vessel, irrespective of the documentation (as documents are explicitly mentioned in (ii) and (iii) following), does not meet international environmental regulations and therefore should undergo a detailed inspection. This matches the wording of article 217(3) (*in fine*), which specifies that other States must have clear grounds for believing that the conditions on board the vessel do not correspond to the information contained in the certificates issued by the flag State before these other (port) States can carry out a detailed inspection.

It should be noted that detailed inspections can also be instigated without there being clear grounds, for instance if a detailed inspection is performed as part of a *concentrated inspection campaign*, where all ships are subject to the same 'limited' detailed inspection. Such concentrated inspection campaigns are occasionally also coordinated between the different MoUs, for instance between the Tokyo and Paris MoUs, with a focus on ensuring compliance with the sulphur limits under regulation 14.[49]

A. What are 'Clear Grounds'?

Nine examples of what 'clear grounds' include are specified in paragraph 2.4.2 of IMO Resolution A.1052(27) on the procedures for Port State Control, which emphasises that a detailed inspection could be justified if: principal equipment is missing, certificates are missing or clearly invalid, the PSCO observes deficiencies, the crew is unfamiliar with essential procedures or crew members are unable to communicate with each other, false distress emissions are not properly cancelled, or the port State receives information from a third party that a ship appears to be substandard. These examples coincide with those set out in article 13(3) of and Annex V to the PSC Directive, although the list in Annex V is somewhat more detailed.

Section 1.6 of the Paris Memorandum refers to Annex 9 of the Memorandum, which also lists several conditions that can constitute clear grounds indicating that there is a need for further, detailed investigations, for example if the ship's *risk profile* demands this.

Section 3.2 of and Annex 7 to the Paris Memorandum specify the many factors and criteria that will be used cumulatively to establish a ship's risk profile, which will 'determine its priority for inspection, the interval between its inspections and the scope of the inspection'.

Annex 7 should be read in conjunction with Annex 8, which lists the different criteria for classifying and selecting ships for PSC inspection, taking into account a ship's cargo, type, age, former deficiencies noted during previous PSCs and the performance of its flag State pertaining to its being white-, grey- or black-listed.

Article 12 of and Annex I to the PSC Directive set out similar conditions for assessing a ship's risk profile by deeming the ship to be a high risk ship (HRS), a standard risk ship (SRS) or a low risk ship (LRS), which then contributes to determining whether the ship is to be categorised as 'Priority I' or 'Priority II' to undergo PSC.

Article 14 and Annex VII determine whether a ship should undergo an *expanded* (detailed) *inspection*. Article 14 also refers to expanded inspections of certain types of ship, irrespective of their risk profile, including passenger ships, oil tankers, gas tankers, chemical tankers and bulk carriers if these ships are more than 12 years old.

[49] The Paris and Tokyo MoUs carried out a concentrated inspection campaign regarding compliance with MARPOL Annex VI from 1 September 2018 to 30 November 2018. See at https://www.parismou.org/launch-joint-concentrated-inspection-campaign-marpol-annex-vi.

B. What are 'Clear Grounds' Relating to Regulation 14 of MARPOL Annex VI?

When focusing on enforcement of the sulphur limits under MARPOL Annex VI, as affirmed in chapter 1, different technologies can provide clear grounds for believing that a ship is non-compliant. For example, if measurements from a sniffer attached to a drone, plane, helicopter, bridge, etc indicate a violation.

This could lead to a detailed PSC inspection in the next port of call, where a fuel sample is taken from all fuel tanks on board,[50] while simultaneously BDNs, the fuel changeover log, etc are checked and the crew questioned. This would also correspond with regulation 10.3 of Annex VI, which refers to the procedures under article 5 of the MARPOL Convention applying to the enforcement of Annex VI. To this article 5(2) adds, inter alia:

> Any such inspection shall be limited to verifying that there is on board a valid certificate, unless there are clear grounds for believing that the condition of the ship or its equipment does not correspond substantially with the particulars of that certificate.

It would also constitute clear grounds if a ship, or ships from the same shipowner, were continuously to claim non-availability and issue a FONAR, or continuously claim breakdown of an EGCS. This should lead to a detailed inspection, as such recurring incidents could indicate a pattern of non-compliance that requires further investigation.

It should be noted, as seen in chapter 1, that sulphur-detection technologies are under constant development, which also could influence how future PSCs check compliance with the sulphur limits under regulation 14 of Annex VI. One advance could be the use of *handheld measuring devices*, which would allow for on-the-spot tests of the sulphur content in fuel samples, thereby enabling on-the-spot detention if the device shows non-compliance.

Further, sniffer technology could become so advanced that it might be able to provide conclusive evidence of non-compliance. This would render the subsequent drawing of fuel samples unnecessary, as the sniffer data, by itself, could prove regulation 14 infringements beyond any reasonable doubt, including violations of the 0.5% limit on the high seas.

VI. Port State Control Sanctions

If violation of one or more international regulations is detected during a PSC, it can result in various sanctions, primarily fines and detention.

[50] The drawing of fuel samples from all fuel tanks on a vessel will become relevant after the carriage ban, also discussed in ch 1, comes into force.

Neither the IMO guidelines nor the Paris Memorandum refers to how detected infringements should be penalised, as decisions on such criminal sanctions are often the national prerogative of each individual State. These various means of penalising breaches of international regulations must nonetheless be within the terms of UNCLOS, including the special provisions of part XII, when PSCs enforce MARPOL Annex VI.

A. Fines

A violation that is detected during a PSC can result in the port State's issuing a fine.

These fines will often be imposed on the shipowner, but can also target the master and, in principle, other members of the crew, for instance the chief engineer or second officer. The responsibility of the crew will nevertheless often require the infringements to be attributable to their actions, while the shipowner's responsibility, and to some extent the master's, is of a more objective nature, as the mere occurrence of a violation can result in a fine.

Some port States empower their national maritime or environmental authorities to enforce PSC by directly imposing *administrative fines* when a violation is detected.[51] Other port States use the criminal system for imposing such fines, referring the case to the police and public prosecutors, who investigate and issue a fine and potentially take the case to court.[52]

A hybrid version is also seen, where the PSC authorities have a legal basis for imposing administrative fines up to a certain amount and any fine exceeding that amount must be issued by the police.

Irrespective of the legal approach adopted, it should lead to the same result – an effective fine that confiscates all savings and imposes a punitive element. This is of special importance when considering effective enforcement of the sulphur regulations, because of the possible savings described in chapter 1. What constitutes an *effective fine* is discussed in chapter 8.[53]

According to article 230 of part XII of UNCLOS, non-flag States, for example port States, are permitted to issue fines for detected violations of environmental regulations, including MARPOL Annex VI and future regulations on GHG emissions.

[51] An example of a State that uses administrative fines for violations of the EU Sulphur Directive and MARPOL Annex VI is Sweden; see at http://www.airclim.org/acidnews/sweden-will-fine-high-sulphur-ships.

[52] An example of State that sends cases of detected violations of the EU Sulphur Directive and MARPOL Annex VI to the police and public prosecutors is Denmark; see at https://mst.dk/service/nyheder/nyhedsarkiv/2017/maj/shipping-company-to-pay-dkk-375-000-for-using-fuel-containing-too-much-sulphur/.

[53] Ch 8 examines what constitutes an effective *flag State fine*. The same principles, originating from art 18 of the EU Sulphur Directive, also apply to what *port State fines* should cover.

B. Imprisonment

Imprisonment is rarely used as a sanction in PSC, except in extreme cases where deliberate infringements result in, or there is a risk of an infringement's resulting in, major damage. This general PSC approach to sanctioning also matches the jurisdictional limitations set out in article 230 for violations relating to the protection of the marine environment.

Article 230(1) of UNCLOS limits the jurisdiction of coastal and port States to impose penalties, as it stipulates that only monetary fines can be imposed for violations committed by foreign ships outside a State's territorial waters. This also applies to such violations that take place inside the territorial sea, unless the offence was *wilful* and of a *serious* nature, according to article 230(2).

A reverse conclusion from the wording of article 230 allows all States to exercise full adjudicatory jurisdiction in internal waters, as this is not mentioned in the article and therefore not covered by its limitations.

It should be noted that some port States penalise violations that seemingly occurred in the territorial sea, in the EEZ or on the high seas with imprisonment. This might seem at first sight to conflict with article 230, but it is important to take note of what exact violation is being sanctioned and where it took place. For example, if the master and/or a crew member were to be questioned by a PSCO during a PSC about a violation that took place in territorial waters, in the EEZ or on the high seas, and the master or crew member provided false information to the PSCO, that information could, by itself, constitute a new violation occurring in internal waters, as it occurs (is consummated) the very second it is committed (uttered), that is, while in port. Some States penalise the giving of such false information to PSCOs with the threat of imprisonment.

The same applies if a ship (ie the shipowner, master and/or the crew) has falsified documentation on board that is presented during an initial PSC to the PSCO. This could be seen as constituting forgery of documents, which in many States is also a crime punishable by imprisonment. The US PSC, enforced by the USCG, has applied this reasoning and interpreted it as being in accordance with article 230[54] – an interpretation with which this author is in full agreement.

Flag States are not subject to the limitations of article 230, as a ship is obviously not a 'foreign vessel' when it flies the flag of that State. The aim of article 230 is to safeguard the crews of *foreign* ships from a coastal or port State's disproportionate threat of prosecution. The flag State is not covered by this provision and can therefore impose stronger penalties, such as imprisonment, for violations committed by a ship under its flag on the high seas, or in the EEZ or territorial sea of another (coastal) State.

[54] See CD Michel, 'Vessel Source Pollution and Protection of the Marine Environment' in Nordquist, Koh and Norton Moore (eds), n 30, 480. The author's views are his own (as a Captain in the USCG), not necessarily those of the USCG itself.

This corresponds not only with the general flag State principle, but also with article 228(3) (part XII) of UNCLOS regarding enforcement of environmental regulations, which stipulates that it is the prerogative of the flag State to 'take any measures, including proceedings to impose penalties, according to its laws irrespective of prior proceedings by another State'. This means that a flag State can supplement a port State's PSC fine for a high seas, EEZ or territorial sea violation with other penalties, including imprisonment.

Article 230, including article 230(3), which refers to the recognised rights of an accused, is discussed in further detail in chapter 12.

C. Detention Due to Safety or Environmental Concerns

Article 19(1) of the PSC Directive requires the competent authority to be satisfied that any (minor) violations found during a PSC are, or will be, rectified. This means that the PSCO can demand that the ship remedy all deficiencies that are easily resolvable, and then may allow the ship to sail with any remaining deficiencies that cannot be easily fixed, provided they do not present an immediate danger and that the ship takes action to resolve these remaining deficiencies within a specified timeframe.

If a PSC reveals a violation that could threaten the safety of the ship or of navigation, or pose a danger to the marine environment, it can result in the PSCO's detaining the ship until these deficiencies are resolved.[55] If this requires repairs to the ship then the PSCO can demand that these be carried out while the ship is at berth in port, or the PSCO can instruct the ship to sail to nearest shipyard to undergo repairs, provided this journey does not present any danger, in accordance with article 21(1) of the PSC Directive.

Annex 2 to Resolution A.1052(27) lists numerous detainable deficiencies, including violations of the global sulphur limit according regulation 14.1 of MARPOL Annex VI and of the 0.1% SECA limit pursuant to regulation 14.4. This would mean that a ship could be required by a PSCO to remedy this deficiency by debunking (unloading) the non-compliant fuel and bunkering (taking on board) compliant fuel. The carriage ban will make it possible to require all non-compliant fuel on board to be debunkered, even if it is situated in fuel tanks that do not feed directly into the engine.

That a violation of regulation 14 of MARPOL Annex VI can lead to detention is in complete alignment with regulation 10.3 of the Annex, which denotes that the procedures under article 5 of the MARPOL Convention apply to the enforcement of Annex VI.[56] Article 5(2) obligates all (port) States party to the

[55] See para 3.5 of Annex 2 to IMO Resolution A.1052(27), art 19(2) of the PSC Directive and sections 3.4–3.13 of the Paris Memorandum.

[56] This also corresponds with the previously mentioned art 1(2) of the MARPOL Convention, which stipulates that all provisions of the Convention itself, unless otherwise stated, also apply to all of the Annexes to the Convention.

MARPOL Convention – and Annex VI, in accordance with regulation 10.3 – to take all steps necessary to 'ensure that the ship shall not sail until it can proceed to sea without presenting an unreasonable threat of harm to the marine environment'.

That a vessel can be detained due to detected infringements during a PSC, and for violations of regulations for the protection of the marine environment, is supported by article 219 and article 226(1)(c) of part XII of UNCLOS. Article 219 of UNCLOS imposes a special duty on *all* States to detain any ship within their ports or offshore terminals if the ship has violated IMO rules pertaining to the *seaworthiness* of the vessel and this may cause a threat to the marine environment. The vessel can be permitted to proceed to the nearest appropriate shipyard. The State must allow the ship to proceed once the threat (deficiency) has been removed or repaired.

Article 219 presumably refers to violation of the SOLAS Convention, etc because of the reference to the *seaworthiness* of a ship. Such conditions would normally not be covered by part XII of UNCLOS, but the provision refers directly to a violation's having a potential impact on the marine environment, which establishes the causal link between the violation, of for example SOLAS, and the enforcement via article 219 of part XII.

Article 219 refers to all States, which includes flag States. The obligation for flag States to ensure that ships under their flag do not leave any port in the State unless they fulfil applicable international environmental requirements is also found in article 217(2), which includes ensuring that the ship has undergone all mandatory surveys and is carrying all mandatory documents and certificates, in accordance with article 217(3). The obligation for (flag) States to detain vessels under their flag while at berth in a port within their jurisdiction is to be found in article 217(2), as this is a specific *lex specialis* flag-State obligation in comparison with the general wording of article 219.

Article 226(1)(c) of UNCLOS also makes provision for the PSC detention of ships that represent a danger to the marine environment, and states that this can be resolved by seeking repair at the nearest shipyard.[57] It specifies that the flag State must be informed of such actions, and that it can challenge such a detention or instruction at one of the dispute settlement institutions mentioned in part XV of UNCLOS. This must be a direct reference to article 292,[58] which allows for an expedited court procedure regarding matters of detention at the International Tribunal for the Law of the Sea (ITLOS) or the ICJ, or one of the other jurisdictional entities mentioned in part XV, discussed in further detail in chapter 12.

[57] H Ringbom, 'Enforcement of the Sulphur in Fuel Requirements: Same, Same but Different', available at https://webcache.googleusercontent.com/search?q=cache:EBk8-0aReA0J:https://www.duo.uio.no/bitstream/handle/10852/61600/SO-Artikel-Ringbom.pdf%3Fsequence%3D4%26isAllowed%3Dy+&cd=1&hl=da&ct=clnk&gl=no, 32.

[58] E Molenaar, *Coastal State Jurisdiction over Vessel-Source Pollution* (Kluwer Law International, 1998) 506.

D. Detention until Financial Security is Posted

Although neither IMO Resolution A.1052(27), nor the PSC Directive or the Paris Memorandum mentions this possibility, PSC can also result in detention with the aim of procuring *financial security* (a bond or bail), posted by the shipowner to cover a fine that could be imposed after the ship has departed. This includes for violations of the sulphur limits under MARPOL Annex VI.

A fine will only be imposed if one or more deficiencies are detected that, by themselves, could form the basis for detention. Yet a minor deficiency, or a deficiency that has been repaired or rectified, would not necessarily constitute sufficient grounds for detaining or continuing the detention of a ship. But if the detected violations, be they a minor violation or violations that have been rectified, will also result in a fine, the ship can be detained until financial security is posted by the shipowner, often by placing the required amount in escrow in a bank account designated by the PSC authorities. For instance, if a ship is found not to be in compliance with regulation 14 of Annex VI, it can of course be detained until it has debunkered the non-compliant fuel and bunkered compliant fuel. It can, however, continue to be detained after the bunkering until economic security is posted by the shipowner.

This arrangement serves to benefit the authorities and shipowners alike. Posting of financial security can be in the interest of shipowners, as it often can take weeks or even months before a PSC authority, perhaps via the police and public prosecutor's office, is ready to impose a fine for a detected violation since investigatory procedures and due process safeguards must be observed beforehand. These procedural steps can also be prolonged if the ship(owner) contests that a violation occurred. It would constitute a disproportionate punishment if a ship were to be detained while awaiting the imposition and payment of a fine, because even the slightest delay to a merchant vessel can come at a high cost for shipowners, with the delivery of cargo being delayed, port slots in the next port of call being missed, etc.

The PSC authorities can also benefit from the imposition of such financial security, as it can be difficult to find and contact the relevant 'shipowner' afterwards. It may be recalled from chapter 1 that term 'shipowner' is used in this book as an umbrella term covering the legal entity (person or company) responsible for the ship's non-compliance. But in the world of charter agreements, shipping agents, etc, a PSC authority can, after a ship has left port, find it difficult to find the right legal entity to question, subpoena and fine. And even if this entity is found, it might still be difficult to execute and enforce a fine imposed on a shipowner situated in another country. It is therefore also in the interests of the PSC authorities to obtain such economic security to cover a subsequently imposed fine.

Article 226(1)(b) of UNCLOS acknowledges the right for port State authorities to require a bond or *other appropriate financial security* if an investigation indicates a violation of rules and regulations for the protection and preservation of the marine environment. The right for port States to obtain a bond or financial security is reiterated in article 218(4) and article 228(1) relating to the procedural steps that apply when a port State's jurisdiction overlaps with another (coastal or flag) State's jurisdiction.[59]

That fact that a port State, pursuant to article 226(1)(b), can detain a vessel for a detected violation of regulation 14 of MARPOL Annex VI until a financial security is posted by the shipowner is of monumental importance, as the subsequent legal procedures of UNCLOS, especially those pertaining to awaiting flag State actions and information in accordance with article 228(1), can be ongoing for a long time. That a port State can await such actions with sufficient financial security 'in the bank' is paramount, if the port State is later to resume any suspended proceedings.

As mentioned, article 226(1)(c) refers to the settling of disputes regarding detention in accordance with part XV. To that end, article 292(4) explicitly regulates detention pertaining to financial security:

> Upon the posting of the bond or other financial security determined by the court or tribunal, the authorities of the detaining State shall comply promptly with the decision of the court or tribunal concerning the release of the vessel or its crew.

This, including how the ICJ, ITLOS, etc can assess the proportionality of the required bail, is discussed further in chapter 12.

E. Refusal of Access

Section 4 of the Paris Memorandum and article 16 of the PSC Directive refer to port States having the right to refuse access to their ports and off-shore terminals. This allows port States to deny a foreign ship entry to its ports due to the potential risk that the ship poses to navigational safety or to the marine environment.

The right to 'refusal of access' is a PSC sanction that is distinct from both detention and the imposition of fines, as the PSC authority in the port State at which the ship intends to call has not yet inspected the ship and detected any violations or deficiencies.[60] A refusal of access therefore often rests on the findings made by other PSCOs during previous PSC inspections in other ports.

The refusal of access is closely linked to the previously described *black*, *grey* and *white lists* made by the different MoUs, to which the PSC Directive also directly refers in article 16 (see section I.D). Section 4.1.1 of the Paris Memorandum and article 16(1) of the PSC Directive provide that a port State may refuse a foreign ship access to its ports if the ship flies the flag of a flag State that is *black-listed* and has been *detained more than twice within the last 36 months* in an EU Member

[59] Part II of this book examines whether port States can exercise jurisdiction over violations of reg 14.1 of Annex VI on the high seas in accordance with art 218 and, if so, how the consequently overlapping flag State jurisdiction is resolved; see art 228(1). Art 218, including art 218(4), is examined in ch 10; and art 228(1) is discussed in ch 11.

[60] 'Refusal of access' is sometimes also referred to as 'refusing of port entry'. See, eg, Ringbom, n 57, 36.

State or Paris MoU party port. The port State is also allowed to refuse access to foreign ships flying flags from a *grey-listed* flag State that have been subject to *detention more than twice within the last 24 months* in an EU Member State or Paris MoU party port.

A port State that is the ship's next intended port of call may impose such a PSC sanction from the moment the ship departs its last port of call where it was subject to its third detention, in accordance with section 4.1.2 of the Paris Memorandum and article 16(1) (*in fine*) of the PSC Directive.

Section 4.1.6 obligates the port State to immediately inform the flag State of any refusal of access to a ship flying its flag. Yet this notification is for information only, as the port State is not required to await any response or acceptance from the flag State before implementing refusal of access.

Section 4.1.3 and article 16(2) establish that the refusal of access shall be lifted only after three months, and if the ship meets the conditions of paragraphs 3 to 9 of Annex VIII to the PSC Directive, including that all deficiencies have been rectified.

Sections 4.1.4 and 4.1.5 and article 16(4) stipulate that if a ship does not rectify or repair the deficiencies detected during a previous PSC, it can result in that particular vessel's being permanently banned from all ports in EU Member States or in (port) States party to the Paris MoU. An example might be if a ship were released from its third PSC detention on condition that the ship sailed immediately to the nearest shipyard for repairs, but it did not comply and continued its voyage as originally planned.

Article 21(6) of the PSC Directive enables a port State to allow a vessel, which has been refused access, to enter its ports in the event of *force majeure* or overriding safety considerations, or to reduce or minimise the risk of pollution or to have deficiencies rectified, provided the shipowner, operator and/or master can ensure the PSC authorities that adequate measures have been implemented to ensure safe entry.

The principle of refusing access also applies to violations of MARPOL Annex VI, including regulation 14, as regulation 10.3, mentioned in section VI.C, refers to the procedures of article 5 of the MARPOL Convention. Article 5(3) allows (port) States party to the MARPOL Convention, and consequently to Annex VI, to deny a foreign ship entry to their ports or offshore terminals if that the ship does not comply with the provisions of the Convention and Annex VI.

UNCLOS does not explicitly mention the possibility of refusing access to ports, but this PSC sanction must nonetheless be in complete conformity with the Convention, as article 219 and article 226(1)(c) make provision for a ship to proceed to the nearest shipyard to undergo repairs. Thus, if the ship does not comply and effect such repairs, it will pose a danger to the marine environment wherever it sails, and to the marine environment in the next port of call. It is therefore in complete alignment with a State's exclusive territorial jurisdiction in internal waters, given its complete sovereignty over this area pursuant to article 2 of UNCLOS, that it can refuse access to a foreign ship that intends to enter a port in these waters.

F. Unlawful Sanctions

It should finally be noted that article 232 of UNCLOS stipulates that all States are liable for damage or loss attributable to them following unlawful or unwarranted enforcement measures pursuant to section 6, which includes articles 219 and 224–227, imposed in the light of available information. Every port State is liable for any unlawful PSC sanctions that result in damage to or economic loss for a shipowner. States shall also, in accordance with article 232, provide for recourse to actions in their courts in respect of such damage or loss. This is especially relevant pertaining to *detentions*, as such a sanction – unlike fines, imprisonment and refusal of access – is a measure that is quickly decided upon by the PSCO based on the information available at the time.

However, the last part of article 232, referring to unlawful measures *in the light of available information*, absorbs and absolves situations where the enforcement measures taken by the PSCO, such as detention, were justified based on the information at hand. Especially when the unlawful measure is assessed in light of what the potential catastrophic consequences could have been if the suspected deficiency had been present. Recalling that detailing inspection and detention would be the natural result if an initial inspection revealed that documentation was missing or flawed, or if the ship's crew or owner by their actions and statements – or the lack of same – contributed to the PSCO's believing that there was a violation allowing for detention.

VII. Different Databases Relating to Port State Control Inspections

Findings made during a PSC, including information on violations, deficiencies and sanctions (detention), are often entered into one or more international databases by the port State authorities. As previously stated in section I.D, the Paris and Tokyo MoUs[61] use these data, inter alia, to create white, grey and black lists of flag States, pertaining to their performance in ensuring compliance by ships sailing under their flags.

There are many reporting databases, but the following subsections will focus on the regional databases of the Paris MoU and the EU, including *THETIS* and *SafeSeaNet*, the quasi-global information system of *Equasis* and the global IMO-administered *GISIS* database. The GISIS database in particular could have several important applications for ensuring enforcement of MARPOL Annex VI on the high seas. The links between UNCLOS and the information entered into these systems are therefore also described.

[61] Tokyo MoU's reporting system, APCIS (the Asia Pacific Computerised Information System) and the Black Sea MoU's BSIS (the Black Sea Information System) are systems hosted by Russia.

A. The Thetis Database – Information relating to PSC Inspections

The principles of the Paris MoU (and the other MoUs) and the PSC Directive seek to establish a unified PSC approach by coordinating PSC inspections between the participating parties to ensure maritime safety and protection of the marine environment.

This coordinated PSC inspection scheme, which includes giving ships a risk profile, conducting follow-up surveys of previous detected deficiencies, and drawing up white, grey and black lists, necessitates the electronic exchange of PSC information between the different parties and PSC authorities.

Annex 3 to the Paris Memorandum refers to violations and deficiencies detected during PSCs being registered in a joint *Information System on Inspections*. The PSC Directive refers to the development of an EU database that also uses data relating to inspections carried out within the Paris MoU regime (see article 2(22) and article 24(1)).

In 2011, a new joint Paris MoU/EU PSC reporting system was established, called THETIS, which serves both PSC regimes and is linked to the general information available within the EU SafeSeaNet system noted in article 24(2) of the PSC Directive. The THETIS system is managed by EMSA.[62]

The THETIS system calculates the different risk factors, that is, establishing a ship's risk profile, thereby indicating which ships have priority for a PSC inspection.[63] The outcome of PSC inspections, including violations, deficiencies and sanctions, are recorded in the system.

The THETIS system also includes several individual modules directly linked to separate EU regulations that are not, as such, encompassed by the Paris MoU PSC inspection scheme, in non-EU countries such as Russia and Canada. THETIS EU (previously THETIS-S) records information on non-compliance with the EU Sulphur Directive,[64] which to a considerable extent implements the relevant regulations of MARPOL Annex VI (mainly articles 14 and 18) pertaining to sulphur limits and document requirements. Also, THETIS MRV records information on ship efficiency following Directive 2015/757[65] on the monitoring, reporting and verification of CO_2 from marine transport.[66]

The THETIS system is therefore a valuable information system for notifying PSC authorities that are party to the Paris MoU of any detected violations of the sulphur regulations in MARPOL Annex VI, as is THETIS-EU pertaining to violations of the Sulphur Directive.

[62] http://emsa.europa.eu/psc-main/thetis.html.

[63] https://www.parismou.org/inspection-search/inspection-search.

[64] http://emsa.europa.eu/ship-inspection-support/thetis-eu.html.

[65] Regulation (EU) 2015/757 of the European Parliament and of the Council of 29 April 2015 on the monitoring, reporting and verification of carbon dioxide emissions from maritime transport, and amending Directive 2009/16/EC, [2015] OJ L123/55.

[66] http://emsa.europa.eu/ship-inspection-support/thetis-mrv.html.

B. SafeSeaNet – General Information on Ships Entering and Leaving EU Ports

SafeSeaNet is an EU-based information system that also monitors vessel traffic. It is administered by EMSA and there are several references to it in the PSC Directive, for instance in article 24(1). It was established by Directive 2002/59/EC,[67] and seeks to enhance maritime and port safety and security as well as protecting the marine environment. It serves a maritime network for data exchange within the EU and for Norway and Iceland.

The information that is entered into the database includes, but is not limited to, information on current and previous vessel positions by the Automatic Identification System (AIS),[68] details of hazardous goods carried on board, ATAs, ATDs, and safety and pollution-related incidents affecting ships that have been refused access to EU ports.

Most of the information entered into SafeSeaNet does not solely focus on PSC, and the system is therefore not a specific PSC-reporting system but a general EU maritime information system, containing all relevant information on ships expected to enter and leave EU ports. Much of this information is nonetheless of relevance for PSC inspections.

The link between SafeSeaNet and the Paris MoU THETIS system is seen in article 24(2) of the 2009 PSC Directive, requiring that information entered into the joint Paris MoU/EU PSC inspection database (ie THETIS) also be entered into the SafeSeaNet system where relevant.

SafeSeaNet therefore serves as a regional information system that has a supporting role in effective PSC enforcement, including under MARPOL Annex VI because of the reference to protection of the marine environment.

C. Equasis – A Public System with General Information on the Global Fleet

The Equasis system is a maritime data collection system that also is managed by EMSA.[69] The system was established in 2000 and collects data relating to ship safety. The work on creating the Equasis system was instigated in 1997 by the European Commission and France.[70]

[67] Directive 2002/59/EC of the European Parliament and of the Council of 27 June 2002 establishing a Community vessel traffic monitoring and information system and repealing Council Directive 93/75/EEC, [2002] OJ L208/10.

[68] The AIS provides near-real-time ship positions by updating every six minutes.

[69] http://emsa.europa.eu/equasis-statistics.html.

[70] http://www.equasis.org/EquasisWeb/public/About?fs=HomePage&P_ABOUT=MainConcern.html.

Equasis sets itself apart from THETIS and SafeSeaNet by collecting and processing information on a global scale on all merchant ships above 100 GT and their owners. This information is made available to all, free of charge. Equasis is therefore not directly linked to PSC inspections and non-disclosed PSC information, but provides factual data for all on the performance of different shipping companies, which potential costumers, etc can rely on and take into account before choosing a shipping provider.

Equasis also sets itself apart by the fact that it is not only PSC authorities from one region that provide information to the system. The IMO, seven PSC MoUs, EMSA and different private classification societies, vetting companies, industry associations and 'P&I Clubs'[71] also enter relevant information into the system.

The Equasis system is a free and public tool for all companies, customers, etc to use to obtain information on shipowners.

D. GISIS – A Global Database

The IMO launched the *Global Integrated Shipping Information System* (GISIS) in 2005 to enable IMO Member States to report directly into a joint database, for example regarding compliance with IMO rules and regulations.

On 26 November 2009, the IMO's General Assembly adopted Resolution A.1029(26),[72] which urges the IMO Member States to use the GISIS database for the reporting and transfer of data to be reviewed by relevant IMO committees, and to 'sustain and, even, enhance compliance with mandatory reporting requirements, as contained in those mandatory instruments to which they are Parties'.[73]

The GISIS database consists of several modules pertaining to different IMO conventions, standards and guidelines, including: Maritime Security, Port Reception Facilities, Marine Casualties and Incidents, Contact Points, Recognised Organizations, Ballast Water Management, Pollution Prevention Equipment and Anti Fouling, Survey and Certification, Piracy and Armed Robberies, Communications and SAR, LRIT Data Distribution Plan, Ship and Company Particulars, MARPOL Annex VI and Port State Control. The last two modules, on MARPOL Annex VI and Port State Control, are especially noteworthy.

The MARPOL Annex VI module enable administrations, that is flag State authorities, to report mandatory IMO notifications to the IMO and other Member States via GISIS. This, inter alia, covers reports made in accordance with

[71] P&I refers to 'protection and indemnity insurance', which is the insurance policy shipowners take out to protect themselves against liability claims from crew, passengers and third parties following incidents such as collisions, pollution accidents and the removal of wrecks. The term 'P&I Clubs' therefore refers to the companies and associations that offer these insurance policies.

[72] Resolution A.1029(26), 'Global Integrated Shipping Information System (GISIS)'.

[73] Para 3 of IMO Resolution A 26/Res.1029 (26) (adopted on 26 November 2009).

regulation 4.2 (approval of EGCSs), regulation 18.2.5 (evidence of non-availability of compliant fuel oil, ie FONARs) and regulation 18.9.6 (failure of fuel oil suppliers to meet the requirements specified in regulations 14 or 18 of Annex VI).

The module also contains a separate link to the IMO's 'Data collection system for fuel oil consumption of ships', often merely referred to as the IMO's 'Data Collection System', where information on ship efficiency (EEDI) can be entered pursuant to the reporting obligation of chapter 4 (regulation 22.A) of Annex VI – information that, inter alia, relates to the reduction of CO_2 emissions.

Yet the most valuable information that can be reported into the Annex VI module of the GISIS database, in the context of this book, is the obligation for flag States to report their bringing of proceedings against detected violations, as stated in regulation 11.4 of Annex VI. This is underlined by (overlaps) the same obligation for flag States to report their effective enforcement to the IMO and other States pursuant to article 217(7) when receiving information on an infringement, as stipulated in article 217(6) of UNCLOS.

The Port State Control module is linked to reports made by different MoUs, including the Paris MoU, Tokyo MoU, Indian Ocean MoU, Mediterranean MoU, Abuja MoU and Black Sea MoU. It should be noted that the Port State Control module is not open to the public.

The GISIS system is, because of its Annex VI and Port State Control modules, of extreme relevance when examining how the 0.5% sulphur limit, and future GHG (CO_2) legislation, can be effectively enforced, as it is a global database in which flag State and PSC proceedings for regulation 14 violations are registered.[74]

E. The Port State Control Reporting Systems Relating to UNCLOS and VCLT

Enforcement of international regulations for the protection of the marine environment is subject to the *lex specialis* provisions of part XII of UNCLOS, including the obligations for flag States to inform the IMO and other States of their effective enforcement according to article 217(7). As already mentioned, this obligation ties into the reporting obligation under regulation 11.4 of MARPOL Annex VI.

The reporting obligation in regulation 11.4 is also of importance when a flag State is required to report to a port State its (the flag State's) effective enforcement, after having invoked the main rule in article 228(1) of UNCLOS enabling it to suspend legal proceedings instigated by a port State.

[74] It has been discussed at MEPC (MEPC 73 and 74) and PPR (PPR 6) whether the GISIS database should be enhanced to include more information pertaining to enforcement of MARPOL Annex VI, eg on FONARs submitted and the collection of data ('data collection') on fuel availability and quality.

The reporting obligations in article 217 are discussed in chapter 8 and the obligations under article 228 in chapter 11.

Some flag States may object to providing such information on enforcement, referring to this as conflicting with its national laws, for example laws on non-disclosure of information in criminal proceedings. Articles 26 and 27 VCLT provide that such objections are without relevance if a flag State has voluntarily become party to UNCLOS or MARPOL Annex VI. Article 26 VCLT codifies the basic customary principle of *pacta sunt servanda,* which requires a State to fulfil its voluntarily accepted treaty obligations. Article 27 VCLT is, however, of special importance in countering such flag State claims that national legislation bars compliance with the reporting obligations in articles 217 and 228 of UNCLOS and regulation 11.4 of Annex VI to the GISIS database.

Article 27 VCLT clearly stipulates that a party to a convention or treaty 'may not invoke the provisions of its internal law as justification for its failure to perform a treaty'. Article 27 thereby 'trumps' such national legislation. The reasoning behind this is that a State should be held accountable for ensuring that its national legislation does not conflict with the terms of a treaty or convention into which the State has voluntarily entered. It must also continuously ensure that any subsequent adopted national legislation does not conflict with existing responsibilities under a treaty or convention.

It should be noted that the information the flag State is obliged to report under articles 217 and 228 of UNCLOS and regulation 11.4 of Annex VI is not necessarily of a sensitive nature as it should not contain personal information. The information that needs to be provided to GISIS should primarily focus on what deficiencies were found on a ship and what sanctions have resulted for the shipowner. If members of the crew are fined for a violation, or a shipowner is a private individual, information that can identify them could in principle be withheld, such as names, social security numbers, etc. Yet the need to anonymise such information is not predominant, as the information in the Port State Control module of GISIS is not open to the public.

The flag State information provided to GISIS should essentially answer two questions: What was the violation? And what was the sanction? The IMO and the relevant port State can then evaluate whether the sanction was effective and dissuasive.

It is important, when looking at enforcement of the sulphur regulations in Annex VI, that the information from the flag State contains a calculation of the economic gain (savings) the shipowner achieved by the violation, as it relates to the 'effectiveness' of the sanction.

Finally, a flag State cannot invoke article 302 of UNCLOS, which stipulates that no State is required to 'supply information the disclosure of which is contrary to the essential interests of its security' as a reason for not entering the required information into GISIS, as information pertaining to enforcement of international environmental regulation, such as MARPOL Annex VI, is *not* information that is essential to a flag State's national security.

VIII. Conclusion on Port State Control

The evolution of the 1978 Hague Memorandum into the Paris MoU in 1982 helped set the scene for the effective and international coordinated PSC inspection and enforcement scheme of international maritime regulations. This includes IMO conventions and guidelines pertaining to protection of the marine environment, such as MARPOL Annex VI.

The nine MoUs, and the independent USCG PSC, will ensure that MARPOL Annex VI, including the 0.5% global sulphur limit, can be enforced effectively, initially by ensuring that all required documentation relating to the sulphur regulations is in order, by checking BDNs, fuel changeover logs, IAPP Certificate and so on. Any FONARs issued are furthermore subject to the approval or rejection of the PSC authorities. The master and crew can also be questioned by the PSCOs during this initial part of the inspection.

This can lead to a detailed PSC inspection if the initial PSC provides *clear grounds* for believing that the ship is not complying with international regulations, for example with regulation 14 of Annex VI. This itself can lead to the drawing of fuel samples for testing, if possible using hand-held measuring devices that can give an instant test result, which can lead to instant detention.

It is noted that if sulphur-detecting methods, for instance sniffers (attached to drones, etc), fuel calculating, Continuous Emission Monitoring Systems (CEMSs), etc, indicate a sulphur violation, this should constitute clear grounds for a detailed inspection.

A global *concentrated inspection campaign* between the different MoUs, especially between the Paris, Tokyo and Indian Ocean MoUs, could, in cooperation with the US, cover most of the high seas and allow for detailed sulphur inspections of all targeted ships.

The principle of NMFT allows all port States parties to MARPOL Annex VI to impose the sulphur rules on all foreign ships that voluntarily call into their ports. The NMFT principle is codified in article 5(4) of the MARPOL Convention, to which regulation 10.3 of Annex VI directly refers.

A deficiency exposed during a PSC, including a violation of regulation 14 of Annex VI, can result in several different PSC sanctions. For example, port State authorities can impose fines, which should be calculated using the same criteria that flag State authorities are obliged to meet, as described in chapter 8. Detention can also be used to penalise regulation 14 violations, in part to ensure compliance by requiring that non-compliant fuel is de-bunkered (and compliant fuel bunkered), but also to ensure that the shipowner posts economic security (bail) that is proportionate to the potential fine.

The initial and detailed PSC inspections, and the different PSC sanctions, are in accordance with articles 219 and 224–227 of UNCLOS.

Violations of MARPOL Annex VI are entered into the Port State Control module of the GISIS system by PSC authorities. Such information can also be entered into regional MoU database systems such as THETIS, which allows violations of MARPOL Annex VI to influence the overall risk profile of ships and shipowners, and whether the violating ship's flag State is deemed to be white-, grey- or black-listed.

The GISIS reporting platform is of immense importance for supporting global PSC enforcement of regulation 14 of MARPOL Annex VI, as it receives PSC reports from several MoUs, including the Paris, Tokyo and Indian Ocean MoUs.

Flag States report their enforcement of MARPOL Annex VI, including the sulphur regulations in regulation 14 (and the required EEDI information of chapter 4), to the MARPOL Annex VI module of GISIS. Flag States must provide such information pursuant to regulation 11.4 of Annex VI and article 217(7) and article 228(1) of UNCLOS.

Flag States that are parties to UNCLOS and MARPOL Annex VI cannot exempt themselves from fulfilling these reporting obligations by referring to their national laws (see article 26 and, in particular, article 27 VCLT).

These conclusions also apply to PSC enforcement of future IMO GHG legislation. Especially as it has been suggested[75] that some of the proposed short-term regulatory measures that the IMO can adopt to reduce GHG emissions encompass a revision (strengthening) of the EEDI regulation set out in chapter 4 of MARPOL Annex VI.

[75] See point 4.7.1 of Resolution MEPC 304(72), 'Initial IMO Strategy on reduction of GHG emissions from ships' (adopted 13 April 2018 at MEPC 72).

5

The EU and the Sulphur Directive Pertaining to UNCLOS

The European Union (EU) consists of 28 European Member States.[1] These States have, by becoming Member States of the Union, also become signatories of the Treaty on the Functioning of the European Union (TFEU) and the Treaty on European Union (TEU).[2]

The scope and range of the EU's powers to adopt legislation that has a binding effect in the Member States within a specific regulatory field depend on whether the Member States have conferred full (*exclusive*) or partial (*shared*) competence on the EU to regulate a given matter. Member States are free to regulate at a national level within areas where there is no exclusive or shared EU competence. The Member States are, on the other hand, completely barred from prescribing national rules within areas of exclusive EU competence.

Shared competence allows Member States to legislate within the regulatory fields, provided the EU has not exercised its right to legislate. If the EU has exercised its right to regulate within areas of shared competence, this bars Member States from doing the same independently at a national level. Article 4(2)(e) and (g) TFEU stipulate that there is shared competence within the regulatory fields of the environment and of transport, which includes shipping.

Many EU Member States have significant shipping and environmental interests, and can be deemed flag, coastal and/or port States, depending on the circumstances. The EU has therefore adopted several legislative acts pertaining to shipping with the aim of ensuring safety at sea and protecting the marine environment. For example, the Port State Control Directive (PSC Directive),[3] which was examined in chapter 4. It is not surprising, given the interests of the EU and its Member States in shipping and environmental protection, that the EU has also used its shared competence to adopt EU legislation regulating sulphur pollution from ships in the Sulphur Directive.[4]

[1] As of February 2019.

[2] Treaty on European Union of 26 October 2012 [2012] OJ C326/13 (TEU); Treaty on the Functioning of the European Union of 26 October 2012 [2012] OJ C326/47 (TFEU).

[3] Directive 2009/16/EC of the European Parliament and of the Council of 23 April 2009 on Port State Control [2009] OJ L131/57.

[4] Directive (EU) 2016/802 of the European Parliament and of the Council of 11 May 2016 relating to a reduction in the sulphur content of certain liquid fuels [2016] OJ L132/58.

This corresponds with the course set out in Directive 2008/50/EC of the European Parliament and of the Council on ambient air quality and cleaner air for Europe, giving priority to inspecting and improving EU regulation of these pollutants so that emissions can be further reduced. The Directive directly refers to sulphur in marine fuels, stating that 'The necessary Community measures to reduce emissions at source … and to address the sulphur content of fuels including marine fuels should be duly examined as a priority by all institutions involved.'[5]

The first version of Sulphur Directive was adopted in 1993,[6] being amended by a new Directive[7] in 1999 and then again in 2005[8] and 2012.[9] Finally, a codified version of the Sulphur Directive entered into force in 2016.[10]

The 2012 Sulphur Directive, which entered into force on 18 June 2014, introduced the 0.1% Sulphur Emission Control Area (SECA) limit applying to many of the waters of many EU Member States, but it also reiterated the 0.5% global limit. The 2012 amendments to the Directive thereby reflected the 2008 revision of MARPOL Annex VI.[11]

Several of the sulphur regulations in the Directive overlap (implement) the provisions of MARPOL Annex VI, for instance the aforementioned sulphur limits. This gives certain regulations under the Directive, in the view of this author, an interpretational worth pertaining to enforcement of Annex VI, especially as to how violations should be sanctioned. This is discussed in section I of this chapter.

Nevertheless, the Sulphur Directive does also go beyond the regulatory frame of MAPROL Annex VI, for example by demanding that ships at berth for more than two hours in an EU port always use 0.1% fuel, outside SECAs too. Or by setting a 3.5% limit for the maximum permissible sulphur content in marine fuels used by ships with an open loop scrubber system. Or by requiring passenger ships in regular service in EU waters[12] to use fuel with a maximum 1.5% sulphur content.

[5] Para 17 of the Preamble to Directive 2008/50/EC of the European Parliament and of the Council of 21 May 2008 on ambient air quality and cleaner air for Europe [2008] OJ L152/1.

[6] Council Directive 93/12/EEC of 23 March 1993 relating to the sulphur content of certain liquid fuels [1993] OJ L074/81.

[7] Council Directive 1999/32/EC of 26 April 1999 relating to a reduction in the sulphur content of certain liquid fuels and amending Directive 93/12/EEC [1999] OJ L121/13.

[8] European Parliament and Council Directive 2005/33/EC of 6 July 2005 amending Directive 1999/32/EC as regards the sulphur content of marine fuels [2005] OJ L191/59.

[9] Directive 2012/33/EU of the European Parliament and of the Council of 21 November 2012 amending Council Directive 1999/32/EC as regards the sulphur content of marine fuels [2012] OJ L327/1.

[10] As stated in ch 1, any references in this book to the 'Sulphur Directive' or the articles therein, unless otherwise specified, are to be considered references to this consolidated version of the Directive from 2016, ie Directive 2016/802 (n 4).

[11] International Convention for the Prevention of Pollution from Ships (adopted 11 February 1973, as modified by the Protocol of 17 February 1978, entered into force 2 October 1983) 1340 UNTS 61 (MARPOL), Annex VI, IMO Publication: IMO-520E.

[12] The term 'EU waters' is used in this context to describe the national and territorial waters and the EEZs of EU Member States; see the wording of art 6(5) of the Sulphur Directive.

These differences, and the European Court of Justice (ECJ) case law following the *Manzi case* relating to the special 1.5% limit for passenger ships,[13] are examined in section II.

Sections III–V respectively study the EU's affiliation with the International Maritime Organization (IMO), the EU's connection to UNCLOS,[14] and the relevant ECJ case law, that is, *Intertanko*[15] and the *Bosphorus Queen*.[16]

I. The Sulphur Directive's Connection to MARPOL Annex VI

It should initially be noted that the EU is not, unlike UNCLOS, an independent party to MARPOL Annex VI. Nonetheless, nearly all EU Member States are themselves independent parties to the Convention, as are several European States that are not members of the EU.[17]

Those EU Member States that are not party to MARPOL Annex VI must respect and implement the Sulphur Directive due to the aforementioned shared competence within this regulatory field.[18] If a EU Member State therefore were to withdraw from its commitment to MARPOL Annex VI, it would still have to follow the regulations under the Directive. The Sulphur Directive is thus an independent piece of legislation, but there are nevertheless strong ties between the Directive and the regulation of sulphur under MARPOL Annex VI.

Furthermore, paragraphs 20–24 of the Preamble to the Sulphur Directive explicitly refer to the work of IMO and MARPOL Annex VI. Inter alia, they refer to the SECA and global limits noted at the start of this chapter, but they also underline that any changes to MARPOL Annex VI should be implemented in the Directive. It is further stated that the designation of new SECA zones should be done in accordance with the IMO process under Annex VI, as provided in article 6(3) of the Directive.

The global sulphur limit laid down in regulation 14.1 of Annex VI is reiterated in article 6(1) of the Directive, while the 0.1% SECA limit under regulation 14.4 is echoed in article 6(2).[19] Both paragraphs refer to those limits applying to all

[13] Case C-537/11 *Manzi and Compagnia Naviera Orchestra*, ECLI:EU:C:2014:19.

[14] United Nations Convention on the Law of the Sea (adopted 10 December 1982, entered into force 16 November 1994) 1833 UNTS 3.

[15] Case C-308/06 *Intertanko, Intercargo, Greek Shipping Co-operation Committee, Lloyd's Register, International Salvage Union v Secretary of State for Transport*, ECLI:EU:C:2008:312.

[16] Case C-15/17 *Bosphorus Queen Shipping Ltd Corp v Rajavartiolaitos*, ECLI:EU:C:2018:557.

[17] Norway, Iceland, Albania and Montenegro are examples of European States that are not members of the EU but are parties to MARPOL Annex VI.

[18] Hungary and Austria are not parties to MARPOL Annex VI.

[19] It should be noted that the European Commission has set special conditions for ships using Liquefied Natural Gas (LNG) as propulsory means in relation to requirements of the Sulphur Directive. See Commission Decision of 13 December 2010 on the establishment of criteria for the use by liquefied natural gas carriers of technological methods as an alternative to using low sulphur marine fuels

vessels of all flags, irrespective of where a vessel's journey began, thereby utilising the principle of 'no more favourable treatment' (NMFT) described in chapter 4.

A. The Importance of Article 18 of the Directive to UNCLOS and MARPOL

Article 18 of the Sulphur Directive is, in the view of this author, one of its most important provisions, in terms of offering clarity not only as to the scope and extent of the Directive, but also as to how the sulphur regulations of MARPOL Annex VI should be enforced in accordance with the MARPOL Convention and UNCLOS.

Article 18 specifies that EU Member States must determine the penalties that shall be applicable to breaches of Directive. It also stipulates that:

> The penalties determined shall be effective, proportionate and dissuasive and may include fines calculated in such a way as to ensure that the fines at least deprive those responsible of the economic benefits derived from the infringement of the national provisions as referred to in the first paragraph and that those fines gradually increase for repeated infringements.

This precise list of criteria setting out what a penalty (fine) for violating the sulphur regulations should comprise is important, as article 4(4) of the MARPOL Convention merely requires violations of its provisions – including, in accordance with article 1(2), violations of the Annexes – to be met with penalties 'adequate in severity to discourage violations', irrespective of where the infringement occurs.[20]

Article 217(1) of UNCLOS simply refers to flag States' being obliged to enforce violations *effectively* irrespective of where these occur, and article 230 merely bestows a right for non-flag States, such as port States, to penalise Port State Control (PSC) violations with fines. It is therefore difficult, using the principles of MARPOL and UNCLOS, to establish what constitutes an *effective* and *dissuasive* penalty (fine) for a violation of regulation 14 of Annex VI.

Article 18 of the Sulphur Directive offers insight into this matter, as it describes what elements a penalty (fine) for violations of the 0.1% or 0.5% sulphur limit should contain, including:

(a) being effective;
(b) being proportionate;
(c) being dissuasive;

meeting the requirements of Art 4b of Council Directive 1999/32/EC relating to a reduction in the sulphur content of certain liquid fuels as amended by Directive 2005/33/EC of the European Parliament and of the Council on the sulphur content of marine fuels (notified under document C(2010) 8753) [2010] OJ L328/15.

[20] H Ringbom, 'Enforcement of the Sulphur in Fuel Requirements: Same, Same But Different', available at https://webcache.googleusercontent.com/search?q=cache:EBk8-0aReA0J:https://www.

(d) depriving those responsible of all economic benefits derived from the infringement; and
(e) that those fines should gradually increase for repeated infringements.

The last two elements in particular offer significant clarity on what constitutes an effective fine for sulphur violations, as it is specified that all savings must be stripped away and that fines should increase in the event of repeated infringements. These are crucial elements for ensuring effective enforcement of regulation 14.1.3 of Annex VI, as violations, as described in chapter 1, are expected to yield high profits for non-compliant shipowners, especially violations on the high seas.

It might, at the outset, seem illogical to use the principles found in regional legislation, such as article 18 of the Directive, to interpret and clarify the scope and content of international legislation, such as the application of article 217(1) of UNCLOS and article 4(4) of the MARPOL Convention pertaining to Annex VI infringements. Nevertheless, there is a certain *lex specialis* element to article 18 that has to be recognised, as it focuses solely on the enforcement of the same sulphur limits as are codified in MARPOL Annex VI. The principles of article 18 of the Sulphur Directive should therefore be recognised as being a *source of international law* when determining the scope of article 217(1) and article 4(4) for enforcement of regulation 14 of Annex VI.

This assumption is based on article 18's representing a *subsidiary means for the determination of rules of law*, as it was developed and presented by a 'highly qualified publicist' in international law, that is the EU, thereby allowing the International Court of Justice (ICJ) to apply the principles in accordance with article 38(1)(d) of the ICJ Statute.[21] This is relevant, as disputes between States regarding the interpretation of UNCLOS, including the scope and extent of article 217(1), can be submitted to the ICJ pursuant to article 287 of UNCLOS, which is discussed in chapter 12 of this book.

The details relating to what an *effective* and *dissuasive* flag State fine should comprise are examined further in chapter 8.

II. Differences in Relation to MARPOL Annex VI

The distinct differences between the Sulphur Directive and MARPOL Annex VI are set out in this section. These differences represent obligations that EU Member States alone have competence to enforce. This also means that all ships calling into ports in EU Member States must respect these strengthened sulphur regulations, compared to Annex VI, as the Directive applies the NMFT principle.

duo.uio.no/bitstream/handle/10852/61600/SO-Artikel-Ringbom.pdf%3Fsequence%3D4%26isAllowed%3Dy+&cd=1&hl=da&ct=clnk&gl=no, 25.

[21] The ICJ Statute is available at https://www.icj-cij.org/en/statute.

As noted in the introduction to this chapter, the Sulphur Directive sets out special requirements for ships at berth for more than two hours in an EU Member State port, special limits for the sulphur content of fuels used by ships with open loop scrubber systems and special conditions for passenger ships in regular service in EU waters.

A. Vessels at Berth for More than Two Hours in an EU Port

Article 7(1) of the Sulphur Directive demands that all ships at berth for more than two hours in a port of an EU Member State must use fuel with a maximum sulphur content of 0.1%, even if the port is located outside of a SECA, for example in the Mediterranean Sea.[22] A similar obligation is not found in MARPOL Annex VI.

The reason for this rule relates to how some ships let their engines, often auxiliary engines, run while at berth in port, as this produces power and electricity for the whole ship, allowing lights, communications, technical equipment, cooling units, electronic devices in residential areas, etc to function on board during the stay in port. But if the engines are in operation whilst a ship is in port, it results in harmful substances, such as sulphur emissions, being released into the immediate vicinity, which sometimes includes highly populated work and/or residential areas, where they can be inhaled and absorbed into the human body, potentially resulting in premature death, asthma and other respiratory disorders.

This is also described in paragraph (9) of the Preamble to the Sulphur Directive as a great problem for many EU Member States, and under paragraph (10), ships at berth are encouraged to use shore-side electricity, also known as 'cold ironing'.[23]

Article 7(2)(b) of the Sulphur Directive explicitly refers to ships being exempt from the 0.1% obligation set out in article 7(1) if they use shore-side electricity to provide power on board, as this obviates the need to run the engines while at berth.

B. Ships that Use Open EGCS Technologies: Open Loop Scrubbers

Another difference between the Sulphur Directive and MARPOL Annex VI is found in article 5 of the Directive. This stipulates that EU Member States shall ensure that marine fuel with a higher sulphur content than 3.50% is not used

[22] There are discussions on whether the Mediterranean, or parts thereof, should be designated as a new SECA under MARPOL Annex VI, and subsequently implemented in the Sulphur Directive. See at http://www.rempec.org/viewNews.asp?NewsID=474&.

[23] When a ship at berth is powered by shore-side electricity, it is sometimes also referred to as 'shore-to-ship power' (SSP) or 'alternative maritime power' (AMP). For more information, see at https://www.marineinsight.com/marine-electrical/what-is-alternate-marine-power-amp-or-cold-ironing/.

by ships unless they have an Exhaust Gas Cleaning System (EGCS) operating in *closed mode*. A reverse conclusion leads to the assumption that EGCSs in *open mode*, that is open loop scrubbers, are not permitted to use marine fuels with a sulphur content exceeding 3.5%.[24]

Neither MARPOL Annex VI nor the IMO scrubber guidelines[25] prescribe such a maximum sulphur limit for fuels used by ships that have an EGCS installed. The guidelines refer to EGCSs' being approved in accordance with *Scheme A* or *Scheme B*, where Scheme A, as described in chapter 2, merely refer to ships' adhering to the specifications and limits set by the EGCS manufacturer.

What the term 'closed mode' in article 5 actually covers has been discussed, as certain closed loop scrubbers also have a *bleed-off* discharge. This is not defined in the Directive, but evidently open loop scrubbers are not covered by the 3.5% exemption under article 5.

All ships with open mode EGCS (open loop scrubbers) must therefore comply with the regional requirement of the Sulphur Directive and not use fuel with a sulphur content exceeding 3.5% while sailing in EU waters, even though the EGCS could clean fuels with a higher sulphur content, such as 4.0% or 4.5%, to the required equivalent emission levels, and could meet the pH discharge limits of the IMO scrubber guidelines.

C. Passenger Ships Operating on a Regular Service in EU Waters until 2020

Passenger ships engaged in regular service to or from EU ports shall not use fuels with a sulphur content exceeding 1.5%, pursuant to article 6(5) of the Sulphur Directive. This expressly provides that this requirement will cease to have effect from 1 January 2020. This is obviously due to the entry into force of the 0.5% limit, also embodied in the Sulphur Directive, rendering the 1.5% limit obsolete, as passenger ships also must comply with the global limit. This is also directly expressed in paragraph (27) of the Preamble to the Directive, which also highlights the reasoning behind this special EU regulation. It explains that passenger ships operating on a regular service to or from EU ports frequently operate in highly populated coastal and port areas. Thus there are vast environmental, that is human health, benefits to requiring a lower sulphur limit for such ships.

Article 6(5) also codifies references to the flag State principle and to the principle of NMFT, as it places a dual obligation on all EU Member States, in their capacity as flag and port States, to ensure that all ships under their flag and all ships at berth in their ports, irrespective of their flag, comply with the requirement.

[24] The possibilities for coastal States to unilaterally adopt national legislation pertaining to the use of open mode EGCS, ie open loop scrubbers, are covered in ch 7.

[25] Resolution MEPC.184(59) (adopted 17 July 2009), amended by Resolution MEPC.259(68) of 2015 (adopted 15 May 2015), 2015 Guidelines for exhaust gas cleaning systems.

The fact that article 6(5) sets special requirements for passenger ships that are not found in Annex VI, and invokes the NMFT principle, has left EU Member States, who also are independent parties to Annex VI, in a difficult position when enforcing this regional legislation over vessels not flying an EU Member State flag, due to the treaty commitments of these EU Member States towards non-EU States that are Annex VI parties. This formed the basis of the so-called *Manzi case* brought before the ECJ.

i. The *Manzi* Case

In the *Manzi case*,[26] the court of Genoa in Italy referred three questions to the ECJ regarding the obligation for EU Member States to ensure that passenger ships used marine fuel with a maximum sulphur content of 1.5%, while other ships in the Mediterranean, including in Italian waters, merely had to adhere to the global 4.5% limit in force at the time (2008).

The case in Genoa was between the Italian Ministry of Transport and the shipping company Compagnia Naviera Orchestra and Captain[27] Matti Manzi, as in July 2008 the port authorities in Genoa had sanctioned the cruise ship MSC *Orchestra* and her captain for the ship's use of non-compliant fuel in breach of the Directive.[28] They had used marine fuel with a sulphur content exceeding 1.5% but not exceeding 4.5%. The ship sailed under a Panamanian flag.

The first question the ECJ took in to consideration was whether the ship fulfilled the criteria of being (i) a passenger ship, and (ii) 'operating on regular services to or from any Union port'. The ECJ found that the vessel and its voyage plan met these criteria, so that the ship was required to use fuel with a sulphur content of 1.5% instead of 4.5%.[29]

The next question the ECJ had to decide, and which is of interest here, related to the obligation for EU Member States – which also were parties to Annex VI – to enforce the 1.5% requirement on ships flying flags of non-EU Member States that were parties to Annex VI. The question was whether this obligation would force the EU Member States to violate the international principle of *pacta sunt servanda*[30] and the principle of 'cooperation in good faith' under article 4(3) TEU towards non-EU States that also were party to Annex VI.[31] That is, if Italian

[26] Case C-537/11, n 13.

[27] Although this book uses the term 'master', the ECJ uses the term 'captain', so this term is adopted in this description of the *Manzi case*.

[28] It should be noted that all references in the case to the Directive, ie the Sulphur Directive, are references to the 2005 version, ie Directive 1999/32 as amended by Directive 2005/33.

[29] *Manzi case*, n 13, paras 17–35.

[30] *Pacta sunt servanda* is a basic principle of international law that dictates that lawful agreements should be kept. This principle is codified in art 26 of the Vienna Convention on the Law of Treaties 1969 (adopted 22 May 1969, entered into force 27 January 1980) 1155 UNTS 331 (VCLT).

[31] *Manzi case*, n 13, para 36.

authorities were to enforce the Sulphur Directive, including a 1.5 % sulphur limit not found in Annex VI, over a ship flying the flag of Panama.

The ECJ held that the validity of the 1.5% limit could not be determined or viewed in light of MARPOL Annex VI, as the EU was not an independent contracting party to Annex VI,[32] nor were the regulations in Annex VI expressions of customary law.[33] The EU was therefore not bound by the regulations under Annex VI and had not violated the principle of *pacta sunt servanda*[34] and the principle of cooperation in good faith by establishing the 1.5% limit.[35]

The last question presented before the ECJ was whether the 1.5% limit did not apply to ships flying the flag of non-European States that were party to MARPOL Annex VI.[36] The Court merely stated that since neither the EU nor *all* EU Member States were contracting parties to Annex VI, the Court was not required to interpret the validity of the 1.5% limit's applying to ships flying non-EU Member State flags.[37] The Court underlined that as it was not required, and since the EU was not party to Annex VI, the ECJ should actually refrain from interpreting the validity of the 1.5% limit under the Annex, as such an interpretation could have consequences for the interpretation of EU law and thereby have a binding effect upon all EU Member States, which would constitute creating obligations for *third States* and thereby violate the principle of *acta tertiis nec nocent nec prosunt* embodied in article 34 VCLT.[38] The Court referred in this respect to paragraphs 49–52 of the *Intertanko* decision, which is reviewed later in this chapter.[39]

Then again, the ECJ did state obiter that:

> Even assuming that the Court could interpret Article 4a(4) of Directive 1999/32 in the light of the sulphur content laid down in Annex VI, it suffices to state that, in the light of the objective pursued by that annex and set out in the title thereof, namely to protect the atmosphere by a reduction in harmful emissions produced by marine transport, that provision, in so far as it fixes a maximum limit on the sulphur content of marine fuel lower than that provided for by that annex, does not appear to be incompatible with such an objective.[40]

This statement clearly indicates that if the ECJ had seen itself capable of answering the third question, it would not have found that the 1.5% limit violated the purpose and objective of MARPOL Annex VI, thereby also impliedly accepting the application of the NMFT principle in the Sulphur Directive.

[32] ibid, para 37.
[33] ibid, para 39.
[34] ibid, para 38.
[35] ibid, para 41.
[36] ibid, para 42.
[37] ibid, para 51.
[38] ibid, para 47.
[39] ibid, para 45; for further discussion of *Intertanko*, see section V.A.
[40] ibid, para 53.

III. The EU's Relationship with the IMO

The EU's close implementation in the Sulphur Directive of the sulphur regulations set out in the IMO's MARPOL Annex VI, shows a link between the EU and the IMO.

Article 220(1) TFEU allows the EU to establish appropriate cooperation with the organs of the UN and its *specialised agencies*, which includes the IMO. This, and the fact that EU Member States have granted the Union shared competence within the maritime and environmental regulatory fields, allows the EU to coordinate a joint EU position on many of the issues discussed and regulated in the Marine Environmental Protection Committee (MEPC) and in the Pollution Prevention Response (PPR) subcommittee, for example regarding the work on *consistent implementation of regulation 14.1.3 of MARPOL Annex VI*, which was an ongoing work within MEPC and PPR leading up the entry into force of the 0.5% sulphur limit.

That work, inter alia, covered adoption of the carriage ban, making a uniform Fuel Oil Non-availability Report (FONAR) template, adopting a non-mandatory Ship Implementation Plan, designation of sampling points, applying a confidence limit on fuel samples, and reviewing relevant guidelines on PSC, fuel sampling and scrubbers. These matters, which primarily relate to *detection* of violations of the sulphur regulations, are discussed in chapter 1.

The joint EU approach towards many of these subjects was often coordinated during one or more meetings in the European Commission in Brussels leading up to MEPC meetings. A final coordinating meeting was often also carried out *sur place* on the first day of these meetings.

Article 220(1) TFEU, and the shared competences, also allows the EU to spearhead coordinated submissions to MEPC and PPR, etc. The content of these submissions was often coordinated and developed within the European Sustainable Shipping Forum (ESSF)[41] and its sub-working groups.[42]

As the EU can coordinate closely between its Member States in such matters, and as the EU can participate directly in IMO negotiations, the Union can afterwards more easily adopt its regional legislation on the same subjects. These regional rules often implement the IMO regulations to a certain degree, thereby giving these a binding effect in all EU Member States.

This IMO-inspired EU work has, besides resulting in the Sulphur Directive, also resulted in the EU's adopting its Regulation on ship recycling,[43] pertaining to

[41] The ESSF group was established by the European Commission by Commission Decision of 24 September 2013 on setting-up the group of experts on maritime transport sustainability – The European Sustainable Shipping Forum (ESSF), Decision C2013 5984 final.

[42] The members of the EESF sub-working group on sulphur implementation were instrumental in securing acceptance of the carriage ban during the plenary discussions at MEPC 72 in 2018.

[43] Regulation (EU) No 1257/2013 of the European Parliament and of the Council of 20 November 2013 on ship recycling and amending Regulation (EC) No 1013/2006 and Directive 2009/16/EC [2013] OJ L330/1.

the Hong Kong Convention,[44] and its Regulation on MRV (Monitoring, Reporting, Verification),[45] relating to CO_2 reduction, which also is the objective of chapter 4 of MARPOL Annex VI.

IV. The EU's Position on UNCLOS

When becoming a member of the EU and entering into the TFEU and TEU, besides surrendering exclusive or shared legislative rights within certain regulatory fields, EU Member States also consent to the Union itself entering into international agreements, treaties and conventions with non-EU Member States or international organisations on behalf of the EU Member States, under article 216 TFEU.

The EU's mandate for entering into such international agreements is limited to regulatory areas in which it has full or shared competence. An example is the EU's[46] signing of UNCLOS on 7 December 1984 and its ratification[47] of the Convention on 1 April 1998.[48]

Upon signing the Convention, and in alignment with article 310[49] of UNCLOS, the EU formulated one of the most comprehensive declarations, particularly regarding part VII of UNCLOS, on the conservation and management of the living resources in the high seas, and section 2 of part XI pertaining to the exploitation of 'the Area', that is, the seabed below the high seas.[50] The EU also stressed in its declaration the extent of its exclusive and shared competences within the regulatory fields relating to UNCLOS. It affirmed that it had full and exclusive competence within the regulatory field of conservation of marine biological resources under the common fisheries policy, in accordance with article 3(1)(d) TFEU.[51]

[44] The Hong Kong International Convention for the Safe and Environmentally Sound Recycling of Ships (adopted 15 May 2009, not yet entered into force).

[45] Regulation (EU) 2015/757 of the European Parliament and of the Council of 29 April 2015 on the monitoring, reporting and verification of carbon dioxide emissions from maritime transport, and amending Directive 2009/16/EC [2015] OJ L123/55.

[46] In 1984 the EU was referred to as the 'EC' (European Community).

[47] The EU was the 124st party to ratify UNCLOS.

[48] M Reuß and J Pichon, 'The European Union Exercise of Jurisdiction Over Classification Societies – An International Law Perspective on the Amendment of the EC Directive on Common Rules and Standards for Ship Inspection and Survey Organisation and for the Relevant Activities of Maritime Administrations' (2007) 67 *Zeitschrift für ausländisches öffentliches Recht und Völkerrecht* 119, 126, available at https://www.zaoerv.de/67_2007/67_2007_1_a_119_144.pdf.

[49] Art 310 of UNCLOS allows for declarations and statements to be made upon signing and ratifying the Convention.

[50] See http://www.un.org/Depts/los/convention_agreements/convention_declarations.htm.

[51] This part of the declaration reads: 'The Community points out that its Member States have transferred competence to it with regard to the conservation and management of sea fishing resources. Hence in this field it is for the Community to adopt the relevant rules and regulations (which are enforced by the Member States) and, within its competence, to enter into external undertakings with third States or competent international organizations. This competence applies to waters under national fisheries jurisdiction and to the high seas.'

This exclusive competence, and the declaration thereof, would, inter alia, allow the EU to fully coordinate how EU Member States should approach the ongoing work in the UN regarding Biodiversity Beyond National Jurisdiction (BBNJ), as described in chapter 3.

The EU also specified in its declaration that it had shared competence within the regulatory fields of maritime transport, safety, shipping and environmental protection, for instance pertaining to part XII of UNCLOS, stating that:

> With regard to the provisions on maritime transport, safety of shipping and the prevention of marine pollution contained inter alia in Parts II, III, V, VII and XII of the Convention, the Community has exclusive competence only to the extent that such provisions of the Convention or legal instruments adopted in implementation thereof affect common rules established by the Community. When Community rules exist but are not affected, in particular in cases of Community provisions establishing only minimum standards, the Member States have competence, without prejudice to the competence of the Community to act in this field. Otherwise competence rests with the Members States.[52]

V. EU Enforcement of IMO Regulations in Accordance with UNCLOS

The EU has, unlike the IMO, a wider discretionary right to determine how Member States should penalise violations of its regulations. This is exemplified by article 18 of the Sulphur Directive, which provides more detailed and specified guidance on how a violation should be sanctioned compared to article 4(4) of the MARPOL Convention and article 217(1) of UNCLOS.

The EU has furthermore adopted legislation exclusively pertaining to the enforcement and sanctioning of specific regulations, including of international regulations. An example is Directive 2005/35/EC (as amended by Directive 2009/123/EC[53]) on ship-source pollution and on the introduction of penalties for infringements. The of purpose this Directive was to ensure uniform implementation and enforcement of MARPOL Annexes I and II in EU Member States.[54] Directive 2009/123/EC specifically defines its purpose, and that of Directive 2005/35/EC, in paragraph (1) of the Preamble, as being to 'approximate the definition of ship-source pollution offences committed by natural or legal persons, the scope of their liability and

[52] See at http://www.un.org/Depts/los/convention_agreements/convention_declarations.htm.

[53] Directive 2009/123/EC of the European Parliament and of the Council of 21 October 2009 amending Directive 2005/35/EC on ship-source pollution and on the introduction of penalties for infringements [2009] OJ L280/52.

[54] Directive 2005/35/EC (as amended by Directive 2009/123/EC) of the European Parliament and of the Council of 7 September 2005 on ship-source pollution and on the introduction of penalties for infringements [2005] OJ L255/11, art 3(1); see also art 2(1)–(3).

the criminal nature of penalties that can be imposed for such criminal offences by natural persons'.

Such regional regulations on penalties and sanctions must nonetheless be enforced by the EU Member States in accordance with the jurisdictional frame of UNCLOS, particularly the provisions of part XII. Following its ratification of the Convention in 1998, the EU also has an explicit obligation to ensure that its regional legislation is in alignment with UNCLOS. This is reflected in article 9 of Directive 2005/35, which refers to the need for sanctions to comply with international law, including the safeguarding provisions in part XII of UNCLOS. Also, paragraph (12) of the Preamble refers to article 220 of UNCLOS on coastal State enforcement.

Directive 2005/35 has, despite these references to ensuring enforcement in alignment with UNCLOS, raised questions regarding its compliance with international law, which were brought before the ECJ in the *Intertanko case*[55] in 2008. It has also, after its amendment by Directive 2009/123/EC, formed the legal basis of another ECJ decision – the *Bosphorus Queen* (2018)[56] – where the ECJ interpreted this EU legislation in accordance with UNCLOS. These cases, particularly pertaining to the ECJ's assessment of the legal implications of the EU's ratification of UNCLOS, are now discussed in further detail.

A. The *Intertanko Case*

A dispute between the British Ministry of Transport and different shipping associations, including the International Association of Independent Tanker Owners (Intertanko), was submitted to the ECJ in 2006 by the High Court of Justice of England and Wales. The case included several questions regarding Directive 2005/35, most of which related to whether articles 4 and 5 of the Directive – which outlined the extent of criminal responsibility for infringements of MARPOL Annexes I and II – were invalid compared to the actual regulations in MARPOL Annexes I and II.[57]

The ECJ refrained from taking a position on these questions, as the EU was not an independent party to the MARPOL Convention or its Annexes,[58] and the relevant provisions of MARPOL Annexes I and II did not represent international customary law.[59] Thus the Court found that it was not 'incumbent upon the Court to review the directive's legality in the light of the Convention'.[60] Yet one question also challenged the compatibility of article 4 of the Directive with UNCLOS.

[55] Case C-308/06, n 15.
[56] Case C-15/17, n 16.
[57] It was specifically the compatibility of reg 11(b) of Annex I and reg 6(b) of Annex II with arts 4 and 5 that was in question.
[58] *Intertanko case*, n 15, paras 47 and 49.
[59] ibid, para 51.
[60] ibid, para 50.

Article 4 required EU Member States to ensure that violations of Annex I or Annex II within their territory, including in straits under their jurisdiction, were to be considered as criminal offences by individuals on board if they were committed with intent, recklessness or serious negligence. The question to which this gave rise was whether the provision required EU Member States to violate their obligations as coastal States (i) to allow foreign ships a right of *innocent passage* through their territorial seas pursuant to articles 17–19 of UNCLOS and (ii) to allow a right of *transit passage* through an international strait located within their territory pursuant to article 38 of UNCLOS.

The ECJ noted that the EU was party to UNCLOS,[61] so it found that the Court *was* able to determine the compatibility of the Directive with that Convention.

The ECJ stated that 'UNCLOS seeks to strike a fair balance between the interests of States as coastal States and the interests of States as flag States, which may conflict'.[62] It also found that:

> It is true that the wording of certain provisions of UNCLOS, such as Articles 17, 110(3) and 111(8), appears to attach rights to ships. It does not, however, follow that those rights are thereby conferred on the individuals linked to those ships, such as their owners, because a ship's international legal status is dependent on the flag State and not on the fact that it belongs to certain natural or legal persons.[63]

The ECJ therefore concluded that:

> In those circumstances, it must be found that UNCLOS does not establish rules intended to apply directly and immediately to individuals and to confer upon them rights or freedoms capable of being relied upon against States, irrespective of the attitude of the ship's flag State.
>
> It follows that the nature and the broad logic of UNCLOS prevent the Court from being able to assess the validity of a Community measure in the light of that Convention.[64]

This view of the ECJ was in contrast with the Advisory Opinion offered by Advocate General Kokott,[65] who did not find that individuals were barred from relying on the rights established under UNCLOS.[66] Nevertheless, the Advisory Opinion did conclude that article 4 of Directive 2005/35 did not violate the right to innocent passage, stating that:

> Restrictions on the sovereignty of coastal States could also affect the enforcement of penalties on vessels at sea. Article 4 of Directive 2005/35, however, does not go that far. In particular, passage is not prohibited and no particular enforcement measures against vessels en route are required. Rather, the Member States are merely to prohibit certain

[61] ibid, para 53.
[62] ibid, para 58.
[63] ibid, para 61.
[64] ibid, paras 64 and 65.
[65] Case C-308/06, n 15, Advisory Opinion by Advocate General Kokott, delivered on 20 November 2007.
[66] Para 66 of the Advisory Opinion.

> acts which are not necessary for passage. In this respect the relatively low standard of care laid down in Marpol 73/78 is raised only slightly. That must be possible as a rule on prevention and reduction of environmental pollution within the meaning of Article 21(1)(f) of the Convention on the Law of the Sea.[67]

One of the conclusions of the *Intertanko case* is that, since the EU is party to UNCLOS, it allows the ECJ to assess whether regional EU legislation, such as Directive 2005/35, is in violation of the principles set out in UNCLOS. And perhaps also where individuals invoke the rights and duties laid down in UNCLOS. At least according to the Advisory Opinion of Advocate General Kokott.

B. The *Bosphorus Queen*

In 2018 the ECJ was presented with another case – the *Bosphorus Queen*[68] – pertaining to the implications of UNCLOS when enforcing EU regulations, again relating to Directive 2005/35/EC, now as amended by Directive 2009/123/EC.[69]

The case concerned the *Bosphorus Queen*, a ship that sailed under the Panamanian flag, which on 11 July 2011 discharged approximately 900 litres of oil into the sea within Finland's EEZ. This was a violation of MARPOL Annex I, and the Finnish Government was required to take action against the violation pursuant to Directive 2005/35/EC, as amended by Directive 2009/123/EC. The ship subsequently did not call into a Finnish port but continued to Saint Petersburg in Russia.

During the return voyage on 23 July 2011, when the ship again made passage through the Finnish EEZ, the Finnish authorities stopped the ship and charged the vessel for its infringement of Annex I. Financial security of €17,112 was demanded, upon payment of which the ship was released two days later.

It was claimed before the Finish court, which referred the claim to the ECJ, that Finland had exceeded its jurisdiction as a coastal State pursuant to article 220(3) and (5)–(6) of UNCLOS, when it stopped and prosecuted the *Bosphorus Queen*, which was a foreign ship engaged in a direct passage of the Finnish EEZ.

It should initially be noted that article 220 is discussed in chapter 9, particularly in connection with enforcement of MARPOL Annex VI. It is, however, underlined that a coastal State has almost unlimited jurisdiction to investigate and prosecute foreign ships for all environmental violations occurring within its waters, including in the EEZ, provided the ship afterwards voluntarily calls into a port in the State, in accordance with article 220(1). Conversely, a coastal State's

[67] Para 121 of the Advisory Opinion.

[68] Case C-15/17, n 16.

[69] The Framework Decision (2005/667/JHA) – which is directly referred to in art 4 of Directive 2005/35 – was annulled by the ECJ on 23 October 2007 in Case C-440/05 *Commission v Council (Ship-Source Pollution)* [2007] ECR I-09097. The legal vacuum in which the annulment resulted led to adoption of amending Directive 2009/123/EC.

right to stop and prosecute a foreign ship in direct transit through the EEZ is very limited under article 220(3) and (5)–(6), as it requires that discharge from the ship is 'causing major damage or threat of major damage to the coastline or related interests of the coastal State, or to any resources of its territorial sea or exclusive economic zone'.

The shipowner claimed that these conditions were not met, so that the prosecution of the *Bosphorus Queen* had been unlawful according to article 220(3) and (5)–(6) of UNCLOS, even though it may have been lawful in accordance with the regional regulations in Directive 2005/35/EC, as amended by Directive 2009/123/EC.[70]

First, it should be noted that the ECJ again placed emphasis on the fact the EU is an independent party to UNCLOS[71] but not to MARPOL or its Annex I.[72] The Court also explicitly referred to the previous ECJ statement on this in paragraphs 42, 47, 52 and 53 of the *Intertanko case*. The ECJ therefore found itself competent to address and determine whether Finland had infringed the principles in article 220(3) and (5)–(6) of UNCLOS while acting against the breach of MARPOL Annex I. This also allowed the ECJ to interpret the applicability and scope of article 220(3) and (5)–(6) of UNCLOS when answering many of the questions put before the Court.

One of the questions in the case was whether the specific geographical and ecological characteristics and sensitivity of the Baltic Sea, which, as described in chapter 3, had been designated a Particularly Sensitive Sea Area (PSSA) since 2006,[73] could have an effect on assessing whether the conditions in article 220(6) relating to the *threat of major damage* were met. The ECJ found that it could have an impact on the overall assessment, but a violation in a sensitive area (PSSA) would not automatically constitute a violation that would allow for the stopping and prosecution of a vessel in accordance with article 220(6) of UNCLOS and article 7(2) of the Directive.[74] An individual assessment of the pollution impact (or threat thereof) should always be made.

This must be interpreted as a demand for a causal link between the violation and the damage sustained (or the threat thereof), which is supported by the principles expressed by Advocate General Wahl's Advisory Opinion in this case.[75]

The ECJ also concluded that it is unnecessary to determine whether 'significant pollution' (or the threat thereof) has occurred according to article 220(5), if it has

[70] The claim against the Finnish authorities was presented before the ECJ as 10 separate questions relating, inter alia, to the different conditions of art 220(6) that must be fulfilled before a coastal State can invoke this provision. See *Bosphorus Queen*, n 16, para 42. The following will only focus on some of the aspects these 10 questions addressed.

[71] ibid, para 44.

[72] ibid, para 45.

[73] See IMO Resolution MEPC. 136 (53).

[74] *Bosphorus Queen*, n 16, para 108.

[75] ibid, Advisory Opinion of Advocate General Wahl, delivered on 28 February 2018, paras 106–108.

been established that 'major damage' (or the threat thereof) has taken place pursuant to article 220(6).[76]

The ECJ took it upon itself to determine the meaning of the term 'related interests' in article 220(6) by interpreting it in accordance with article 31 VCLT.[77] The Court also emphasised that the term 'any resources', also used in article 220(6), meant that the word 'resources' should be interpreted in the broadest possible sense.[78] These deductions are of importance when interpreting article 218, discussed in chapter 10.

Finally, it should be noted that although the Court referenced the approach applied in the *Intertanko case*, in the *Bosphorus Queen*, a decade later, the ECJ does seem to be more open to interpreting the wording and scope of the provisions of UNCLOS, such as article 220. This is an approach more aligned with the Advisory Opinion offered by Advocate General Kokott in the *Intertanko case*.

It should further be noted that the Advisory Opinion of Advocate General Wahl in the *Bosphorus Queen* also refers to the principles set out in the *Intertanko* and *Manzi cases* pertaining to determining whether or not an IMO regulation represents a customary principle of international law and therefore could be binding upon the EU and the ECJ.[79]

VI. Conclusions

The EU Member States have, by their membership of the EU and their accession to the TEU and TFEU, accepted that the EU has shared competence within the regulatory fields of transport, including shipping and the environment, according to article 4(2)(e) and (g) TFEU. They also accept, in accordance with articles 220 and 216 TFEU, that the EU can cooperate with specialised UN agencies, such as the IMO, and enter into international agreements, such as UNCLOS.

This shared competence has resulted in the EU's adopting the Sulphur Directive, which implemented many of the regulations and principles found in MARPOL Annex VI relating to the regulation of sulphur emissions, including the global and SECA sulphur limits in regulation 14.

Cooperation with the IMO, and the fact that the Sulphur Directive has implemented much of Annex VI, has led to the EU's – through the EU Commission and the ESSF – participating actively in the ongoing MEPC and PPR work on consistent implementation of regulation 14.1.3 of MARPOL Annex VI.

The Sulphur Directive does go beyond Annex VI in certain respects, with provisions relating to ships at berth for over two hours, and setting an implicit

[76] *Bosphorus Queen*, n 16, para 92.
[77] ibid, para 67.
[78] ibid, para 81.
[79] ibid, Advisory Opinion of Advocate General Wahl, delivered on 28 February 2018, paras 46–47.

maximum sulphur limit for fuel used on board vessels with open loop scrubber systems and a 1.5% sulphur limit applying to passenger ships in regular service between EU ports.

The discrepancies between the Directive and MARPOL Annex VI regarding passenger ships formed the basis of the *Manzi case*, where the ECJ, inter alia, found that the EU had not violated the principles of *pacta sunt servanda* and of cooperation in good faith by establishing the 1.5% limit, as the EU was not an independent contracting party to Annex VI and the regulations of the Annex did not express customary law.

The facts that the EU is not party to MARPOL Annexes I and II but is a party to UNCLOS were also key elements in the rulings of the ECJ in the *Intertanko case* and the *Bosphorus Queen*, as the Court refused to determine whether Directive 2005/35/EC, as amended by Directive 2009/123/EC, was compatible with MARPOL Annexes I and II, but was willing to scrutinise whether the Directive was compatible with the jurisdictional frame of UNCLOS.

In particular, the ECJ's clear and probing examination and interpretation of UNCLOS article 220(3) and (5)–(6) in the *Bosphorus Queen* leave one with the impression that the Court has become more inclined to assess the extent of the jurisdictions under UNCLOS than it was 10 years earlier in the *Intertanko case*.

Conversely, following the principles of the *Intertanko* and *Manzi cases* – which were reiterated in the Advisory Opinion of Advocate General Wahl in the *Bosphorus Queen* as still representing applicable law – the ECJ refrains from interpreting EU legislation, such as the Sulphur Directive and Directive 2005/35, in accordance with international regulations, such as MARPOL Annexes I, II and VI, if the EU is not an independent party to these regulations. The Court has explained this position by referring to ECJ rulings' having a binding effect on all EU Member States, including those not party to the international rules in question.

In conclusion, the ECJ places great emphasis on whether or not the EU is party to an international convention (that does not represent international customary law) when establishing which international legislative boundaries the EU must respect when adopting its own regional regulations or directives.

This would mean, for example, that the ECJ would be able to determine whether EU Member States can enforce the 0.5% sulphur limit, also implemented in the Sulphur Directive, in accordance with article 218 of UNCLOS, as the EU is party to that Convention.[80]

[80] The ECJ would presumably not see itself as competent to test the compatibility of the Regulation on ship recycling (n 43) with the underlying international IMO Convention on Ship Recycling (the Hong Kong International Convention, n 44), as the EU is not party to this Convention, nor does this IMO Convention represent international customary law.

A. Future EU Legal Measures on GHG and the Role of the ECJ

These conclusions would also apply if the EU in the future were to adopt legislative measures to reduce greenhouse gas (GHG) emissions from ships, if these measures basically were implementations of IMO measures. The ECJ would therefore not, at the outset, be able to test the compatibility of the EU's GHG legislation with the IMO's regulations unless the EU was an independent party to the latter. The ECJ would be able to establish if any EU legislation on GHG complied with the jurisdictions under UNCLOS.

However, it should be recalled that Part IV of this book examines whether it is conceivable that future IMO regulations on GHG could be seen as protecting peremptory (jus cogens) norms. If this is conceivable, it would enable the ECJ to ensure that any EU-adopted GHG measures are in accordance with the underlying IMO regulations, regardless of whether the EU is party to these or not, as these IMO regulations would represent *international customary law.*

6

Basic Jurisdictional Principles of International Law

Before the specific regulations of part XII of UNCLOS[1] for the protection and preservation of the marine environment are analysed in the next part of this book (Part II), some basic principles of international law need clarifying, regarding how States can legislate and enforce. Some of these jurisdictional principles will be discussed in section I, including a brief introduction to the *Lotus case*,[2] as several cross-references will be made to the decision and its findings throughout this chapter and others.

This will be followed by a closer examination of the principles for prescribing, enforcing and adjudicating legislation outside a State's territory (section II), and how this relates to the law of the sea and the provisions of part XII of UNCLOS (section III), especially relating to the flag State principle and how other States can exercise extraterritorial jurisdiction over violations committed by foreign vessels on the high seas. In section IV some basic principles of customary international law are examined as to whether these can provide a legal basis for exercising extraterritorial jurisdiction.

I. Basic Principles of Jurisdiction

Jurisdiction is a legal term used to describe a State's right to exercise its legal authority over events constituting a violation of its laws and legislation. Several basic forms of jurisdiction govern to what extent a State may prescribe and – without any overlapping claims of jurisdiction from other States – enforce its legislation, including maritime legislation.

Different forms of jurisdiction can be invoked by a State depending on the different circumstances in play, for example where the infringement took place, where it had – or threatened to have – an effect, and the nationality of those who committed the violation or of any victims. Also, some violations are by themselves

[1] United Nations Convention on the Law of the Sea (adopted 10 December 1982, entered into force 16 November 1994) 1833 UNTS 3.

[2] *SS Lotus (France v Turkey)* PCIJ Rep Ser A No 10.

deemed to be of such a grave and unacceptable nature that they inevitably give all States jurisdiction over them.

An attempt was made in 1935 to unify and codify these basic jurisdictional principles recognised within international law.[3] This resulted in the listing of the following principles:

- *Territorial jurisdiction* is the most fundamental and well-recognised jurisdictional principle, as it embodies the principal right in international law for all States to legislate and enforce without restriction within their own territory. This is based on a State's having exclusive sovereignty within its own territory, where no other State can, without consent, exercise its jurisdiction. It should be recalled that article 2(1) and (2) of UNCLOS stipulate that the sovereignty – and thereby the (territorial) jurisdiction – of a coastal State extends from its internal waters to the territorial sea, which includes the air space above and the subsoil beneath that sea.[4]
- *Nationality jurisdiction*, also referred to as the 'active personality principle', links a State's jurisdiction to a person's nationality, requiring nationals to comply with the laws of their country, even when abroad. This jurisdiction allows a State to exercise jurisdiction over its nationals who commit an offence outside its territory.
- The *passive personality principle* also allows a State to assert jurisdiction over crimes committed outside its territory based on the nationality of a person. But this jurisdiction pertains to the nationality of the victim. This form of jurisdiction is not as well-recognised as the other jurisdictional principles, and its scope will often overlap with these.
- *Protective jurisdiction* allows a State to establish and enforce legislation that protects various interests of a State, including interests of national security, etc. The US has in some cases relied on this jurisdictional principle to justify enforcement of anti-terror legislation.
- *Universal jurisdiction* is a special form of jurisdiction that confers a right for a State to claim jurisdiction over certain violations due to their grave nature, irrespective of their occurring outside the State's territory and not having any effect on, nor links to, nationals of the State.[5] The acts that are covered by the

[3] See the Harvard Draft Convention on Jurisdiction with Respect to Crime (1935) 29 *American Journal of International Law* 439.

[4] M Reuß and J Pichon, 'The European Union Exercise of Jurisdiction Over Classification Societies – An International Law Perspective on the Amendment of the EC Directive on Common Rules and Standards for Ship Inspection and Survey Organisation and for the Relevant Activities of Maritime Administrations' (2007) 67 *Zeitschrift für ausländisches öffentliches Recht und Völkerrecht* 119, 132, available at https://www.zaoerv.de/67_2007/67_2007_1_a_119_144.pdf. See also D Bodansky, 'Protecting the Marine Environment from Vessel Source Pollution: UNCLOS III and Beyond' (1991) 18(4) *Ecology Law Quarterly* 737.

[5] Section IV of this chapter discusses the application of the universal jurisdictional principle regarding the principles of jus cogens and erga omnes.

scope of universal jurisdiction are those deemed as being the most damaging to and heinous – and therefore unacceptable – by the international community as a whole. These acts include, but are not limited to, crimes against humanity, genocide, torture, war crimes and others.

Some of these forms of jurisdiction that allow States to exercise jurisdiction outside their territory, for example for controlling or protecting nationals abroad, are often described as being expressions of *extraterritorial jurisdiction*.

The aim of this book is to establish whether a non-flag State, under certain conditions, can assert jurisdiction over certain violations taking place outside the State's jurisdiction, for instance on the high seas. This is therefore a question of whether a non-flag State can exercise extraterritorial jurisdiction over foreign ships. The right for States to exercise extraterritorial jurisdiction over vessels for violations occurring on the high seas formed the base of the *Lotus case* in 1927.

A. The *Lotus Case* – Establishing Principles that are Still Applicable Today

The *Lotus case* concerned the collision of the French steamer SS *Lotus* with the Turkish steamship *Boz-Kourt* on 2 August 1926. The collision took place on the high seas, and eight people on board the Turkish ship lost their lives.

The SS *Lotus* called at a Turkish port after the incident, carrying 10 survivors from the *Boz-Kourt*. The Turkish authorities charged the master of the ship (captain) and the officer of the watch on duty at the time of the accident with involuntary manslaughter, in accordance with the Turkish Penal Code. The national Turkish Penal Code allowed Turkey to assume extraterritorial jurisdiction over offences committed against Turkish citizens, irrespective of where this took place, including on the high seas. A Turkish court sentenced the French master – by virtue of its extraterritorial legislation – to 80 days' imprisonment and also imposed a fine.

France protested, as it claimed that such enforcement violated international law, and brought the case before the Permanent International Court of Justice (PCIJ).[6] France supported its claim by referring to the flag State principle, which gave flag States (here, France) exclusive jurisdiction over their ships and crew on the high seas.

The PCIJ recognised the right of France to exercise jurisdiction over the ship under its flag, but it also found that Turkey had not violated international law by exercising its extraterritorial jurisdiction on the high seas as the case concerned a

[6] The PCIJ was established by the League of Nations. After the establishment of the United Nations (UN) it was replaced by the International Court of Justice (ICJ). Many of the legal precedents set by PCIJ case law still apply today, eg the *Lotus cas*, n 2, and *SS Wimbledon (UK v Japan)* 1923 PCIJ Rep Ser A No 1.

collision between two ships that affected a Turkish interest, meaning the Turkish flagged vessel (the *Boz-Kourt*).[7] The Court thereby on the one hand cemented the flag State principle by recognising the rights of both France and Turkey as flag States over their respective vessels, but on the other hand it also recognised that the flag State principle is not exclusive when the jurisdictions of a flag State and another State overlap on the high seas.[8]

The PCIJ viewed the jurisdiction of a flag State as being closely linked to the principle of territorial jurisdiction, by basically perceiving the ship as being a part of the flag State's territory whilst it sailed on the high seas. Although this view, that flag State jurisdiction represents a territorial jurisdiction, might not be fully accepted by all, the principles of the case relating to how States can exercise jurisdiction are regarded as principles representing accepted international law.[9]

Two principles in particular have been derived from the case, pertaining to the possibilities for States to exercise jurisdiction and limitations on that exercise, which are still considered relevant and applicable today. These are often referred to as 'the *Lotus* principles'. The first *Lotus* principle established that a State cannot exercise its jurisdiction outside its territory unless an international treaty or customary law permits this. The PCIJ expressed this as follows:

> Now the first and foremost restriction imposed by international law upon a State is that – failing the existence of a permissive rule to the contrary – it may not exercise its power in any form in the territory of another State. In this sense jurisdiction is certainly territorial; it cannot be exercised by a State outside its territory except by virtue of a permissive rule derived from international custom or from a convention.[10]

The second *Lotus* principle, conversely, established that a State may exercise its jurisdiction unrestricted within its own territory, provided no rule under an international treaty or customary law prevents this.[11]

II. Extraterritorial Jurisdiction for Prescribing, Enforcing and Adjudicating

The previously described jurisdictional principles and those set out in the *Lotus case* show the differing jurisdictions of a State, on a theoretical level, to

[7] This principle – that a collision at sea can result in other States asserting jurisdiction because the effect of the collision has an impact on them (be it their ships or their coastline) – is to an extent represented in art 221 of UNCLOS, art 11 of the Convention on the High Seas (Convention on the High Seas (adopted 29 April 1958, entered into force 30 September 1962) 450 UNTS 11) and in the Intervention Convention (Convention Relating to Intervention on the High Seas in Cases of Oil Pollution Casualties (adopted 29 November 1969, entered into force 6 May 1975) 970 UNTS 211).

[8] *Lotus case*, n 2, paras 71–84.

[9] See A Henriksen, *International Law* (Oxford University Press, 2017) 96.

[10] *Lotus case*, n 2, para 45.

[11] ibid, paras 46–47.

legislate without restriction inside and outside its territory over all matters concerning the State, curbed, on a practical level, by limitations on enforcing such legislation outside its territory. These differences between legislating and enforcing are often described as the right of States to *prescribe, enforce* and *adjudicate on* legislation.

It is the right of every State to *prescribe* legislation that applies and governs in territory under the State's sovereignty, as established by the second *Lotus* principle (see section I). This right includes the adoption of national laws and the implementation of international rules.

A State can, in principle, also prescribe legislation applying outside its territory, for instance protecting national interests abroad. This right is curtailed by the right of States – or the lack of it – to *enforce* and *adjudicate on* these rules. For example, the right to arrest, detain and prosecute a ship can be seen as the right to *enforce*, whereas the right to impose and execute a sanction (fine) relates to the right to *adjudicate*.

A State must have an *extraterritorial jurisdictional basis* for enforcing and adjudicating on a State's prescribed rules outside of territory under its sovereignty. According to the first *Lotus* principle (see section I), the exercise of such extraterritorial jurisdiction must be based on an international accepted convention or on recognised principles of international customary law. UNCLOS represents such an international accepted convention that, in certain respects, bestows extraterritorial rights on its participating States. This is discussed further in section III of this chapter.

Principles of customary international law can also provide States with an extraterritorial jurisdictional basis for enforcing international (International Maritime Organization (IMO)) regulations outside of their territory. This is examined in section IV.

III. UNCLOS Provides a Legal Basis for Extraterritorial Jurisdiction

UNCLOS constitutes an international convention that, pursuant to the second *Lotus* principle, requires all participating States to respect the limitations of the Convention when exercising their otherwise unrestricted right of enforcement within their territory. For instance, article 2(3) of UNCLOS, which refers to the exceptions of the Convention and international law for limiting a coastal State's territorial jurisdiction in the territorial sea. These exceptions include the right to *innocent passage* (articles 17–19) and the right to make a *transit passage* of a strait used for international navigation (article 38).[12] See further chapter 3.

[12] See Bodansky, n 4, 745–48.

The first *Lotus* principle is also embodied in UNCLOS, as the Convention bestows on participating States limited and clearly defined extraterritorial rights to enforce and adjudicate on prescribed legislation. One example of an UNCLOS provision that provides an extraterritorial basis for exercising jurisdiction is the aforementioned codification of the flag State principle in article 92, which allows a flag State to exercise its prescribed legislation over vessels flying its flag on the high seas outside the territory of the flag State.

Article 217 also codifies the flag State principle and is, compared to article 92, *lex specialis* concerning the flag State's right and obligation to enforce regulations for the protection of the marine environment. Article 217 thereby also confers an extraterritorial legal basis for flag States to enforce environmental protections, for example in MARPOL Annex VI, irrespective of where they occur, including outside the flag State's territory. And just as article 92 refers to exceptions to this flag State principle, so does article 217(4), which refers to the flag State's jurisdiction being without prejudice to articles 218, 220 and 228.

These exceptions (articles 218, 220 and 228) provide non-flag States, such as port and coastal States, with extraterritorial jurisdiction for enforcing regulations for the protection of the marine environment outside of their own territory, thereby also expressing the first *Lotus* principle. Article 218(1) provides a port State with an extraterritorial jurisdiction to penalise discharge violations taking place outside its territory, for example on the high seas.

Article 220 provides a coastal State with a legal basis for penalising violations that take place in waters under its full or partial jurisdiction, including in the Exclusive Economic Zone (EEZ). As the EEZ is not an area where the coastal State enjoys full sovereignty and unlimited territorial jurisdiction, any jurisdiction conferred to the coastal State for regulating violations in the EEZ could be seen as conferring extraterritorial jurisdiction. Article 220 grants such extraterritorial jurisdiction to coastal States for enforcement in the EEZ.

These extraterritorial jurisdictions of port and coastal States according to articles 218 and 220 consequently overlap with the extraterritorial jurisdiction of the flag State pursuant to article 217 to regulate such (discharge) violations on the high seas and violations that take place in another State's EEZ.

Article 228(1) determines whether the coastal/port State or the flag State can assert primary jurisdiction over these violations and bring the legal proceedings and impose sanctions.[13] In the view of this author, it therefore provides the coastal, port or flag State with the final extraterritorial jurisdiction to adjucate, in accordance with the first *Lotus* principle.

Article 217 is analysed further in chapter 8, article 218 in chapter 10, article 220 in chapter 9 and article 228 in chapter 11, as these provisions form the cornerstones of Part II of this book.

[13] The overlapping extraterritorial jurisdiction between coastal and port States pursuant to art 218(1) and art 220, is resolved through art 218(2)–(4), discussed in ch 10.

A. The Perceived Lack of Extraterritorial Jurisdiction Bars Effective Enforcement of Annex VI

As noted in chapter 1, many non-flag States have penalised violations of the sulphur limits in Annex VI, predominantly violations of the 0.1% SECA limit, with fines ranging on average from $10,000 to $50,000.[14] These Port State Control (PSC) fines often primarily focus on punishing the shipowner (in the same way that non-compliance with, for instance, the safety regulations of the SOLAS Convention[15] is punished) and not on removing (confiscating) illegal savings achieved outside the State's territory.

These fines may nonetheless have been adequate when sanctioning violations of the 0.1% limit that entered into force in 2015, as the 0.1% requirement is geographically limited to the SECA zones coinciding with the 200-nautical-mile EEZs of the affected States, thereby limiting the potential economic savings for shipowners by not complying with this requirement.[16] These States have thus also 'only' penalised violations occurring within areas (internal and territorial waters and in the EEZ) where they have a clear – territorial/extraterritorial – jurisdictional basis for imposing such fines according to article 220(1). However, the potential savings for not complying with the 0.5% sulphur limit will skyrocket, as shown in chapter 1, with a calculated $750,000 saving for one violation on a normal commercial route between Asia and Europe. Needless to say, a $50,000 fine is hardly likely to be dissuasive if a ship can save $750,000 from one violation, in particular when one considers that not all violations are necessarily detected.

This means that, in order to be effective, the level of fines must increase to cover all savings achieved, and must also entail a punitive element that will increase in the event of aggravating circumstances, such as repeated violations, to counter any temptation not to comply.[17]

Many States have argued that this would require them to have a jurisdictional extraterritorial right to *enforce* and *adjudicate* the part of the violation of regulation 14.1.3 of MARPOL Annex VI that occurs on the high seas. An extraterritorial right that must be found within UNCLOS.

This author believes that the analysis set out in Part II of this book will show that there is such an extraterritorial right for (port) States, pursuant to article 218,

[14] H Ringbom, 'Enforcement of the Sulphur in Fuel Requirements: Same, Same But Different', available at https://webcache.googleusercontent.com/search?q=cache:EBk8-0aReA0J:https://www.duo.uio.no/bitstream/handle/10852/61600/SO-Artikel-Ringbom.pdf%3Fsequence%3D4%26isAllowed%3Dy+&cd=1&hl=da&ct=clnk&gl=no, 21, fn 74.

[15] International Convention for the Safety of Life at Sea (SOLAS) 1974 (adopted 1 November 1974, entered into force 25 May 1980) 1184, 1185 UNTS 2.

[16] Ringbom, n 14, 28–29.

[17] An in-depth discussion of what constitutes an effective fine for violations of the 0.5% sulphur limit is found in ch 8, inter alia building on the principles set out in art 18 of the EU Sulphur Directive (Directive (EU) 2016/802 of the European Parliament and of the Council of 11 May 2016 relating to a reduction in the sulphur content of certain liquid fuels [2016] OJ L132/58), described in ch 5.

to proceed against such sulphur violations irrespective of where these occur, including on the high seas, thus allowing port States to impose fines that confiscate all savings achieved from the entire trip and also entailing a punitive, dissuasive element.

IV. Customary Law Provides a Legal Basis for Extraterritorial Jurisdiction

Customary international law can, in accordance with the second *Lotus* principle (see section I), limit the right of all States to carry out enforcement within their territory. For example, the right for foreign ships to make *innocent passage* through a territorial sea must be recognised, thereby requiring all (coastal) States to abide by this rule even if they are not party to UNCLOS.

The first *Lotus* principle bars States from exercising extraterritorial jurisdiction outside their territory unless this is explicitly permitted in accordance with a convention or by customary law. Some of the extraterritorial principles of UNCLOS, for example under article 218(1), can be seen as representing customary principles of international law, therefore also allowing States that are not party to UNCLOS to exercise this jurisdiction.[18]

Other principles of customary law, not necessarily relating to law of the sea or UNCLOS, can – in the view of this author – also effect the extraterritorial enforcement of IMO regulations for the protection of the marine environment. This assumption forms the basis of the analysis in Part IV of this book, but some of these principles should, for the sake of clarity, be briefly described here.

A. Customary Principles of Jus Cogens and Erga Omnes

The principles of jus cogens and erga omnes represent principles of recognised international customary law that allow and obligate States to protect certain basic recognised rights and seek to stop and penalise violations of them. Some basic rights are viewed as being rights that have the status of *peremptory norms*, also referred to as 'jus cogens norms'. No derogation is allowed from the protection afforded by these norms. Any agreement or treaty violating a jus cogens norm is void pursuant to article 53 or article 64 of the Vienna Convention on the Law of Treaties (VCLT).[19] These norms, inter alia, aim to protect humans from crimes that are so serious and damaging that they are deemed unacceptable

[18] The reason for assuming that art 218(1) reflects customary law is set out in ch 10.

[19] Vienna Convention on the Law of Treaties 1969 (adopted 22 May 1969, entered into force 27 January 1980) 1155 UNTS 331 (VCLT).

by the international community as a whole. Jus cogens norms therefore include prohibitions on genocide, torture and crimes against humanity. The right of all States to protect against such grave offences is also – in the view of this author – rather closely linked to the erga omnes principle. This principle confirms that any State has an obligation, owed towards the rest of the international community, to ensure the preservation of these basic human rights and prosecute any violation thereof. The erga omnes principle was accepted by the ICJ in the *Barcelona Traction case*.[20]

That such a linkage exists between recognised jus cogens norms and recognised erga omnes obligations has also been expressed by other authors, for example Anders Henriksen, who states that 'In practice, erga omnes obligations will include the norms of a peremptory/jus cogens character.'[21]

Thus, a State's right, by virtue of the principle of erga omnes, to exercise jurisdiction over such violations has ties to the principle of *universal jurisdiction*, as this principle, as described previously in section I, covers violations that often are deemed to be the gravest and most damaging to human life. This allows all States to assert jurisdiction over such violations, despite their failing to have any impact on, or other tie to, the State in question.

One of the common denominators underlying all these principles is that they are subject to constant change and development due to the transformation of the international community and which crimes are deemed to be the most atrocious or damaging. The principles of erga omnes and universal jurisdiction originally encompassed piracy, but they have since evolved to include other, more serious crimes, such as those mentioned previously.[22] This is also reflected in an UN report from July 2018, where it is discussed whether unlawful acts in the form of espionage, human trafficking, etc should also give rise to application of the principle of universal jurisdiction.[23] Article 64 VCLT therefore explicitly refers to the development of new jus cogens norms, making any existing treaty void if it contradicts these new norms.

Part IV of this this book seeks to clarify whether the effects of global warming, leading to continual deterioration of the environment and having adverse effects on human health and living conditions, may result in future IMO regulations on greenhouse gas (GHG) emissions being considered international rules protecting peremptory norms, that is jus cogens norms. And if so, whether this would consequently result in all States having an erga omnes obligation to prevent infringements of these norms by exercising extraterritorial jurisdiction over any violation of them through the assertion of universal jurisdiction.

[20] *Barcelona Traction (Belgium v Spain)* [1970] ICJ Rep 3 (Judgment of 5 February 1970) para 33.
[21] See Henriksen, n 9, 36.
[22] See D Haas, *International Law* (Oxford University Press, 2017) 93.
[23] See United Nations General Assembly, *The Scope and Application of the Principle of Universal Jurisdiction*, Report A/73/123 (July 2018) 10.

V. Conclusion

The fundamental principles of jurisdiction include territorial, nationality, passive personality, protective and universal jurisdiction. They allow a State to *prescribe* legislation that regulates matters within the State's own territory but also outside it, pertaining to nationals abroad or acts that threaten the State or which are of a particularly appalling nature.

States also have the right to *enforce* measures and to *adjudicate* on violations taking place within their territory, given their sovereignty and territorial jurisdiction there. The right to enforce within a State's own territory is only limited by regulations of a convention to which the State is party, or by basic principles of international law, as stated by the second *Lotus* principle.

The almost unrestricted right of States to prescribe legislation applying outside their territory is, however, curtailed by a lack of extraterritorial jurisdiction for States. Only where the regulations of a convention or basic principles of international law confer such extraterritorial rights on a State can it enforce and adjudicate outside its territory, as stated by the first *Lotus* principle.

UNCLOS is a convention containing provisions that include the first and the second *Lotus* principles, the latter as it imposes obligations on participating States to respect certain limitations on their territorial jurisdiction, for example to respect the right of foreign ships to exercise innocent passage or transit passage through a State's territorial sea.

UNCLOS also encompasses the first *Lotus* principle, as several of its provisions in part XII confer extraterritorial jurisdiction on participating States, especially in relation to their status as flag States, articles 92 and 217 codifying the flag State principle and bestowing those States with extraterritorial jurisdiction over ships flying their flag on the high seas too.

However, UNCLOS also bestows extraterritorial jurisdiction on non-flag States, such as coastal and port States, which therefore are codified exceptions to the flag State principle.

Article 217(1) relates to flag States' penalisation of environmental violations irrespective of where these occur, including on the high seas, recalling that any such enforcement measure must be 'effective'.

Exceptions to this basic principle are found in article 217(4), which explicitly refers to articles 218, 220 and 228. These articles grant port States (pursuant to article 218) and coastal States (pursuant to article 220) extraterritorial jurisdiction to punish violations of environmental measures, such as those in MARPOL Annex VI, outside their territorial sea, that is on the high seas and in the EEZ.

Article 228 (specifically article 228(1)) determines the overlapping extraterritorial jurisdictions between a flag State and a port or a coastal State.

These provisions concerning extraterritorial jurisdictions pursuant to the first *Lotus* principle – articles 217, 218, 220 and 228 – are analysed in chapters 8–11 of this book. Part II of the book will prove that non-flag States, especially port States pursuant to article 218, have an extraterritorial jurisdictional basis for enforcing and adjudicating violations of the global sulphur limit pursuant to regulation 14.1.3 of MARPOL Annex VI on the high seas too. This enables port States to impose fines irrespective of where the violation took place. Those fines can confiscate all savings achieved and entail a punitive, dissuasive element.

Finally, the extraterritorial jurisdictional basis of universal jurisdiction has strong ties to the principles of jus cogens and erga omnes. Those principles, and the application of universal jurisdiction, form the basis of Part IV of this book, which examines whether it is possible that future IMO regulations on GHG could attain the status of rules protecting norms of a jus cogens nature.

It is noted that such principles are considered basic principles of customary international law, so that they are encompassed by the first and second *Lotus* principles, as they provide an exterritorial basis for enforcing and adjudicating outside a State's territory, but also provide limitations regarding enforcement and adjudication inside that territory.

PART II

Special Obligations and Jurisdictions under UNCLOS Part XII to Enforce Air Pollution Regulations

7

Protecting the Environment: UNCLOS Part XII

Part XII of UNCLOS[1] contains 46 articles (articles 192–237) in 11 sections. It is primarily the provisions of section 5 (articles 207–212) on 'International rules and national legislation to prevent, reduce and control pollution of the marine environment', section 6 (articles 213–222) on 'Enforcement' and section 7 (articles 223–233) on 'Safeguards' that will be analysed in this part (Part II) of the book.[2]

The main focus in Part II is clarification of the rights and obligations for the different State entities (flag, coastal and port) in implementing and enforcing MARPOL Annex VI[3] through these provisions of part XII of UNCLOS.

This chapter will concentrate on the general obligations for all States to cooperate and to implement International Maritime Organization (IMO) regulations. The right to adopt national legislation in environmental matters will also be clarified, as will the scope of articles 212 and 222, which specifically relate to air pollution.

The increased obligations for flag States to investigate, enforce and inform pursuant to article 217 of UNCLOS, and the special extraterritorial jurisdictions for coastal and port States according to article 220 and article 218,[4] will form the basis of the analysis in chapters 8–10. Chapter 11 studies article 228(1) relating to overlapping jurisdiction between flag States and port or coastal States, while chapter 12 describes the safeguarding provisions of section 7 and the procedural dispute-settlement possibilities of part XV, before chapter 13 offers a conclusion on Part II.

[1] United Nations Convention on the Law of the Sea (adopted 10 December 1982, entered into force 16 November 1994) 1833 UNTS 3.

[2] Some of the provisions of pt XII, eg art 219 and arts 224–227, have already been studied in ch 4 relating to investigation and detention during Part State Control (PSC).

[3] International Convention for the Prevention of Pollution from Ships (adopted on 11 February 1973, as modified by the Protocol of 17 February 1978, entered into force 2 October 1983) 1340 UNTS 61 (MARPOL), Annex VI, IMO Publication: IMO-520E.

[4] A Kanehara, 'Environmental Protection of Ocean and Flag-State Jurisdiction' Rikkyo University, Faculty of Law, paper delivered at the 8th Conference of SCA Qingdao, China, 27–30 May 2008, 4, where it is stated that 'In respect to vessel-source pollution, the relevant provisions of the UNCLOS are mainly Articles 211, 217, 218, 219 and 220. Among them the most important provisions are Articles 218 and 220 from the perspective of the flag state principle and its compromise with other states' concurrent jurisdiction.'

I. Introduction to Part XII of UNCLOS

Many provisions of part XII of UNCLOS refer to the implementation and enforcement of 'international rules and standards ... for the prevention, reduction and control of pollution of the marine environment from vessels'. This book will use the term 'international rules for the protection of the marine environment' as a shorter reference to these regulations.

All such references in the provisions of part XII to international rules for the protection of the marine environment encompass the regulations in MARPOL Annex VI.[5] This is based on article 1(1)(4) of UNCLOS, which defines 'pollution of the marine environment' as manmade pollution that poses a danger to living resources, marine life and human health. Sulphur pollution fits this description. For further details, see chapter 3.

That MARPOL is covered by the provisions of part XII of UNCLOS is supported by the following statement made by the IMO's Legal Committee:

> The power to impose sanctions conferred by IMO regulations on the port State (notably in the MARPOL Convention) should be related to the rights and obligations provided in part XII of UNCLOS.[6]

The references in part XII to international rules protecting the marine environment cover a number of other IMO conventions, such as other annexes of the MARPOL Convention, the London Convention, the Anti-fouling Convention,[7] the BWM Convention,[8] etc.[9] Future IMO legal measures on greenhouse gases (GHGs) will undoubtedly also be embraced by the enforcement principles of part XII.[10] All references in part XII to the 'competent international organization' are direct references to the IMO.[11]

Finally, it is noted that article 236 stipulates that the provisions of part XII do not apply to any State-operated warship or naval vessel (or aircraft) in government service.

[5] Some provisions of pt XII have explicitly narrow scope to cover certain specific forms of violation, eg art 216, referring to 'dumping', which obviously has direct ties to the Convention for the Prevention of Marine Pollution by Dumping of Wastes and Other Matter (adopted 13 November 1972, entered into force 30 August 1975) 1046 UNTS 120 ('the London Convention'), and not MARPOL Annex VI.

[6] IMO Legal Committee, 'Implications of the United Nations Convention on the Law of the Sea for the International Maritime Organization', LEG/MISC. 8 (30 January 2014) 13.

[7] International Convention on the Control of Harmful Anti-fouling Systems on Ships (adopted 5 October 2001, entered into force 17 September 2008) ('the Anti-Fouling Convention' or 'AFS Convention').

[8] The International Convention for the control and management of ship's ballast water and sediments, 2004 (adopted 13 February 2004, entered into force 8 September 2017) ('the BWM Convention').

[9] This is specified by IMO's Legal Committee in 'Implications of the United Nations Convention on the Law of the Sea for the International Maritime Organization', n 6, 56–57.

[10] Y Tanaka, 'Regulation of Greenhouse Gas Emissions from International Shipping and Jurisdiction of States' (2016) 25 *Review of European, Comparative and International Environmental Law* 337.

[11] IMO's Legal Committee, n 9, 56.

II. The Development of Part XII of UNCLOS

As described in chapter 3, UNCLOS, and particularly part XII of the Convention, reflects an evolution of international law, and consequently of the law of the sea, pertaining to protection of the marine environment. The statement made by Chairman Tommy Koh at the adoption of UNCLOS in 1982 may be recalled, in which he underlined that 'the Convention contains important new rules for the protection and preservation of the marine environment from pollution'. This evolution was made necessary following the technological progress of the international community, and its increased need for the transport of passengers and for the shipping of commodities and cargo between States, which led to more and bigger ships sailing the seas, leading to more pollution, including pollution of the atmosphere.

The special regulations under part XII reflect the shift in focus within the international community, compared to the 1958 Conventions (see chapter 3, section I), to address concerns regarding pollution from ships. Similar concerns can be traced back to the Action Plan adopted at the Stockholm Conference in 1972 (see chapter 1, section IV), where several of the Plan's 109 points have clear links to the broadened jurisdictions found in part XII of UNCLOS. For example, the general appeal for cooperation between States and for participating in the development of international rules for the protection of the marine environment, an appeal echoed in section 2 of part XII of UNCLOS, predominantly in article 197. Also, the specific reference in recommendation 86(b) of the Action Plan is to a considerable extent reiterated in article 217 of UNCLOS, concerning the extended obligations of flag States to ensure that ships flying their flags comply with such newly developed international rules for the protection of the marine environment. The precise wording of recommendation 86(b) was:

> *It is recommended* that Governments …:
>
> …
>
> (b) Ensure that the provisions of such instruments are compiled with by ships flying their flags and by ships operating in areas under their jurisdiction and that adequate provisions are made for reviewing the effectiveness of, and revising, existing and proposed international measures for control of marine pollution …

This work to adapt international maritime law towards having an increased focus on ensuring flag State implementation and enforcement of international environmental rules resulted in the development of part XII of UNCLOS and the new legal principles rooted therein, including article 217.[12] Daniel Bodansky elegantly critiques this development, as follows:

> Despite these legal requirements, questions remain about the adequacy of flag state implementation. In part, this can be attributed to the development of flags of

[12] Y Tanaka, *The International Law of the Sea*, 2nd edn (Cambridge University Press, 2015) 276.

convenience, which may not accept international conventions such as MARPOL or be willing or able to enforce these standards adequately. But enforcement of international pollution standards has been limited, even by flag states that do accept international environmental conventions. Reporting to the International Maritime Organization on enforcement measures has been spotty at best. Even when flag states take enforcement action, the penalties imposed have often been insufficient to serve as a deterrent. UNCLOS III attempts to remedy these problems by making the legal obligations of flag states universal.[13]

III. Obligations to Cooperate in Protecting the Marine Environment

The first provision of part XII of UNCLOS, article 192, asserts that all States must protect and preserve the marine environment. Article 194 specifies that States must, individually or jointly, seek to prevent and reduce pollution from all sources affecting the marine environment.

States should also, pursuant to article 197, cooperate on a global and regional basis to develop international rules, standards, procedures and recommended practices for the protection of the marine environment. The reference to global cooperation could be to cooperation within the IMO, whereas cooperation on a regional basis could be within the EU.

Article 197 also calls on States to cooperate directly with one another, which may be done through the adoption of bilateral or multilateral agreements. An example of such direct cooperation is the north European cooperation within the OSPAR Commission,[14] where 15 countries[15] are parties to the OSPAR Convention,[16] which requires the parties to work together to protect the marine environment of the North-East Atlantic.[17] The Preamble to the OSPAR Convention denotes article 197 of UNCLOS as the basis for this cooperation.

[13] D Bodansky, 'Protecting the Marine Environment from Vessel Source Pollution: UNCLOS III and Beyond' (1991) 18(4) *Ecology Law Quarterly* 719, 742–43.

[14] See at https://www.ospar.org/.

[15] The 15 nations cooperating in the OSPAR forum are Denmark, Belgium, Finland, France, Ireland, Iceland, Luxembourg, Netherlands, Norway, Portugal, Spain, Sweden, Switzerland, the United Kingdom (UK) and Germany.

[16] The Convention for the Protection of the Marine Environment of the North-East Atlantic 1992 (adopted 22 September 1992, entered into force 25 March 1998) 2354 UNTS 67 (OSPAR Convention) is a codification of the Oslo Convention of 1972 on dumping (Convention for the prevention of marine pollution by dumping from ships and aircraft (adopted 15 February 1972, entered into force 7 April 1974, 13268 UNTS 5)) and the Paris Convention of 1974 on prevention of marine pollution from land based sources (adopted 4 June 1974, entered into force 6 May 1978) 1546 UNTS 103. Hence 'OSPAR'(Oslo/Paris).

[17] S Cole, MJ Ortiz and C Schwarte, *Protecting the Marine Environment in Areas Beyond National Jurisdiction – A Guide to the Legal Framework for the Conservation and Management of Biodiversity in Marine Areas Beyond National Jurisdiction* (Foundation for International Environmental Law and Development (FIELD), 2012) 22–23.

The duties for all States to cooperate in such environmental matters are repeated in article 198 and article 199 of UNCLOS. Article 198 compel States that becomes aware of a risk of pollution to inform all other States deemed likely to be affected by it.

Article 199 encourages all (coastal) States affected to work together to address and eliminate such pollution. It also suggests that (coastal) States should attempt to pre-empt such situations by making joint contingency plans for responding to any pollution incidents, likely in the form of agreements between neighbouring coastal States.

Article 211(1) and article 212(3) also stipulate that States should cooperate through the IMO[18] in establishing international rules and standards to prevent, reduce and control pollution of the marine environment from ships.

IV. Implementing International Rules and Adopting National Laws for the Protection of the Marine Environment

The general requirements of UNCLOS article 197, article 211(1) and article 212(3) for all States to participate in the development of international rules for the protection of the marine environment, for example through the IMO, are elaborated in article 211(2)–(6), under which flag States (article 211(2)) and coastal States (article 211(3)–(5)) are required subsequently to adopt national laws, including the implementation of international rules.

Article 211(7) obligates all States to inform coastal States whose coastlines or related interests may be affected by a pollution incident.[19]

Articles 212 and article 222 specifically relate to the implementation, adoption and enforcement of rules regulating *air pollution*, and are therefore discussed separately from article 211, in section V of this chapter.

A. Flag State Rules under Article 211(2)

A flag State is, in accordance with article 211(2), required to adopt national laws that *at least* offer the same level of protection as the international rules issued by the IMO.[20] The conclusion must be that flag States are therefore able to draft national laws containing environmental protection standards stronger than those under the international rules.

[18] Recalling that the reference to the 'international organization' is a direct reference to the IMO and the conventions and guidelines issued by that body. See IMO Legal Committee, n 6, 56.

[19] Art 217(7) will, due to its scope, not be examined further in this chapter.

[20] Tanaka, n 12, 292.

While article 211(2) enables a flag State to *prescribe* national legislation that applies to all ships under its flag, article 217 allows this legislation to be extraterritorially *enforced* and *adjudicated on* wherever these ships sail.[21]

Violations of such exclusive national flag State laws are not the subject of enforcement measures by other non-flag States.[22] If another State (eg a coastal State) actually has the same legislation applying within its territory as the flag State has imposed on ships under its flag, for example an implemented IMO regulation, then the other State will enforce that legislation in accordance with its own rules and not those adopted by the flag State.

It should be noted that part XII of UNCLOS also contains other obligations for flag States to implement very specific regulations, for instance on *dumping*, according to article 210(1), and on *reporting formalities* for ships sailing in another State's EEZ, pursuant to article 220(4). These implementation obligations are *lex specialis* compared to the general obligation concerning implementation in article 211(2).

B. Coastal State Rules in Internal and Territorial Waters under Article 211(3)–(4)

Coastal States have the same rights to create national laws for the protection of the marine environment in areas under their sovereignty, according to article 211(3)–(4) of UNCLOS. Article 211(3) covers national regulation in the State's *internal waters and ports*[23] and article 211(4) relates to national legislation in its *territorial waters*.

Article 211(3) reiterates the principle of a coastal State's having full sovereignty within its internal waters and ports pursuant to article 2 of UNCLOS, which gives it full territorial jurisdiction to prescribe legislation and to subsequently enforce and adjudicate on it. This includes penalising violations in this area with non-monetary penalties such as imprisonment (conversely, see article 230). The only precondition the coastal State must meet is to give *due publicity* to such rules.

Article 211(4) also refers to the codified exceptions for the territorial sea, which – as described in chapter 3 – mean that a coastal State's enforcement of its rules and regulations cannot hinder a foreign ship's privilege to exercise its

[21] The principles of prescribing, enforcing and adjudicating legislation are discussed in ch 6.

[22] It could be seen as conflicting with art 227 of UNCLOS (requiring equal treatment of all foreign ships) if a (coastal) State were to enforce foreign flag State legislation over foreign ships flying that flag within the (coastal) State's own waters, as it would amount to unequal treatment of foreign ships. Other (coastal) States can of course assist flag States in determining whether a violation of the flag State's exclusive national legislation occurred while the ship was sailing in its (the coastal State's) waters, eg by investigating. The prosecution and penalisation of the violation must, however, be carried out by the flag State.

[23] The reference to adopting national regulations in a port could in principle lead to use of the term 'port State'. Yet as art 211(3) also refers to the regulation of internal waters, the term 'coastal State' will be used when referring to that paragraph.

rights of *innocent passage* or *transit passage* in the territorial sea. To this end, the discussion of article 21(1)(f) and article 42(1)(b) in chapter 3 should be brought to mind.

Article 21(1)(f) allows coastal States to implement and adopt national environmental legislation applying to innocent passage, while article 42(1)(b) allows coastal States that border an international strait to adopt regulations pertaining to the prevention of pollution in the form of the discharge of oil, oily wastes and other noxious substances in the strait. However, unlike article 21(1)(f), article 42(1)(b) refers exclusively to the *implementation* of international regulations, such as MARPOL Annexes I and II. Article 42(1)(b) therefore does not allow coastal States to adopt national environmental legislation.

A coastal State's legislation adopted in accordance with article 211(3)–(4) can be enforced in accordance with article 220(1). Environmental legislation adopted pursuant to article 21(1)(f) can be enforced in accordance with article 220(2), while the international regulations implemented following article 42(1)(b) are enforced pursuant to article 233.

C. Coastal State Rules in the EEZ under Article 211(5)–(6)

Just as article 211(3)–(4) present coastal States with the basis for implementing international regulations and adopting national laws with higher standards of protection in their *internal waters* and *territorial waters*, article 211(5)–(6) regulate the possibility for coastal States to do the same in their *EEZs*.

Article 211(5) repeats the possibility of implementing *international* rules, such as MARPOL Annex VI, to give these effect in the EEZ.

Article 211(6) allows coastal States to establish *national* environmental legislation that applies to foreign ships in their EEZ. As this zone extends up to 200 nautical miles (nm) from the baseline, the requirements that coastal States must meet in order to adopt national rules in such a vast area are much more detailed, in accordance with article 211(6)(a)–(c), compared with those under article 211(3)–(4) regulating internal and territorial waters. For example, article 211(6)(a) requires that international rules for the protection of the marine environment developed pursuant to article 211(1) (and implemented in accordance with article 211(5)) must be *inadequate* to meet the *special oceanographical and ecological conditions* present in a *clearly defined area* of the EEZ.

The coastal State must submit scientific and technical evidence directly to the IMO to support and prove that there are such special conditions in a clearly defined area of the EEZ. The IMO must then within 12 months evaluate whether it agrees with the scientific assessments made by the coastal State relating to the need for additional national regulations.

If the IMO approves the adoption of national laws in that area of the EEZ the coastal State must implement these by using standards made available through the IMO, and must ensure that the rules do not become applicable to foreign vessels until 15 months after the submission to the IMO.

The reference to using IMO standards for designating such distinct areas in an EEZ links to the establishment of Particularly Sensitive Sea Areas (PSSAs). The criteria and standards for creating a PSSA are based on IMO Resolution A.982(24), 'Revised guidelines for the identification and designation of Particularly Sensitive Sea Areas', from 2005 (see also chapter 3). Section 7.4.2(1)(a)(*bis* iii) of the Resolution refers directly to article 211(6) of UNCLOS, as it requires coastal States intending to have an area designated as a PSSA to apply the same criteria as required by article 211(6). The legal basis in UNCLOS for establishing a PSSA is thus found in article 211(6).

Even so, the opposite conclusion cannot be drawn from this, that is, to automatically assume that all national pollution areas created in accordance with article 211(6) are inevitably designated as PSSAs. There is a possibility – albeit a somewhat theoretical possibility – that a coastal State may wish to invoke the article 211(6) procedure without necessarily deeming such an area a PSSA in accordance with IMO Resolution A.982(24). This book will therefore, when referring to national rules issued pursuant to article 211(6), not automatically refer to these rules as applying to a PSSA but merely as comprising 'national regulation'.

D. Coastal State Rules in an Ice-covered Area of the EEZ under Article 234

When clarifying the possibilities for coastal States to prescribe and enforce national legislation in their EEZs, attention must also be given to article 234 of UNCLOS.

This article provides a legal basis for coastal States that have 'ice-covered areas within the limits of the exclusive economic zone' to adopt non-discriminatory laws that aim to offer special protection of the marine environment in such areas of the EEZ.

Article 234 is applicable where:

> particularly severe climatic conditions and the presence of ice covering such areas for most of the year create obstructions or exceptional hazards to navigation, and pollution of the marine environment could cause major harm to or irreversible disturbance of the ecological balance. …

Consequently, the term 'ice-covered area' is defined as an area that is covered by ice 'for most of the year'. It is unclear exactly how many months this encompasses, but a strict interpretation of the wording leaves the impression that the term *most* implies more than half, meaning that the condition is fulfilled if an area is covered by ice for more than six months a year.

The applicability of article 234 also depends on the ice's hindering and presenting an exceptional danger to navigation, and that pollution could potentially cause major harm to or irreversible disturbance of the ecological balance in the area.

The consideration for the *ecological balance* repeats the concerns addressed in article 211(6) regarding *ecological conditions*. Both articles furthermore underline that these assessments of the ecological conditions should be based on scientific data.

Yet article 234 sets itself apart from article 211(6) by not requiring coastal States to present the scientific evidence to the IMO, nor requiring IMO approval of the intended national regulations. There is also no specification of an implementation period during which foreign ships are exempt from complying with the national rules.

The legal basis for coastal States' prescribing and enforcing national laws for marine protection in their EEZ is therefore wider in article 234 than in article 211(6), provided the physical requirements are met with regard to there being ice for most of the year that presents a danger in a sensitive ecological area.

Some ice-covered areas, such as the polar regions (Arctic/Antarctic), have been deemed regions that need special protection, which the IMO seeks to provide.[24] For example, regulation 43 of MARPOL Annex I prohibits the use and transport of oil with certain characteristics (viscosity, etc) in the Antarctic; and the mandatory provisions of the International Code for Ships Operating in Polar Waters (the Polar Code) require ships sailing in polar waters to adhere to the additional protective safety and environmental measures of the Code.[25]

V. Special Regulations on Air Pollution: Article 212 and Article 222

UNCLOS part XII contains two distinct provisions pertaining to air pollution from ships, which should be read and reviewed in conjunction with each other.

Article 212 and article 222 require States not only to implement international rules on air pollution, but also to adopt national rules regulating such damaging emissions, and to enforce these in areas under their sovereignty.

A. The Obligation to Adopt Regulations on Air Pollution from Ships under Article 212

Article 212(1) of UNCLOS permits all States, similar to article 211(2)–(5), to adopt national laws on reducing air pollution, making these 'applicable to the air

[24] Other intergovernmental organisations also work on preserving the Artic environment, such as the Arctic Council – see at https://www.arctic-council.org/index.php/en/.

[25] The Polar Code was implemented in the SOLAS Convention (International Convention for the Safety of Life at Sea (SOLAS) 1974 (adopted 1 November 1974, entered into force 25 May 1980), 1184, 1185 UNTS 2) and the MARPOL Convention, and entered into force on 1 January 2017.

space under their sovereignty and to vessels flying their flag or vessels or aircraft of their registry'. This is supplemented in article 212(2) with a reference to 'other measures' for achieving such reductions, which could comprise voluntary national guidelines and standards.

Article 212(3) encourages all States, like article 197 and article 211(1), to participate in the establishment of global rules and standards, including through the IMO, for preventing and controlling air pollution from ships.

B. The Obligation for all States to Implement MARPOL Annex VI under Article 222

Article 222 of UNCLOS refers directly to the enforcement of the national rules and regulations adopted by coastal and flag States in accordance with article 212(1). The provision also requires those States to implement international IMO rules and standards applicable to areas under their sovereignty and ships under their flags.

Article 222 therefore represents a *lex specialis* obligation on all States (flag, port and coastal) to *implement* MARPOL Annex VI. It is also the legal basis for requiring these States to implement the IMO's future rules and regulations for reducing GHG emissions from ships.

Article 222 also reads, at the outset, as a *lex specialis* obligation for all States to *enforce* regulations on air pollution, including those adopted according to article 212 and those implementing regulation 14 of Annex VI.[26] Obviously, the scope of article 222 therefore needs be examined in further detail, as the purpose of this book is to determine how international regulation of air pollution (primarily sulphur and GHG emissions) can be enforced. To this end a distinction must be made between the legal basis the regulations provide for *flag States* and for *non-flag States* (ie port and coastal States).

C. No Clear Guidance on Flag State Enforcement of MARPOL Annex VI under Article 222

In accordance with article 212, article 222 is the *lex specialis* obligation for flag States to *prescribe* (implement) international regulations on air pollution such as MARPOL Annex VI. Article 222 must, in principle, also be considered *lex specialis* regarding the obligation for flag States to *enforce* these rules, including regulation 14 of Annex VI.

Although this article does not directly refer to enforcement by the flag State irrespective of where a ship under its flag is sailing, an implicit extraterritorial

[26] IMO Legal Committee (IMO LEG) argues that art 212 – and thereby art 222 – includes air pollution as regulated in MARPOL Annex VI, including SOx and NOx pollution; see reg 14 and reg 13 of MARPOL Annex VI. See IMO Legal Committee, n 6, 80.

basis is found within this provision, given the simple reference in it to enforcement with regard to 'vessels flying their flag'.

That being said, neither article 222 nor article 212 includes any specifications on how flag States should enforce and what constitutes effective flag State enforcement of such air pollution regulations. These non-descriptive articles should therefore, in the view of this author, be applied in conjunction with the overall, yet detailed, requirements of article 217 for flag State enforcement of all legislation for the protection of the marine environment, also encompassing MARPOL Annex VI. Article 217 contains, for example, specific obligations for effective extraterritorial enforcement and for furnishing information thereon to the IMO and all States, etc.[27]

Article 217 should thus be the *de facto measuring scale* for determining whether a flag State's enforcement of international or national rules on air pollution, including regulation 14 of Annex Vi, has been effective.

Flag States are consequently required to effectively enforce MARPOL Annex VI in accordance with article 222 pursuant to article 217, irrespective of where these violations occur. Article 217 is analysed in the next chapter.

D. No Extraterritorial Jurisdiction for Non-flag States to Enforce Annex VI under Article 222

Article 222 of UNCLOS stipulates that all States, including port and coastal States, are obliged to *prescribe* (implement) MARPOL Annex VI. The geographical scope of article 222 is, however, rather limited regarding *enforcement* by non-flag States, as it explicitly refers to such regulation's being enforced 'within the air space under [the] sovereignty' of a State.

This sovereignty encompasses, according to article 2(2) of UNCLOS, 'the air space over the territorial sea as well as … its bed and subsoil', as under article 2(1) (coastal) States have sovereignty in internal and territorial waters, with the exceptions described in chapter 3 applying to the latter (territorial sea) according to article 2(3).

In other words, article 222 provides *no* extraterritorial jurisdictional for port or coastal States to enforce air pollution regulations, such as MARPOL Annex VI, outside areas under their sovereignty, for example in an EEZ or on the high seas. Article 222 simply reiterates basic principles of international law, as it confirms that all (port and coastal) States can enforce and adjudicate on prescribed laws in

[27] Dr Erik Molenaar and Professor Henrik Ringbom have expressed similar views regarding the limitations of art 212 and art 222. E Molenaar, *Coastal State Jurisdiction over Vessel-Source Pollution* (Kluwer Law International, 1998) 507 and 512; H Ringbom, 'Enforcement of the Sulphur in Fuel Requirements: Same, Same But Different', available at https://webcache.googleusercontent.com/search?q=cache:EBk8-0aReA0J:https://www.duo.uio.no/bitstream/handle/10852/61600/SO-Artikel-Ringbom.pdf%3Fsequence%3D4%26isAllowed%3Dy+&cd=1&hl=da&ct=clnk&gl=no, 10–11.

areas under their sovereignty. Article 222 therefore does *not* offer an extraterritorial jurisdictional solution for ensuring that violations of regulation 14.1 can be effectively enforced on the high seas too.

The legal basis for coastal States to enforce MARPOL Annex VI in the EEZ must be found in article 220(1), which is discussed in chapter 9. The legal basis for port States to enforce regulation 14.1.3 of Annex VI on the high seas must be found in article 218(1), as analysed in chapter 10.

E. Summary of the Applicability of Articles 212 and 222 to Implement and Enforce Annex VI

Article 212 allows all States to adopt national regulations on air pollution from ships, although article 212(3) also refers to developing such rules on an international scale through cooperation at the IMO.

Article 222 obligates all States to implement international regulations on air pollution, such as MARPOL Annex VI.

Article 222 also refers to all (flag, coastal and port) States enforcing these nationally adopted and internally implemented regulations.

However, article 222 contains no specific references as to how flag States should enforce such regulations, so that a flag State's enforcement of MARPOL Annex VI should be supplemented by application of the general and always applicable, but also very detailed, requirements of article 217.

Also, article 222 does not grant coastal or port States any broadened extraterritorial jurisdiction to enforce MARPOL Annex VI, and other regulations on air pollution, outside the territorial sea. Articles 220 and 218 are therefore key for ensuring enforcement by non-flag States of regulation 14 of Annex VI in EEZs and on the high seas.

VI. Legal Basis for Adopting National Sulphur Limits and Scrubber Rules

Putting aside the implementation and enforcement of MARPOL Annex VI for a brief moment, the previously discussed article 211 and article 212 (with article 222) should be studied to determine the applicability of these provisions to allow all States to establish national sulphur regulations pertaining to sulphur limits and limitations on the use of Exhaust Gas Cleaning Systems (EGCSs), particularly regarding the use of *open loop scrubbers*.[28]

[28] See ch 2 for a description of EGCSs such as open loop scrubbers.

A. Flag States Adopting National Sulphur and Scrubber Rules

Flag States can, according to article 212 of UNCLOS, adopt national sulphur limits applying to ships under their flags. Article 212 is *lex specialis* compared to article 211(2) when it comes to adopting rules regulating *air pollution* such as sulphur emissions from ships. Nevertheless, only the flag State would enforce these regulations, as the principle of 'no more favourable treatment' only applies to international rules, such as MARPOL Annex VI. Obviously, a flag State's national sulphur limits should therefore not be enforced during PSC, while the limits set out in regulation 14 of Annex VI always should. The flag State must therefore, in principle, wait until a ship under its flag returns to the flag State and its jurisdiction before any violations of the national sulphur limit can be penalised.

Should a flag State wish to regulate how ships under its flag use an EGCS then this must be effected in accordance with article 211(2). It will be recalled from chapter 2 that flag States, pursuant to regulation 4 of Annex VI, must approve all EGCSs that are installed in ships under their flag, which limits the need to adopt separate flag State rules on the use of these systems.

B. Coastal States Adopting National Sulphur and Scrubber Rules in Internal Waters and in the Territorial Sea

Coastal States can, in accordance with article 212(1), adopt national sulphur limits applicable in ports, internal waters and in the territorial sea – with general exceptions applied to the last of these. Article 212 is *lex specialis* compared to article 211(3)–(4), as it concerns pollution of the atmosphere.

Taiwan[29] and several Chinese port areas around Shanghai, Ningbo, Zhoushan and the Yangtze River Delta[30] were among the first to opt for this solution by creating national 0.5% sulphur zones in internal and territorial waters. That is, areas where the general 3.5% limit would normally be applied in accordance with regulation 14.1.2 of MARPOL Annex VI.

On 1 January 2019, China implemented regulation 14.1.3 of Annex VI, 12 months ahead of the IMO's deadline, by setting a 0.5% sulphur limit in all its territorial waters.[31]

Article 21(1)(f) of UNCLOS allows the coastal State to adopt national rules and implement international rules for the protection of the marine environment

[29] See at https://www.ukpandi.com/knowledge-publications/article/taiwan-to-implement-0-5-sulphur-cap-from-1-january-2019-142785/.

[30] See at http://www.nepia.com/insights/industry-news/china-emission-control-areas-starupdatestar/.

[31] See at https://lloydslist.maritimeintelligence.informa.com/LL1123230/China-to-expand-ECA-to-cover-all-territorial-waters.

applying to foreign ships making an *innocent passage* in accordance with articles 17–19, which could encompass national sulphur limits adopted according to article 212 and thereby already applicable to the territorial sea for non-innocent sailing, such as to and from a port. As mentioned in chapter 3, such national regulations must contain an explicit reference to its also applying to innocent passage, to allow it to be enforced in accordance with article 21(4). A coastal State cannot adopt national sulphur rules applicable in an international strait, as article 42(1)(b) only refers to *implementation of international rules* and only refers to the discharge of *oil and other noxious substances*. If a coastal State wishes to set stronger limits on the use of EGCSs in internal and territorial waters, for example pertaining to discharge from open loop scrubbers, this must be done in accordance with article 211(3)–(4). Singapore,[32] China, the US (in California and Massachusetts) and several European countries, including Belgium and Germany,[33] have used the legal basis of article 211(3) for making stricter regulations on the use of open loop scrubbers in their ports and internal waters.[34]

National laws on the use of open loop scrubbers in internal waters can provide for the complete banning of the use of such systems, as well as applying stricter criteria and (pH) limits for the wash water continuously discharged from such systems, provided the coastal States give 'due publicity to such requirements' and inform the IMO.

Coastal States cannot adopt regulations that de facto ban the use of the open loop scrubber technology for ships making an innocent passage in the territorial sea (see article 21(2)).

C. Coastal States Adopting National Sulphur and Scrubber Rules in the EEZ

Initially, the geographical applicability of article 212(1) and article 222 must be recalled, as it is limited to prescribing and enforcing national laws on air pollution in areas under the sovereignty of the coastal State such as internal and territorial waters in accordance with article 2 of UNCLOS which, of course, excludes the EEZ.

Any national regulations on sulphur, be they lower (ie stricter) sulphur limits or strengthened open loop scrubber standards, must therefore be adopted in accordance with article 211(6). Or in accordance with article 234 if parts of the EEZ are ice-covered (see section IV.D).

[32] See at https://www.mpropulsion.com/news/view,singapore-bans-use-of-openloop-scrubbers-in-port_56074.htm.

[33] A Proelß and VJ Schatz, 'Rechtliche Vorgaben zum Umgang mit Schiffsabwasser Völker-, unions- und nationalrechtliche Anforderungen an Einleitungen von Scrubber-Abwasser, Ballastwasser und häuslichem Abwasser durch Schiffe', Project no 99836 (University of Hamburg, 2019) 31.

[34] See at https://splash247.com/china-to-ban-open-loop-scrubbers-along-its-rivers/.

Article 211(6)(a) allows a coastal State to adopt national laws in its EEZ if international rules are inadequate to meet the special oceanographical and ecological conditions present in a clearly defined area of the EEZ. This must be accepted by the IMO through the submission of scientific and technical evidence.

This means that, in theory, a coastal State could adopt national laws stipulating a lower sulphur limit in the EEZ.[35] However, because of the characteristics and interactions of sulphur pollution as described in chapter 1, it might be difficult to provide scientific data that prove that sulphur emissions from ships in an EEZ have a verifiable negative impact on the oceanographical and ecological conditions in that area.[36]

The applicability of article 234 for adopting a national sulphur emission limit must rest on the same presumptions, as it should be considered doubtful that scientific evidence will prove that the release of SOx particles into the atmosphere has a particularly adverse effect on ice-covered areas. Although IMO approval is not needed, the relevant scientific data and evidence must still be presented.

Regarding potential national regulations on EGCSs, article 211(6)(c) emphasises that the national laws may relate to 'discharges or navigational practices but shall not require foreign vessels to observe design, construction, manning or equipment standards other than generally accepted international rules and standards'. The wording of article 211(6)(c) leaves a knife-edge legal basis for coastal States to balance on if they wish to prohibit the use of *open loop scrubbers* in an EEZ, as many States already have done in their ports and internal waters. On one hand, coastal States may not adopt special national technical or design rules that ban the open loop scrubber technology, or its use, in the EEZ. On the other hand, coastal States could in principle, if they can present the necessary scientific data, obtain IMO approval for setting more stringent discharge (wash water) limits in an area of the EEZ where special oceanographical and ecological conditions make it necessary.[37]

This would also be supported by the general principle embodied in article 195 of UNCLOS, stating that all States have a duty, when adopting rules for the protection of the marine environment, *not to transform one type of pollution into another type of pollution*. This covers not transforming sulphur pollution of the atmosphere in to sulphur pollution of the sea by setting open loop scrubber discharge criteria that are too relaxed.

It must therefore be assumed that if a coastal State can provide the compulsory scientific data that unequivocally shows the need for setting more strict wash-water

[35] See also Dr Erik Molenaar, who points out that the IMO would presumably advise the State in question to become party to MARPOL Annex VI to allow for designation of a (Sulphur) Emission Control Area ((S)ECA) zone, rather than an art 211(6) area (PSSA): Molenaar, n 27, 509.

[36] See the similar discussion in ch 9, section I.D, regarding whether air pollution (eg sulphur pollution) can result in – or threaten to cause – significant pollution of the marine environment in a coastal State's EEZ as required under art 220(5).

[37] See at http://www.egcsa.com/technical-reference/what-is-the-ph-of-the-water-discharge(d)-from-an-exhaust-gas cleaning system-into-the-sea/.

limits in a clearly defined area, then this could result in IMO-approved national regulations on the use of open loop scrubbers while sailing in the EEZ.

It should be noted that any national laws *prescribed* (adopted) in accordance with article 211(6) can be *enforced* by coastal States in their EEZ in accordance with article 220(3)–(7), pursuant to article 220(8).[38]

Article 234 may, like article 211(6), have a key role to play when determining coastal States' jurisdiction for legislating on discharges from open loop scrubbers. A coastal State with ice-covered parts in its EEZ should, according to article 234, be able to adopt more stringent wash water discharge limits, without IMO approval, provided the best available scientific evidence deems it necessary.

This could also encompass a complete ban or prohibition on the use of open loop scrubbers, as article 234 does not include the same restriction as under article 211(6)(c) pertaining to the adoption of national technical and design requirements. Article 234 merely demands that *due regard to navigation* must be ensured.

The opposite scenario could in theory also be envisioned, as some studies show that scrubber systems remove not only high percentages of SOx and NOx, but also particle matter (PM) and, perhaps, also Black Carbon particles, as these could be particular harmful in ice-covered areas.[39] If this is the case, a coastal State may actually require the use of scrubber systems, most likely *closed loop scrubbers*, by ships when sailing in these areas.

VII. Unlawful for States Parties to Annex VI to Lower Protection Standards

It should, on a final note, be stressed that the regulations in articles 211, 212, 222 and 234 of UNCLOS do not allow States that are parties to MARPOL Annex VI to set sulphur limits that are higher than those under regulation 14. The same principle applies to the adoption of national discharge limits on open loop scrubbers that allow more pollutants from the scrubber system to be released into the sea.

A State that is 'merely' a signatory of MARPOL Annex VI also cannot adopt national rules that lower the protective scheme of Annex VI, for example by relaxing the sulphur limits or discharge standards, as this 'would defeat the object and purpose of a treaty' (see article 18(a) of the Vienna Convention on the Law of Treaties).[40]

[38] Art 220 is analysed in chapter 9.

[39] See IMO, *Investigation of appropriate control measures (abatement technologies) to reduce Black Carbon emissions from international shipping* (2015), IMO study report by Litehauz in cooperation with DA Lack, Boulder University, Colorado/J Thuesen and R Elliot, ERRIA, DK. The report was finalised by the authors on 20 November 2012 and published by the IMO in 2015. Available at http://www.imo.org/en/OurWork/Environment/PollutionPrevention/AirPollution/Documents/Air%20pollution/Report%20IMO%20Black%20Carbon%20Final%20Report%2020%20November%202012.pdf.

[40] Vienna Convention on the Law of Treaties 1969 (adopted 22 May 1969, entered into force 27 January 1980) 1155 UNTS 331 (VCLT).

VIII. National Regulations on Greenhouse Gases

The principles of articles 211, 212 and 222 of UNCLOS already discussed would apply not only to the implementation and adoption of sulphur regulations, but also to regulations on GHG emissions.

More specifically, a ship's emission of GHGs would also constitute *air pollution* covered by articles 222 and 212, which would be the legal basis for requiring coastal States and flag States to implement any IMO regulations on GHG. This would also allow all States to adopt stronger GHG (CO_2) emission limits for ships under their flag and in internal and territorial waters. A State could also unilaterally adopt more ambitious overall GHG targets than those accepted by the IMO at MEPC 72 in April 2018.

Coastal States could, regarding the EEZ, implement international GHG rules according to article 211(5), and adopt more stringent national limits pursuant to article 211(6), and enforce these in accordance with article 220(8). Adopting national GHG regulations pursuant to article 211(6) would, however, require scientific data that prove the negative impact from GHGs on the ecological conditions of the EEZ, which might be difficult to achieve. It would also necessitate IMO approval, unless the EEZ is ice-covered in accordance with article 234.

The previously described distinction between adopting national regulations on sulphur *limits* and sulphur *technology* (EGCSs) would also, if relevant, apply to GHGs.

IX. Conclusion

Part XII of UNCLOS encourages all States to cooperate, particularly through the IMO, in establishing international rules for the protection of the marine environment.

Section 5 of part XII contains obligations for all States to *implement* international rules and regulations for the protection of the marine environment, such as MARPOL Annex VI and future IMO regulations for the reduction of GHGs.

Articles 211 and 212 of section 5 allow all (flag, port and coastal) States to *adopt* national laws for the protection of the marine environment, provided these include a higher protection standard than the international regulations, such as lower sulphur limits than those set out in regulation 14 of MARPOL Annex VI.

The coastal States' possibilities for adopting national regulations on air pollution are geographically limited to internal and territorial waters according to article 212. National regulations in the EEZ can be adopted in accordance with the procedure set out in article 211(6) or, if applicable, article 234. This could, in theory, include national regulations on air pollution, but it would require scientific evidence that proves a negative impact from air pollution (eg sulphur pollution) on the ecological conditions of the EEZ, which is considered difficult to achieve.

The *enforcement* of these international implemented – and national adopted – regulations and laws on air pollution is at the outset enforced through article 222, found in section 6 of part XII. This means that article 222, in principle, provides the legal basis for flag, coastal and port States to penalise violations of the 0.1% and 0.5% sulphur limits in Annex VI and any regulations adopted at a national level in accordance with article 212(1).

However, although article 222 refers to the implementation and enforcement of regulations, such as regulation 14 of MARPOL Annex VI, it sets out certain clear limitations on enforcing this effectively. First, article 222 provides no extra-territorial jurisdiction for *non-flag States* to penalise violations outside of internal and territorial waters, such as in the EEZ and on the high seas. The regulations applicable in these areas under MARPOL Annex VI, including the 0.5% limit on the high seas, must therefore be enforced in accordance with the general jurisdiction for coastal States pursuant to article 220 (see chapter 9) and the special extended jurisdiction for port States according to article 218 (see chapter 10).

Second, even though article 222 also refers to *flag States'* enforcement of rules on air pollution, it includes no details on what this enforcement should comprise. Article 222 must therefore, in the view of this author, be used in conjunction with article 217 when determining how a flag State should proceed against MARPOL Annex VI violations, including violations of the 0.5% sulphur limit on the high seas. Article 217 is a general provision that refers to enforcement of all legislation for the protection of the marine environment, including air pollution. It also contains specific obligations that a flag State must meet when enforcing this legislation. Article 217, and its eight paragraphs, is therefore analysed in the next chapter.

8

Special Obligations of the Flag State: Article 217

As concluded in the last chapter, flag States are under a general obligation to implement international and national rules for the protection of the marine environment pursuant to article 211(2) of UNCLOS.[1] The *lex specialis* right for flag States to adopt national legislation to prevent air pollution from ships is found in article 212(1), and the legal basis for enforcing this, and for implementing and enforcing international rules on air pollution, is established in article 222.[2] Flag States must therefore implement and – in principle – enforce MARPOL Annex VI in accordance with article 222 of UNCLOS when becoming party to the Annex.

Although the obligation to implement and prescribe the international regulations in national law is very clear, the reference in article 222 to enforcing these rules is non-specific, merely stating that flag – and other – States *shall enforce* these rules.

To avoid a discrepancy between flag State enforcement of International Maritime Organization (IMO) regulations on air pollution (such as MARPOL Annex VI[3]) and non-air pollution regulations (such as MARPOL Annexes I–V), all IMO regulations for the protection of the marine environment[4] should, as discussed in chapter 7, be subject to the same detailed demands for effective flag State enforcement imbedded in the eight paragraphs of article 217.

Article 217 thereby occupies an unusual position as *lex specialis* compared to the overall principles requiring flag State intervention pursuant to articles 92 and 94 of UNCLOS when dealing with enforcement of the IMO's environmental regulations in general. And at the same time, article 217 is also *lex generalis* when

[1] United Nations Convention on the Law of the Sea (adopted 10 December 1982, entered into force 16 November 1994) 1833 UNTS 3. See Y Tanaka, *The International Law of the Sea*, 2nd edn (Cambridge University Press, 2015) 292.

[2] Special duties to implement *lex specialis* regulations on dumping or reporting formalities are also imposed on the flag State in accordance with art 210 on dumping and art 220(4) on reporting formalities in an EEZ.

[3] International Convention for the Prevention of Pollution from Ships (adopted on 11 February 1973, as modified by the Protocol of 17 February 1978, entered into force 2 October 1983), 1340 UNTS 61 (MARPOL), Annex VI, IMO Publication: IMO-520E.

[4] Recalling from ch 3 that art 1(1)(4) defines pollution of the marine environment as including pollution that has an adverse effect on human health, eg air pollution.

dealing with IMO rules on air pollution, filling in the legal vacuum left by the wording of article 222 when determining flag State responsibilities.

Article 217(1)–(8) therefore form the legal basis by which a flag State must abide when bringing enforcement proceedings following violations of the sulphur limits under MARPOL Annex VI, regulation 14, committed by ships under its flag. The provision also serves as a scale of measurement for determining if a flag State's enforcement has been effective.

Article 217 will also be the basis for determining flag State enforcement of future IMO legislative measures on greenhouse gases (GHGs), regardless of whether these focus on reducing air pollution (GHG/CO_2) by using zero-carbon or fossil-free fuels, or focus on reducing GHGs by performance-based measures such as speed optimisation or revising the Energy Efficiency Design Index (EEDI) for certain ships.[5] The scope of article 217 embraces all of these possible regulatory adoptions.

I. Obligation to Enforce Effectively: Article 217(1)

In accordance with article 217(1), flag States must ensure that ships under their flags comply with national and international rules for the protection of the marine environment, which includes MARPOL Annex VI. Article 217(1) further specifies that 'Flag States shall provide for the effective enforcement of such rules, standards, laws and regulations, irrespective of where a violation occurs.'

This establishes two clear requirements that flag States must fulfil when enforcing legislation for the protection of the marine environment:

(a) the flag State enforcement must be *effective*;
(b) a flag State must enforce the regulations, etc *wherever* a ship under its flag commits a violation, including on the high seas.

As a result, pursuant to article 217(1), flag States are required to effectively enforce the stricter 0.5% sulfur limit under regulation 14.1.3 of Annex VI for violations on the high seas too.

The *second requirement* does not bar other (coastal or port) States from also assuming jurisdiction over a violation committed by a – in their view – foreign ship. It merely reflects that flag States will always have overlapping (competing) jurisdiction with these States over such violations. The legal basis for determining which State can assume primary (final) jurisdiction is provided for in article 228(1), which is analysed in chapter 11.

[5] The proposed measures for future IMO regulation of GHG reduction – following the IMO's 2018-adopted GHG reduction strategy – are described in Part III of this book, which also considers how these may be enforced through UNCLOS.

The *first requirement* of article 217(1) is worth examining further regarding what constitutes *effective* enforcement.

It should be noted that flag States are not subject to the restrictions under article 230 of UNCLOS. An article which specifies that only monetary penalties can be imposed for violations taking place outside territorial waters, as the scope of the provision is limited to *foreign* ships, therefore not encompassing the exercise of flag State jurisdiction. Reaffirming that flag States can – in principle – take action against violations that includes criminal penalties such as imprisonment, although this is seldom seen. Except in cases where violations cause major damage to the marine environment and it can be attributed to a deliberate act or, at minimum, gross negligence. Nonetheless, the possibility for flag States to impose criminal sanctions exists, and article 228(3) allows flag States to impose their own penalties, such as imprisonment, even though the shipowner has already been penalised by the imposition of a fine, detention, etc by a coastal or flag State.[6]

Despite the theoretical possibility of imposing a prison sentence, violations of the IMO's environmental regulations are normally met with monetary sanctions (fines) by flag States. Such flag States have a natural extraterritorial jurisdictional *right* to impose fines for violations, irrespective of where these occur, pursuant to the flag State principle codified in article 217(1). Evidently, the wording of the provision, requiring that flag States 'shall ensure compliance by vessels flying their flag' and 'shall provide for the effective enforcement', clearly stipulates that it is also an *obligation* for flag States to ensure effective enforcement.

Section I.A will study what *effective flag State enforcement* comprises, with a focus on determining which minimum criteria must be met to establish that a fine, imposed by a flag State for a violation of the sulphur limits under regulation 14 of Annex VI, can be deemed *effective*, thereby allowing the flag State to fulfil the first requirement in article 217(1).

A. What is an *Effective* Flag State Fine under Article 217(1)?

Effective is a very open-ended term, but the principles of article 18 of the EU Sulphur Directive[7] (described in chapter 5) offer, in the opinion of this author, clear and unequivocal guidance on the basic principles that should be relied on for fines imposed for violations of the MARPOL Annex VI sulphur limits. This is (also reiterating what is said in chapter 5) based on article 18's representing a *subsidiary means for the determination of rules of law*, as it implements the MARPOL Annex VI limits and was developed by the EU, which could be considered a *highly qualified publicist* within international law. The International Court of Justice (ICJ)

[6] Art 228(3) and art 230 are both examined in ch 12.

[7] Directive (EU) 2016/802 of the European Parliament and of the Council of 11 May 2016 relating to a reduction in the sulphur content of certain liquid fuels [2016] OJ L132/58.

could therefore, in accordance with article 38(1)(d) of the ICJ Statute,[8] apply the principles of article 18 of the Sulphur Directive when determining whether a flag State's enforcement against a violation of regulation 14 of Annex VI was to be considered effective pursuant to article 217(1) of UNCLOS.[9]

Another, and more simple way, of viewing this is that the principles of article 18 represent common principles of law – and of common sense – as they dictate that it should not be possible to benefit from committing a crime, or plotting to do so.

The principles of article 18 of the Sulphur Directive are therefore applicable to determine the effectiveness of non-EU States' enforcement of the 0.1% and 0.5% sulphur limits in regulation 14 of MARPOL Annex VI.

It should be recalled that article 18 requires sanctions to: be effective; be proportionate; be dissuasive; at least strip the economic gains from the infringement; and gradually increase in the event of repeated infringements. One of the most important of these criteria is that *all economic gains made by the infringement are removed*, especially considering the enormous potential savings to a shipowner from a violation of the sulphur limits (eg a $750,000 saving on a single trip from Asia to Europe).

This part of the fine – sometimes addressed separately as *confiscation* – needs to be supplemented with a *punitive element* as a punishment for the violation itself. A punitive element must always embody the remaining criteria of article 18, being effective, proportionate, dissuasive and gradually increasing in the case of recidivism. Such a punitive element is important, as it removes the incentive for shipowners who might otherwise dare to be non-compliant. If a fine 'merely' removes the savings from a violation, or the supplementing punitive element is not high enough, it could lead to shipowners' weighing the risks of getting caught during a Port State Control (PSC) (and the potential fine) against the potential savings. It is therefore necessary for the fine to be *dissuasive*, which should ensure that it is also *effective*.

Ensuring that a fine is dissuasive is also in alignment with article 4(4) of the MARPOL Convention, which requires that all parties, including flag States, prescribe penalties for violations of the Convention that are 'adequate in severity to discourage violations of the present Convention and shall be equally severe irrespective of where the violations occur'. This also applies to enforcement of the Annexes to the MARPOL, including Annex VI (see article 1(2) of the Convention).

Article 217(8) of UNCLOS further underlines that the flag State's national laws and regulations for sanctioning violations should allow for penalties that are 'adequate in severity to discourage violations wherever they occur', emphasising that a flag State's *effective enforcement* pursuant to article 217(1) includes imposing dissuasive (discouraging) penalties.

[8] The ICJ Statute is available at https://www.icj-cij.org/en/statute.

[9] The ICJ is one the judicial entities that, according to art 287(1), can settle a dispute regarding interpretation of UNCLOS, eg on the applicability and scope of art 217(1). Art 287, and the other dispute settling mechanisms of pt XV of UNCLOS, is discussed in ch 12.

The requirement in article 18 of the Sulphur Directive for the flag State authorities to exercise *proportionality* when sanctioning could be seen as a double-edged sword. On the one hand, a procedural safeguard that bars the authorities from imposing an unjustly (disproportionate) high fine for the offence, but on the other hand requiring the authorities to impose a fine that removes all savings and includes a punitive element since otherwise it would be an unjustly (disproportionate) low punishment given the crime. Authorities are often skilled in exercising such proportionate sanctioning, as this requirement also applies to the imposition of all penalties, whether criminal sanctions or PSC sanctions.

The principles of effectiveness, dissuasiveness and proportionality also require that fines should always take into account any aggravating or mitigating circumstances. *Repeated infringements* are always seen as an aggravating circumstance, which naturally leads to an increased fine as it reflects a *mens rea* (a guilty mind) that should be penalised more harshly than a first-time offender who exceeds the limit by only a small amount, who could more easily plead mitigating circumstances.

The severe penalties for recurring infringements are extremely important when enforcing the 0.5% sulphur limit, as the savings from infringements can be gargantuan, and any indication that shipowners are likely to reoffend should be met with equally dissuasive punishment.

Many States account for *aggravating circumstances* when calculating a fine by simply multiplying the normal (effective, proportionate and dissuasive) fine by a predetermined factor.

It should be noted that the concept of repeated infringements should *not* be interpreted simply as referring to the same ship's repeating the same violation of either the 0.1% or the 0.5% sulphur limit. If different ships under the same shipowner repeatedly violate any of the sulphur limits under regulation 14 of Annex VI with a 'guilty mind' (ie not involving minor exceedances), this must be viewed as a *repeated infringement* by that shipowner. This is based on the accountability of shipowners for their ships, as the owners bear the legal responsibility[10] for ensuring that all ships in their fleet abide by all relevant rules, including those implemented by the flag State in accordance with article 222. The shipowner also has the power and responsibility to instruct the masters and crews on board the ships to comply with all regulations, and to ensure that they are able do to so by laying down relevant procedures and instructions, for example pertaining to the planning of fuel purchases or fuel changeovers when entering a SECA zone.

Finally, the shipowner is normally the entity that stands to gain from a ship's non-compliance, as the shipowner pays for the fuel purchased and therefore reaps the savings from buying non-compliant fuel; savings that can be used to generate

[10] There are numerous different ways of changing the legal responsibilities of shipowners by entering into different agreements on chartering, etc. These different possibilities are within the realm of private law and are not covered in this book, the reference to the 'shipowner' being used as a term for the legal entity responsible for any non-compliance by the ship.

greater profits, enabling the company (shipowner) to offer lower freight rates than compliant competitors, thereby creating an uneven playing field.

When it comes to determining how many violations the word 'repeated' covers, a strict interpretation indicates that it is more than one, making two violations 'repeated violations'. Then again, the principle of proportionality dictates that the higher the number of repeated violations, the greater the fine (eg using the aforementioned predetermined multiplication factor, which increases in accordance with the number of violations).

The principle of proportionality also dictates that the size of the violation (the amount of sulphur) should be of relevance when calculating a fine, and when discussing recidivism. A substantial violation, for example by using 3.5% fuel instead of 0.5% fuel, is a clear indication of intent to violate, while the use of 0.53% fuel could indicate that the violation was not intentional, as such small instances of non-compliance can be the result of contaminated fuels or contaminated fuel systems. Minor violations should also be sanctioned, of course, but the principle of proportionality dictates that the sanction should be less severe, as the shipowner cannot achieve an economic benefit from such a violation.

Minor violations will in general not generate a profit, as the ship has bought expensive fuel that was labelled as compliant. And 0.53% fuel and those with similar specifications are not the most common fuel products. That being said, any leniency towards violations involving fuels of 0.60% and more should be subject to careful consideration, as shipowners could become astute blenders of different fuel types, which allows for savings and small-scale violations.[11]

Thus recidivism involving substantial breaches of the sulphur limit should be penalised more severely than recurring small-scale violations, as the former suggests the *mens rea* for non-compliance for profitable economic gains.

It should be noted that the term 'aggravating circumstances' can include circumstances other than repeated violations. For example, *wilful* and *serious* violations of the sulphur limit. These circumstances were just described in connection with the scale of sanctions for recidivism. If any such aggravating circumstances are present, the principle of proportionality dictates that these should be met with higher fines, for a first-time offender too. This means that a first-time offender that uses 3.5% fuel should be subject to a higher fine than a first-time offender that uses 0.53% fuel.

Finally, the principles of *effective flag State enforcement* according to article 217(1) of UNCLOS – harmonised through the explicit principles in article 18 of the Sulphur Directive – also apply to other aspects of sulphur enforcement, such as repeated violations of the regulations by use of a faulty or non-functioning EGCS system, falsified Bunker Delivery Notes (BDNs) or recurring

[11] Discussions at PPR5 and the intersessional PPR meeting in 2018 and PPR6 in 2019 revolved around applying a '95% confidence limit' by using ISO Standard 4259 when testing fuel samples, as this would allow a small measure of leniency towards such minor infringements.

misuse of the 'Fuel Oil Non-availability Report' (FONAR) system (including seeking out ports with no compliant fuel), etc. These principles of effective enforcement also apply to enforcement of all IMO regulations for the protection of the marine environment, and will apply to future IMO rules on measures for the reduction of GHG emissions.

Where a flag State fails to meet the requirements in article 217(1), it must automatically be considered to have 'disregarded its obligation to enforce effectively', which is relevant to the discussion of the two exceptions to final flag State jurisdiction laid down in article 228(1) (see chapter 11).

Proceeding with the dissection of article 217 of UNCLOS, the onus on flag States to enforce international environmental regulations effectively may start with article 217(1) but it does not end there, as article 217(2)–(8) also impose different enforcement obligations on the flag State that must be fulfilled. Thus, when assessing whether a flag State has enforced MARPOL Annex VI effectively, all the obligations set out in article 217 must be met, not just those in article 217(1).

II. Obligation to Detain Ships until they Comply with IMO Regulations: Article 217(2)

Flag States must, in accordance with article 217(2) of UNCLOS, take *appropriate measures* to ensure that ships under their flag are prohibited from sailing until they comply with the IMO's rules on environmental protection, including requirements in respect of the design, construction, equipment and manning of vessels.[12] This requires flag States to detain ships under their flag until they comply with all relevant regulations, such as MARPOL Annex VI.

The reference to *equipment* can also encompass EGCSs, thereby requiring the flag State to ensure that a ship equipped with such a system complies with all rules, including that the EGCS is fully functional and has been approved in accordance with regulation 4.1 of Annex VI and the Guidelines for Exhaust Gas Cleaning Systems (MEPC.184(59)).

The geographical scope of article 217(2) must nonetheless be somewhat limited, as a flag State cannot detain ships under its flag when those are sailing outside the flag State's own waters. A flag State could, in principle, pursue, stop and detain ships under its flag on the high seas, but the practical implications

[12] The reference to *design* could be relevant when ensuring flag State enforcement of the EEDI requirements of ch 4 of MARPOL Annex VI. The references to *construction* and *manning* could be seen as references to the International Convention for the Safety of Life at Sea (SOLAS) 1974 (adopted 1 November 1974, entered into force 25 May 1980) 1184, 1185 UNTS 2 and the International Convention on Standards of Training, Certification and Watchkeeping for Seafarers 1978 (adopted 7 July 1978, entered into force 28 April 1984) 1361, 1362 UNTS 2, as violations of these could result in accidents at sea, which again could lead to pollution of the marine environment.

seem insurmountable, and this option cannot be considered a realistic approach to enforcement.

The possibility to detain a ship in accordance with article 217(2) must be when it is in a port, or in the internal or territorial waters of a flag State. Exceptions to this could entail a flag State's entering into bilateral or multilateral agreements with other (port) States, allowing the flag State to carry out inspections of ships under its flag while these are at berth in a foreign port (ie 'flag State inspections'). Any exercise of enforcement measures by a flag State in a foreign port, such as detention according to article 217(2), would nonetheless likely require active assistance from the port authorities, etc in that port State.

A flag State can also simply request another (port) State to detain a certain ship under its flag, and to inspect it on behalf of the flag State.

III. Obligation to Comply with Requirements for Certificates and Surveys: Article 217(3)

Article 217(3) of UNCLOS imposes a duty on flag States to ensure that ships flying their flag have on board all applicable certificates that are mandatory in accordance with international rules for the protection of the marine environment. This includes the certificates required pursuant to MARPOL Annex VI. Many of these are described in chapter 2, including the obligations for ships of 400 GT or above to carry an International Air Pollution Prevention Certificate (IAPP Certificate) and an International Energy Efficiency Certificate (IEE Certificate) (see regulations 5–9 of Annex VI).

The IAPP and IEE Certificates, inter alia, are issued to ensure that all ships, in accordance with regulation 5 of Annex VI, undergo a periodic survey to confirm that all installations, equipment, etc on board comply with the relevant provisions of the Annex.

This is relevant to compliance with article 217(3), as the paragraph also demands that flag States must ensure that ships under their flags are periodically inspected to verify that the certificates are up-to-date and in conformity with the actual condition of the ship.[13] So, to determine whether a flag State has met the conditions in article 217(3), it must prove that ships under its flag carry IAPP and IEE Certificates, and that the required survey intervals have been met pursuant to regulations 5–9 of Annex VI. Such mandatory periodic surveys are often carried out as part of the IMO's specialised Harmonised System of

[13] Art 217(3) does not affect the right of port States to conduct PSC and ensure that all ships are carrying the required documents and certificates. These certificates are, as described in ch 4, among those presented during the initial PSC document inspection; see art 226(1)(a) of UNCLOS.

Survey and Certification (HSSC),[14] which follows a five-year interval scheme of assessment.

It must be underlined that article 217(3) does not apply to BDNs, as these are issued by the fuel supplier, not by the flag State. A missing or non-compliant BDN[15] is therefore not attributable to the flag State's lack of enforcement of Annex VI.

A. Flag State Use of Classification Societies

The meeting of the flag State's obligations to assess and certify compliance on board ships under its flag is often outsourced from the State to private companies, so-called *classification societies*, that are authorised to carry out these mandatory surveys and the subsequent issuing of certificates on behalf of the flag State.[16] The responsibilities and liabilities of a classification society are often established in an agreement between the flag State authorities and the company.[17]

A flag State cannot evade its external responsibilities pursuant to article 217(3) by concluding such inter partes agreements with a private company. This means that if a classification society makes a mistake, for instance not ensuring compliance with MARPOL Annex VI, then this mistake should be directly attributable to the flag State and will be deemed an infringement of article 217(3), albeit that the flag State may have a private law claim against the classification society.

IV. Obligation Ex Officio to Investigate and Initiate Proceedings: Article 217(4)

If a flag State finds, or suspects, that a ship flying its flag has violated IMO regulations for the protection of the environment, that State must immediately investigate this and, where appropriate, initiate proceedings (legal proceedings) in regard to the infringement, irrespective of where the violation occurred.

The scope of article 217(4) must be seen in conjunction with article 217(6), as the latter provision concerns flag State obligations to investigate and instigate

[14] See Resolution A.1104(29) 2015.

[15] A BDN will be deemed non-compliant if it does not meet the standards set out in app 5 to MARPOL Annex VI; see reg 18.5.

[16] For a detailed review of the conditions applying to classification societies, see NM Hosanee, 'A Critical Analysis of flag State duties as laid down under article 94 of the 1982 United Nations Convention on the Law of the Sea', The United Nations-Nippon Foundation Fellowship Programme, 2009–10, 43–52, section 2.4.

[17] The surveys and certification carried out by classification societies on behalf of flag States relating to art 217(3) should not be confused with the safety assessments made by, or for, insurance companies to determine the ship's risk profile prior to insuring it and setting an insurance premium. The latter procedure is often referred to as 'vetting'.

proceedings following official information on a violation from another State. Article 217(4) is therefore relevant where flag States instigate an ex officio investigation based on information provided by States themselves or by non-State actors, for example by employees in the shipping sector using official *whistleblower systems*, or by other maritime sources, such as other ships, pilots, etc. 'Unofficial' (ie unwritten) information from another State will also be information encompassed by article 217(4), as article 217(6) concerns *written requests* by another State.

Consequently, a flag State is, according to article 217(4), required to investigate and, where the information so warrants, institute proceedings, if it has knowledge of, or suspects, that a violation of MARPOL Annex VI has taken place.

Article 217(4) (*in fine*) also includes a very important point, which this book has cited numerous times. It provides that a flag State's obligation to investigate and prosecute is 'without prejudice to articles 218, 220 and 228'. This is – in the view of this author – an important cross-reference, as it is made explicit that flag State jurisdiction pursuant to article 217 must respect the exceptional extraterritorial jurisdictions allowing port States to enforce on the high seas (article 218) and for coastal States to enforce in their own waters and EEZs (article 220). It must also respect how article 228(1) delimits the overlapping jurisdiction between a flag State and a coastal or port State. The provisions mentioned in article 217(4), articles 218, 220 and 228, can therefore be regarded as some of the *codified exceptions to the flag State principle* to which article 92 of UNCLOS refers.

V. Right to Request Assistance from Other States: Article 217(5)

When a flag State is investigating an alleged violation in accordance with Article 217(4) of UNCLOS, article 217(5) allows the flag State to call upon other States for further information to help clarify the circumstances of a case. The other States should endeavour to comply with such *appropriate* requests.

Article 217(5) thereby sets itself apart from the other paragraphs of the article, as it does not place an obligation on the flag State but grants it a legal basis for requesting any help and assistance it may need to fulfil the obligations under article 217.

VI. Obligation to Investigate and Prosecute Alleged Violations: Article 217(6)

As previously specified, under article 217(6), a flag State is obliged to start an investigation of an alleged violation committed by a ship under its flag if it receives a written request to do so from another State.[18] This provision also explicitly calls for flag States, without delay, to institute proceedings in respect of the alleged violation if the evidence provided or obtained so warrants.[19]

A foreign State could send such a request to a flag State following any findings made during a PSC or – when exclusively focusing on sulphur limits enforcement – following a sulphur measurement from a Continuous Emission Monitoring System (CEMS), or measurements made by a sniffer attached to a drone, bridge, helicopter, etc. For example, if excess sulphur was measured by a sniffer while the ship made an innocent or transit passage through a coastal State's territorial sea.

It should be noted that article 217(6) uses the term 'alleged violation', meaning that other States are not required to present solid evidence at this stage. Mere indications of an infringement, such as sniffer measurements, are sufficient to oblige the flag State to act by investigating and, if warranted, prosecuting.

The obligation in article 217(6) is repeated in regulation 11.4 of MARPOL Annex VI, which requires flag States to respond to any notification from a port State regarding any possible infringements of the Annex, including regulation 14. It also clearly demands that the flag State instigate legal proceedings 'as soon as possible' if a violation has been proved. Regulation 11.4 reads:

> Upon receiving such evidence, the Administration so informed shall investigate the matter, and may request the other Party to furnish further or better evidence of the alleged contravention. If the Administration is satisfied that sufficient evidence is available to enable proceedings to be brought in respect of the alleged violation, it shall cause such proceedings to be taken in accordance with its law as soon as possible. The Administration shall promptly inform the Party which has reported the alleged violation, as well as the Organization, of the action taken.

This requirement is also found in article 4(1) of the MARPOL Convention, which provides:

> If the Administration is informed of such a violation and is satisfied that sufficient evidence is available to enable proceedings to be brought in respect of the alleged violation, it shall cause such proceedings to be taken as soon as possible, in accordance with its law.

And under article 4(2)(b), any party to the Convention should 'furnish to the Administration of the ship such information and evidence as may be in its possession that a violation has occurred'.

There is a consistency in the responsibility of the flag State to investigate and proceed against violations of the sulphur limits under regulation 14 pursuant to article 217(6) of UNCLOS, article 4(1) and (2)(b) of the MARPOL Convention and regulation 11.4 of MARPOL Annex VI.

[18] The same principle is found in the second indent of art 94(6).

[19] See JA Zerk, *Extraterritorial Jurisdiction: Lessons for the Business and Human Rights Sphere from Six Regulatory Areas*, Report of The Harvard Corporate Social Responsibility Initiative (June 2010) 195.

The requirement for flag States to initiate proceedings when a violation is proved is highly dependent on the subjective assessment made by the flag State authorities regarding the strength of the evidence. In spite of that, the authorities in a (open registry) flag State should not automatically be able to dismiss all information and evidence from other States by officially referring to the lack of sufficient evidence, while unofficially looking to offer leniency to ships sailing under their flag. This conclusion is supported by the underlying principle of article 223, which requires all States, including flag States, to 'take measures to facilitate the hearing of witnesses and the admission of evidence submitted by authorities of another State, or by the competent international organization'.

Thus, a flag State must investigate, under article 217(6), if another State forwards objective and credible information that indicates that a violation of regulation 14 of Annex VI has occurred. Article 217(6) also requires the flag State to institute legal proceedings if another State can forward indisputable proof of a violation of the sulphur limits of Annex VI. This might occur, for example, if a CEMS or a sniffer attached to a drone shows a violation of the global 0.5% limit on the high seas, and a subsequent fuel sample in the next port of call confirms this, as the carriage ban prohibits the presence of non-compliant fuel in all fuel tanks.[20] This should constitute sufficient evidence to require the State to sanction the violation pursuant to article 217(6). The sniffer and CEMS technology might also be developed to offer such precise and accurate results as those achieved by laboratory-tested fuel sampling. This would obligate flag States to institute proceedings on the sole basis of sniffer detection.

On the other hand, should a flag State fail to commence legal proceedings in such circumstances, this would automatically be considered an instance of the State's disregarding its duty to enforce effectively under article 217(6). See chapter 11 on article 228(1) for further details.[21]

VII. Obligation to Inform the IMO and All States of All Enforcement: Article 217(7)

Article 217(7) of UNCLOS obliges the flag State to inform the 'requesting State' and the IMO promptly of the action it has taken after receiving information on an alleged violation pursuant to article 217(6). The flag State must also subsequently make this information available to all States.

[20] See ch 1 for discussion of the carriage ban, sniffer technology and fuel sampling, etc.

[21] A flag State's failure to comply therewith could, to some extent, also be deemed a violation of its duty to other contracting States that are parties to UNCLOS or MARPOL Annex VI, ie violating the *pacta sunt servanda* principle embodied in art 26 of the Vienna Convention on the Law of Treaties 1969 (adopted 22 May 1969, entered into force 27 January 1980) 1155 UNTS 331 (VCLT). The flag State would also breach art 300 of UNCLOS, which, inter alia, provides that parties to UNCLOS 'shall fulfil in good faith the obligations assumed under this Convention'.

The information that must be provided pertains to the action taken and the outcome of the investigation and legal proceedings. This means that the flag State must disclose the size of any fine and any other sanctions imposed (imprisonment, etc).

This information must be forwarded *promptly*, which, in the view of this author, means immediately after a final verdict has been rendered or the deadline for appeals has expired. The obligation to make the information available to all States is not subject to the same demand for timely notification.

Obviously, article 217(7) plays a key role when determining whether a flag State has fulfilled its responsibilities pursuant to article 217 to take action against violations of regulation 14 of Annex VI.

The fact that the flag State is under a legal obligation to disclose information on the outcome of its investigation and legal proceedings, including information on its sanctions, enables other States and the IMO to determine if a flag State's fine was *effective* in accordance with article 217(1). This includes assessing if a fine fulfils the criteria of being effective, proportionate, dissuasive, stripping (confiscating) all economic gains from the infringement and increasing in the event of aggravating circumstances, such as repeated violations.

The obligation to inform laid down in article 217(7) is also found in article 4(3) of the MARPOL Convention, which states:

> Where information or evidence with respect to any violation of the present Convention by a ship is furnished to the Administration of that ship, the Administration shall promptly inform the Party which has furnished the information or evidence, and the Organization, of the action taken.

The information provided in accordance with article 217(7) of UNCLOS allows other States and the IMO to evaluate whether the flag State has used the information and proof provided by the requesting State, in accordance with article 217(6)[22] and the principles in article 4(1) and (2)(b) of the MARPOL Convention and regulation 11.4 of Annex VI, to proceed against a violation of the 0.1% or 0.5% sulphur limit effectively.

The clarity of the wording in article 217(7) leaves no room for interpretation by flag States to avoid providing the mandatory information, and any failure to so amounts to *disregarding* its obligations, with the result set out in article 228(1) (see chapter 11).

Some flag States may oppose providing information pursuant to article 217(7) by referring to national legislation that prevents or restricts the authorities from disclosing information or the outcomes of criminal proceedings. To this end article 27 VCLT dictates that 'A party may not invoke the provisions of its internal law as justification for its failure to perform a treaty.' This means that if a flag State voluntarily has become a party to UNCLOS, it must comply with article 217(7)

[22] And the principle in art 223 of UNCLOS.

and cannot use its national public access legislation as an excuse for not promptly furnishing the requesting State and the IMO with the required information, and subsequently making it accessible to all States.[23]

It should also be noted that the information that the flag State must provide in accordance with article 217(7) need not contain sensitive *personal* information, as the information required merely relates to the sanction (fine) imposed on the shipowner, which often is a *legal person* (a company).

Any information on sanctions (fines and/or imprisonment) imposed on *physical persons*, whether the master, chief engineer, other crew members, the director of the company, etc, need not include any information that might reveal the identity of the sanctioned person but only information on the sanction itself, to determine if the enforcement has been *effective*. The information provided pursuant to article 217(7) will also not be made available to the public or the press, as only States (ie competent authorities) should have access to this information, and employees working at the authorities are normally subject to a duty of confidentiality.

To reiterate the statement made in chapter 4, the information provided by the flag State should essentially answer two questions: What was the violation? And what was the sanction? Information that flag States must enter into the GISIS system, discussed next.

A. Using the GISIS Module for Flag State Reporting

As described previously, the flag State reporting obligation in article 217(7) corresponds to the reporting obligation in regulation 11.4 of MARPOL Annex VI, for a flag State to inform the IMO and other States of the action taken by the flag State when responding to a violation of the Annex committed by a vessel under its flag. This might include, for instance, action taken for violations of the sulphur limits in regulation 14.

The information provided to the IMO could be registered in the GISIS database, as it would then be accessible by all IMO Member States, thus simultaneously fulfilling the obligations to inform the IMO and the reporting State, *and* make the information available to all States.

The GISIS database, which has a designated MARPOL Annex VI module, could be used for collecting and evaluating the mandatory enforcement reports from flag States pursuant to article 217(7) and regulation 11.4 of Annex VI, as this is valuable when examining the applicability of the second exception in article 228(1) (discussed in chapter 11).

[23] Art 302 of UNCLOS establishes that a State, including a flag State, is never required to provide information that discloses information contrary to essential security interests. The information that needs to be provided in accordance with art 217(7) can never be exempt under art 302, though, as information on enforcement can never be deemed information contrary to essential security interests.

VIII. Obligation to Ensure National Legislation Can Enforce Effectively: Article 217(8)

As briefly touched upon during the clarification of the scope of effective enforcement pursuant to article 217(1) (see section I), article 217(8) establishes that the laws and regulations of the flag State must not present a bar to imposing sanctions (fines) that are severe enough to 'discourage violations wherever they occur'.

This provision requires flag States to ensure that their national laws allow for the sanctioning of violations of environmental legislation, including implemented international regulations in accordance with article 211(2) or article 222, with dissuasive penalties comprising *effective* fines.

The principle under article 217(8) is also in alignment with article 4(4) of the MARPOL Convention, which stipulates:

> The penalties specified under the law of a Party pursuant to the present article shall be adequate in severity to discourage violations of the present Convention and shall be equally severe irrespective of where the violations occur.

Pertaining to enforcement of the sulphur regulations in MARPOL Annex VI, article 217(8) of UNCLOS and article 4(4) of the MARPOL Convention thus task flag States with ensuring that their national legislation does not present a bar to exercising *effective enforcement* in accordance with article 217(1) of UNCLOS (and article 18 of the Sulphur Directive).

This also means that flag States are required to guarantee that their national laws can take effective action against violations of the global 0.5% limit wherever an infringement occurs, including on the high seas. This includes imposing fines that remove (confiscate) economic gains and contain a punitive (dissuasive) element. Such fines could potentially be in the range of, and above, \$1 million, given the calculations set out in chapter 1 regarding the possible savings from infringing the regulations (eg \$750,000 from one non-compliant trip from Asia to Europe). A flag State's laws must allow for such substantial fines.

The national laws that could limit the possibility for effective sanctioning, and thereby violate article 217(8), could be a flag State's national Criminal Code or the relevant environmental or maritime laws, especially if these allow the relevant (environmental or maritime) authorities to proceed against violations directly with administrative fines.

A. Conflict with National Laws on Administrative Fines

Some States issue so-called *administrative fines*, which enable the responsible authority, often an environmental or maritime authority, to prescribe fines without involving the police, prosecutors and the judicial system.

Some national laws, which provide the legal basis for imposing such administrative fines, also set a safeguarding maximum for an administrative fine.

If a flag State's national law provides for such a maximum on fines for infringements of implemented IMO environmental regulations, such as MARPOL Annex VI, and that maximum is too low to remove economic savings and impose a dissuasive punitive element, the national law is explicitly in breach of article 217(8), but also in breach of article 217(1), as the State's enforcement will be ineffective.[24]

Further, as was the case with article 217(7), a flag State that is party to UNCLOS and/or MARPOL cannot, under article 27 VCLT, use its national legislation on administrative fines as an excuse for not adhering to its obligations to comply with article 217(1) and (8) of UNCLOS and article 4(4) of the MARPOL Convention.

IX. Flag State Obligations under Article 223

On a final note, it should be mentioned that article 223 of UNCLOS consists of obligations that all States, including flag States, must fulfil – obligations that are worth mentioning when examining the legal basis for enforcing MARPOL Annex VI.

The requirement laid down in article 223, to facilitate the hearing of witnesses and the admission of evidence, was briefly mentioned when examining article 217(6) (see section VI). Article 223 also requires all States to 'facilitate the attendance at such proceedings of official representatives of the competent international organization, the flag State and any State affected by pollution arising out of any violation'.

The article also stresses that the official representatives attending such proceedings shall have such rights and duties as may be provided under national and international law. These are presumed to be some of the rights that are normally considered diplomatic rights (see chapter 12).

This means that flag States are not only required to investigate and initiate proceedings under article 217(6) and inform the IMO and other States of the outcome of these proceedings under article 217(7); the flag State must also allow representatives from the IMO to attend such proceedings.[25]

[24] If the national law that provides the legal basis for imposing administrative fines also allowed certain cases to be referred to a court so that it might impose fines higher than the legislative limit, the national law would not be in breach of art 217(8).

[25] Art 223 is studied further in ch 12.

X. Conclusion on Flag State Obligations Pursuant under Article 217

It can be concluded that article 217 of UNCLOS – compared to articles 92 and 94 – is the *lex specialis* provision for determining flag State responsibilities in taking action against violations of all environmental legislation, including MARPOL Annex VI.

Article 217 therefore forms the basis for determining whether flag State enforcement of regulation 14 of Annex VI is adequate or inadequate, even though article 222 specifically regulates air pollution. However, article 222 does not set any specific requirements, while article 217 fills that legal void and is applied in conjunction with article 222.

Seven of the eight paragraphs of article 217 set out obligations that must be fulfilled by a flag State to establish that it has lived up to its duty of effective enforcement. (Article 217(5) does not impose an obligation on the flag State but instead allows it to reach out to other States to request their assistance.)

Pertaining to enforcement of the sulphur limits in MARPOL Annex VI – and other emission rules, including future GHG regulations – a flag State must ensure:

- That the enforcement is *effective* pursuant to article 217(1). This means that any sanction (fine) must be effective, proportionate, dissuasive, deprive the shipowner of the economic gains from the infringement, and increase in light of aggravating circumstances such as repeated infringements, that is, as the principles of article 18 of the EU Sulphur Directive are applied.
- That any ship under its flag that is not complying with MARPOL Annex VI is detained when sailing in areas under the flag State's jurisdiction, until the ship complies (see article 217(2)). This includes ensuring that the ship uses compliant fuel and that any EGCS is approved, functioning and certified in the IAPP Certificate.
- That ships flying its flag have all mandatory certificates, such as the IAPP and IEE Certificates, on board and are periodically surveyed in accordance with article 217(3).
- That all suspected violations, brought unofficially to the attention of the flag State or observed by the flag State itself, are investigated, and that all infringements are proceeded against (prosecuted) pursuant to article 217(4).
- That all suspected violations brought to the attention of the flag State by an official written request from another State are investigated and proceeded against in accordance with article 217(6) and regulation 11.4 of MARPOL Annex VI. The flag State can rely on any evidence and proof supplied by the other State, such as CEMS or sniffer measurements, etc. (See also article 223.) The flag State can request assistance from other States when investigating such alleged violations, as laid down by article 217(5).

- That the State that provided the information under article 217(6) and the IMO are immediately informed by the flag State of its enforcement action. This information must afterwards also be made available to other States pursuant to article 217(7) and regulation 11.4 of Annex VI. The IMO could make such received information available to all States in the GISIS database. It is noted that flag States cannot excuse themselves from complying with this obligation by referring to its being incompatible with their national law (see article 27 VCLT).
- That the national law of the flag State does not present a bar to sanctioning violations of regulation 14 of Annex VI by the imposition of dissuasive penalties (fines) (see article 217(8)). Sanctioning by the use of penalties that discourage further offending is also a requirement of the obligation for effective enforcement under article 217(1) of UNCLOS and article 4(4) of the MARPOL Convention.
- That flag States must, in accordance with article 223, allow authorised official representatives from the IMO to attend the proceedings.

These obligations must all be fulfilled if a flag State is to show that it has taken effective enforcement action against violations of regulation 14 of Annex VI, including violations of the 0.5% sulphur limit occurring on the high seas.

If these criteria are *not* all met, it must be presumed that the flag State has 'disregarded its obligation to enforce effectively', as set out in the second exception in article 228(1), which is analysed in chapter 11.

9

Special Jurisdiction for Coastal States: Article 220

A coastal State's jurisdiction historically has extended to areas under its sovereignty covering internal and territorial waters, which is codified in article 2 of UNCLOS.[1] Certain exceptions – some of which are described in chapter 3 – curtail the jurisdiction in the territorial sea pertaining to the rights to *innocent passage* and *transit passage*.

UNCLOS also represented a development of the international law of the sea for coastal States by, in article 3, empowering those States to claim a 12 nautical mile (nm) territorial sea (instead of the previously accepted 3 nm) and a 200 nm Exclusive Economic Zone (EEZ) in accordance with part V of the Convention.

Coastal States do not have an unlimited sovereign claim over the EEZ, which is reflected in the limited jurisdiction that can be exercised in this area, exemplified by the noticeable differences in article 211,[2] between the provisions permitting coastal States to adopt national legislation for the protection of the environment in internal and territorial waters (article 211(3)–(4)) and those covering the adoption of national laws governing the EEZ (article 211(6)).[3]

Article 220 is a *lex specialis* provision for coastal State enforcement against violations of the International Maritime Organization's (IMO's) environmental regulation in the different areas (waters and zones) described in chapter 3. The article grants the coastal State different jurisdictions for proceeding against such violations committed by foreign ships. Under article 220(1), there is almost unlimited jurisdiction over all violations in all areas and zones, provided the ship subsequently calls into a port or at an off-shore terminal in the coastal State.

Special jurisdiction exists over violations taking place during passage of the territorial sea (see article 220(2)).

Article 220(3)–(8) set out the jurisdictional basis for enforcing all environmental legislation in the EEZ, providing for the possibility of contacting,

[1] United Nations Convention on the Law of the Sea (adopted 10 December 1982, entered into force 16 November 1994) 1833 UNTS 3.

[2] See ch 7 for an analysis of art 211.

[3] Y Tanaka, *The International Law of the Sea*, 2nd edn (Cambridge University Press, 2015) 293.

stopping, investigating and prosecuting ships that have violated such regulations in the EEZ.[4]

The extension of a coastal State's jurisdiction to cover violations of environmental rules in the EEZ outside areas under its sovereignty (as defined in article 2 of UNCLOS) represents *extraterritorial jurisdiction* that has historical roots dating back to before the adoption of UNCLOS in 1982.

Coastal States, their coastlines and the marine life living therein have always been vulnerable to ships discharging or spilling harmful substances, such as oil and chemicals, into the sea, whether such pollution is accidental or the result of collisions at sea. This was – as described in previous chapters – one of the reasons and one of the concerns that led to the enacting of some of the earliest international regulations for the protection of the marine environment, such as the International Convention for the Prevention of Pollution of the Sea by Oil (OILPOL) in 1954[5] and the International Convention Relating to Intervention on the High Seas in Cases of Oil Pollution Casualties ('the Intervention Convention') in 1969.[6] Yet when examining the coastal State's enforcement, one must distinguish between IMO regulations on *air pollution* and other environmental IMO regulations.

As concluded in chapter 7, article 222 is in principle the primary *lex specialis* provision for a coastal State's enforcement of rules on air pollution, including MARPOL Annex VI[7] and future greenhouse gas (GHG) legislation. The geographical scope of article 222 is, however, limited to *internal and territorial* waters. This means that article 220(1) and (3)–(8) provide the *lex specialis* jurisdictional basis for enforcing air pollution regulations in the *EEZ*, including the 0.1% or 0.5% sulphur limits in Annex VI.

As was the case with article 222 in relation to article 217, article 222 does not provide any clear insight as to how a coastal State should enforce these rules on air pollution in its internal and territorial waters. The conclusion as regards 'article 222 v article 220' must therefore be the same as that in 'article 222 v article 217', so that the rather toothless legal content of article 222 must be supplemented by the more specific provisions of article 220(1)–(2), applicable to the enforcement of all international environmental regulations, including MARPOL Annex VI, in internal and territorial waters.

[4] NM Hosanee, 'A critical analysis of flag State duties as laid down under article 94 of the 1982 United Nations Convention on the Law of the Sea', The United Nations-Nippon Foundation Fellowship Programme, 2009–10, 69.

[5] International Convention for the Prevention of Pollution of the Sea by Oil (OILPOL) (adopted 12 May 1954, entered into force 26 July 1958) UNTS 4714.

[6] The Convention Relating to Intervention on the High Seas in Cases of Oil Pollution Casualties (adopted 29 November 1969 and entered into force 6 May 1975). 970 UNTS 211 ('the Intervention Convention').

[7] International Convention for the Prevention of Pollution from Ships (adopted on 11 February 1973, as modified by the Protocol of 17 February 1978, entered into force 2 October 1983) 1340 UNTS 61 (MARPOL), Annex VI, IMO Publication: IMO-520E.

Such enforcement in internal and territorial waters can be very important to some coastal States, as air pollution is something that can have an immense adverse impact on the population of the State, especially those living in cities near larger ports and shipping lanes. It may be recalled that the environmental reports referenced in chapter 1 showed that a clear majority of the 137,000 early deaths and 7.6 million asthma cases among children each year, attributable to sulphur pollution from ships, occur in such areas.[8]

I. Jurisdiction under Article 220

A. Full Jurisdiction Over Ships Voluntarily Calling into a Port or at an Off-shore Terminal: Article 220(1)

Article 220(1) of UNCLOS provides a coastal State with the legal basis for enforcing the IMO's rules and regulations for the protection of the marine environment in all areas and zones under its full or partial jurisdiction, that is, in internal waters, the territorial sea and the EEZ.

For the coastal State to exercise jurisdiction under this provision over a foreign ship, that ship must, after the violation, voluntarily call into a port or at an off-shore terminal in the State. The safeguards in section 7 of part XII of UNCLOS must also be observed by the coastal State (see chapter 12).

This means that a coastal State can proceed against (including prosecute) infringements of MARPOL Annex VI (and future GHG legislation) in its internal and territorial waters and in the EEZ pursuant to article 220(1) and in accordance with article 222,[9] provided the foreign ship afterwards calls into a port or at an off-shore terminal in the State.

There has been a general misconception that the right for coastal States to enforce environmental rules in the EEZ is solely found in article 220(3)–(8), as these paragraphs explicitly refer to the EEZ and lay down detailed demands for coastal State enforcement in this zone. Nevertheless, it must be observed that these regulations pertain to the right of coastal States to question, stop, inspect, detain and prosecute a foreign ship that sails straight through the EEZ and thereby exercises its right to the freedom of transit through the zone pursuant to article 58(1). But if a ship violates an environmental regulation, such as regulation 14 of MARPOL Annex VI, in the EEZ and then voluntary sails into a port

[8] M Sofiev et al, 'Cleaner fuels for ships provide public health benefits with climate tradeoffs' (2018) 9 (article no 406) *Nature Communication* 6, available at https://www.nature.com/articles/s41467-017-02774-9#ref-link-section-d1456e583.

[9] The correct legal reference when referring to coastal State enforcement of air pollution regulations, such as Annex VI, in internal or territorial waters should also include a reference to art 222, eg 'pursuant to art 220(1) in accordance with art 222'.

in the coastal State, then that State has full jurisdiction over the infringement pursuant to article 220(1). This is jurisdiction that overlaps with the flag State's jurisdiction under article 217(1), a matter that can be resolved in accordance with article 228(1) (see chapter 11).

B. Enforcing Violations Committed by Ships Making an Innocent Passage: Article 220(2)

As described in chapter 3, article 21(1)(f) of UNCLOS permits coastal States to adopt national laws and implement international legislation for the protection of the marine environment, by which foreign ships making an innocent passage through the territorial sea must abide (see article 21(4)).

If a ship violates the rules and regulations found in part II, section 3 of UNCLOS (including article 21) during its innocent passage through territorial waters, the coastal State may, in accordance with article 220(2), inspect the ship and, where the evidence so warrants, detain the vessel and institute legal proceedings. Under article 220(2) the coastal State must have *clear grounds* for believing that such an infringement has occurred before carrying out these enforcement measures. The procedural safeguards of section 7 of part XII must also be adhered to (see chapter 12).

When discussing enforcement of air pollution regulations in MARPOL Annex VI, such as SOx (regulation 14), or NOx (regulation 13) or future GHG/CO_2 measures, violations can potentially be detected – or at least suspected – in numerous ways without the ship's having to call into a port and undergo Port State Control (PSC). Some of the different emission surveillance systems referred to in chapter 1 include sniffer systems (attached to drones, bridges, helicopters, other ships, etc) and Continuous Emission Monitoring Systems (CEMSs), which are installed in the funnel of a ship. These systems can all indicate, and perhaps even unequivocally prove, a violation of such emission regulations.

If such a system shows that a ship is non-compliant, for instance with regulation 14, during its innocent passage through the territorial sea, and thereby is in violation of UNCLOS article 21(1)(1) and (4), it would constitute *clear grounds* for the coastal State to believe that such an infringement has taken place, enabling the State to stop, investigate, detain and prosecute the ship(owner) in accordance with article 220(2) and pursuant to article 222.[10]

It would also constitute *clear grounds* if a coastal State were to receive information from the IMO (through the GISIS system) or another State (perhaps the flag State) of a foreign merchant ship's suspected non-compliance while passing

[10] Indicating and/or proving violations of certain non-visible discharge limits, such as infringements of wash water discharge limits for open loop scrubbers, could be difficult. Coastal States would therefore often not have *clear grounds* for believing that a violation of the wash water limits has taken place, unless a Continuous Monitoring System (CMS) shows such a violation of these limits.

through the territorial sea. The coastal State could take measures, including investigating and imposing penalties, against such a violation, as article 27(5) of UNCLOS, as discussed in chapter 3, refers to the applicability of part XII of UNCLOS, which includes article 220(2), thereby allowing coastal State to disregard the usual restrictions on hindering a foreign ship during its *innocent passage*.

C. Requiring Information from Ships in the EEZ: Article 220(3)

Article 220(3) stipulates that if a coastal State has *clear grounds* for believing that a foreign vessel has violated *international* rules for the protection of the environment whilst sailing in the EEZ, that State 'may require the vessel to give information regarding its identity and port of registry, its last and its next port of call and other relevant information required to establish whether a violation has occurred'.

Before analysing the content of this paragraph, it should first be noted that although it directly refers to *international* rules, article 220(8) stipulates that article 220(3) (and article 220(4)–(7)) also applies to enforcement of a coastal State's *national* environmental laws, for example regarding a Particularly Sensitive Sea Area (PSSA), adopted in accordance with article 211(6), as described in chapter 7.

Second, article 220(3) lays down the same requirement as in article 220(2), for the coastal State to have *clear grounds* before invoking the provision. The standard used for determining clear grounds under article 220(2), in situations in territorial waters, could therefore, in principle, also be used in article 220(3) situations in the EEZ, for instance if a sniffer or CEMS points towards a violation of the sulphur limits under MARPOL Annex VI, or if a CMS indicates a discharge violation from a ship using an open loop scrubber system. Information from the IMO or another State on non-compliance would also constitute such clear grounds.

The coastal State can require the necessary information from a foreign ship pursuant to article 220(3) if the ship is still sailing in the EEZ, or if it has continued into territorial waters where it is making an innocent passage.[11] The information the ship must provide is to disclose its identity (name and IMO number), port of registry (flag State) and its last and its next port of call. Also, it must furnish 'other relevant information required to establish whether a violation has occurred', which, when determining whether a violation of regulation 14 of Annex VI has taken place, could include information on what fuel is being used, where and when it was purchased, and by which fuel provider.

[11] If the ship is not making an innocent passage, eg is at anchor in territorial waters at an off-shore terminal, or if it continues into a port in internal waters, the jurisdictional basis for the coastal State to investigate and proceed against the infringement will shift to the wider jurisdiction under art 220(1), as this provision will then become applicable.

Article 220(3) clearly shows how the coastal State's jurisdiction is limited in the EEZ compared to its jurisdiction in internal and territorial waters, as it merely allows the questioning of ships, while article 220(1) and (2), inter alia, allow for investigation and the commencement of proceedings.

D. Obligation for Flag States to Implement Rules on Providing Information: Article 220(4)

Article 220(4) sets itself apart from the other paragraphs of article 220 as it does not impose obligations on, or bestow rights on, the coastal State. Instead, it requires a flag State to implement rules in its legislation ensuring that ships under its flag provide the information a coastal State may request in accordance with article 220(3). Pursuant to article 220(8), this includes information needed to determine if a ship has violated national legislation for the EEZ adopted by the coastal State in accordance with article 211(6).

Article 220(4) is not studied in further detail, but it should be noted that flag States are also required to enforce such legislation pursuant to article 217, which means that they must assist the coastal State in any way to ensure that a ship under their flag provides the information requested.

E. Proceeding against Violations in the EEZ Causing Significant Pollution: Article 220(5)

Article 220(5) also refers to the aforementioned 'article 220(3) situations' where a coastal State has *clear grounds* for believing that a violation has occurred in the EEZ and the ship is present in the EEZ or in the territorial sea.

Compared to article 220(3), article 220(5) expands the jurisdiction for a coastal State to act following such violations, as it offers a legal basis for such a State to stop and inspect a ship sailing in the EEZ or territorial sea. This wider jurisdictional basis for enforcement is, however, curtailed in comparison with article 220(3), by the setting of further requirements for its applicability.

First, the stopping and investigation of the foreign ship must be a result of the ship's not providing information in accordance with article 220(3), or the information provided being insufficient. The coastal State may also undertake an inspection of the ship if it has reason to believe that the information is inconsistent with the factual and actual circumstances of the case, meaning that the coastal State has reason to believe the information provided by the ship is false or erroneous.

Second, the scope of article 220(5) is narrower than that of article 220(3) as it refers to 'a substantial discharge [in the EEZ] causing or threatening significant pollution of the marine environment'. Chapter 10 of this book, relating to port

State jurisdiction pursuant to article 218 of UNCLOS, concludes that the term 'discharge' includes *emissions*, such as sulphur emissions. The reasoning behind this conclusion will not be elaborated upon in this chapter, but it should be noted that while article 218(1) refers to *any* discharge – that is, the broadest possible application of the term 'discharge' – article 220(5) narrows the scope of the term, as seen from the wording of the paragraph quoted above.

Although air pollution is extremely damaging to the environment in general, including to human health, this source of pollution, especially when considering the release of SOx, NOx and CO_2 (GHG), does not create specific, tangible damage to a coastal State.[12] Reputedly, it instead contributes to the overall accumulation of air pollution, thereby reducing the air quality, with adverse effects on human health and the environment. Emissions of air pollution could, from an overall global perspective, be seen as being more damaging than other pollution, for instance more damaging than discharge of oil into the sea, which the term 'discharge' of course encompasses. Nonetheless, as article 220(5) refers to the discharge's being substantial and possibly causing significant pollution, etc, there must be a direct causal link between a specific ship's infringement of the discharge regulation and the specific pollution of the marine environment that results. Distinguishing between the different forms of discharge is necessary as not all discharge violations are covered by article 220(5), since they must result in tangible pollution damage.

Specific violations by a foreign ship of MARPOL Annex VI, should not – although also being *discharge violations* (see chapter 10) – be regarded as violations that potentially can cause specific, perceptible pollution damage to the marine environment of the coastal State. Every infringement contributes instead to the accumulating air pollution, as a long-range transboundary pollutant can travel great distances before it undergoes a chemical reaction or is inhaled by humans.

F. Enforcement against Violations in the EEZ Causing Major Damage: Article 220(6)

Article 220(6) expands the jurisdictional basis of article 220(5), so that coastal States, rather than merely taking measures against violations by stopping and

[12] See Professor Henrik Ringbom, who comes to the same conclusion regarding art 220(5) and (6) of UNCLOS on sulphur pollution: H Ringbom, 'Enforcement of the Sulphur in Fuel Requirements: Same, Same But Different', available at https://webcache.googleusercontent.com/search?q=cache:EBk8-0aReA0J:https://www.duo.uio.no/bitstream/handle/10852/61600/SO-Artikel-Ringbom.pdf%3Fsequence%3D4%26isAllowed%3Dy+&cd=1&hl=da&ct=clnk&gl=no, 10–11. See also Yoshifumi Tanaka, who asserts the same regarding Energy Efficiency Measures and art 220(5) and (6) of UNCLOS: Y Tanaka, 'Regulation of Greenhouse Gas Emissions from International Shipping and Jurisdiction of States' (2016) 25 *Review of European, Comparative and International Environmental Law* 339.

investigating a foreign ship, may – if the evidence warrants it – detain the vessel and institute proceedings.

The scope of article 220(6) is to a certain extent the same as that of article 220(5), as it also refers to violations encompassed by article 220(3) and to the fact that a discharge has taken place in the EEZ and that the ship is subsequently sailing in the EEZ or in the territorial sea.[13]

The wider coastal State jurisdiction for enforcement under article 220(6) is limited, though, by stricter requirements for invoking the paragraph than those applying to article 220(5).[14] While article 220(5) (and article 220(3)) refer to the coastal State's having 'clear grounds' for presuming that a violation has occurred, article 220(6) sets the evidence bar somewhat higher, requiring that the State has 'clear objective evidence' of a foreign ship's (discharge) violation in the EEZ.

This criterion requires the coastal State to prove that a certain discharge is attributable to a specific ship. For example, an inspection carried out in accordance with article 220(5) may prove that such a violation has taken place, or aerial or satellite surveillance may clearly prove a ship's discharge of oil.

Another criterion is introduced in article 220(6) that narrows the applicability of the paragraph, in comparison with article 220(5), as article 220(6) refers to the discharge's 'causing major damage or threat of major damage to the coastline or related interests of the coastal State', it being recalled that article 220(5) simply refers to 'significant pollution of the marine environment'.

This must also be seen as setting a more demanding standard for what sort of damage may result in the detention and prosecution of a foreign ship using its freedom to transit the EEZ pursuant to article 58. Consequently, if an infringement of article 220(6) has been established, it must ipso facto be assumed that a breach of article 220(5) has also taken place.

This conclusion is supported by the ECJ in the *Bosphorus Queen*:

> It is unnecessary, in principle, to take account of the concept of 'significant pollution' referred to in Article 220(5) of the United Nations Convention on the Law of the Sea when applying Article 220(6) …[15]

When it comes to defining exactly what the term 'related interests' means in article 220(6), UNCLOS is of no assistance, even though that term is also used in articles 142(3), 211(1) and (7), and 221(1). However, both the Intervention Convention (in article II(4)) and the Nairobi Convention (in article 1(6))[16] define

[13] It should be recalled that the regulation referred to in art 220(3) can encompass nationally adopted (discharge) laws. See also art 211(6) and art 220(8). Also recalling that the ship, if it continues into internal waters, is subject to the coastal State's full jurisdiction to enforce pursuant to art 220(1).

[14] As noted by the European Court of Justice (ECJ) in Case C-15/17 *Bosphorus Queen Shipping Ltd Corp v Rajavartiolaitos*, ECLI:EU:C:2018:557, para 91.

[15] ibid para 119, point 4 of the conclusions See the Court's ruling on the fourth question in the *Bosphorus Queen* case, which is supported by the conclusion reached by Advocate General Wahl in his Advisory Opinion in the case at para 98.

[16] The Nairobi International Convention on the Removal of Wrecks (adopted 18 May 2007, entered into force 14 April 2015) ('the Nairobi Convention').

it identically,[17] suggesting that the definition could be used for the purposes of interpretation and clarification to determine the scope and applicability of article 220(6). Those Conventions define a coastal State's 'related interests' as including:

(a) maritime coastal, port and estuarine activities, including fisheries activities, constituting an essential means of livelihood of the persons concerned;
(b) tourist attractions and other economic interests of the area concerned;
(c) the health of the coastal population and the wellbeing of the area concerned, including conservation of marine living resources and of wildlife.

Major damage to the coastline or related interests – following these definitions – could therefore be caused by a significant oil spill (or the threat thereof), which also was determined in the *Bosphorus Queen*, where the ECJ scrutinised the term 'related interests' in article 220(6). The Court concluded – also applying the definition in article II(4) of the Intervention Convention[18] and interpreting the term in accordance with article 31 of the Vienna Convention on the Law of Treaties[19] – that Finnish interests had been endangered by a foreign ship's violation of MARPOL Annex I while passing through the Finnish EEZ.[20]

The ECJ also found that it could have an impact on the overall assessment that a discharge violation occurred in a sensitive area, such as a PSSA, but that it would not automatically constitute a violation covered by article 220(6).[21] That being said, national regulations adopted pursuant to article 211(6) – which also forms the basis for designating PSSAs, as described in chapter 3 – are also subject to the coastal State's EEZ jurisdiction under article 220(6), as stipulated by article 220(8).

As the interests of the coastal State encompass the 'health of the coastal population' ((c) above), an argument could be made that article 220(6) should cover violations of regulation 14 of MARPOL Annex VI, as sulphur emissions from ships statistically cause more damage (given the adverse effects on human health) when released near densely populated coastal areas located close to ports and shipping lanes.[22] Albeit a commendable approach, the causality between the violation of

[17] The definitions in subparas (a)–(c) of their respective provisions are identical in both Conventions, but the Intervention Convention does not, unlike the Nairobi Convention, include subpara (d) referring to the term's including 'offshore and underwater infrastructure'.

[18] *Bosphorus Queen*, n 14, para 87.

[19] ibid para 67; Vienna Convention on the Law of Treaties 1969 (adopted 22 May 1969, entered into force 27 January 1980) 1155 UNTS 331 (VCLT).

[20] The ECJ notes in *Bosphorus Queen*, n 14, para 107, that even though the Baltic Sea is deemed a Special Area in accordance with MARPOL Annex I, a violation of Annex I does not automatically result in major damage covered by art 220(6) but it could influence that assessment. The same conclusion is reached in Advocate General Wahl's Advisory Opinion in the case (para 107 of the Advisory Opinion).

[21] *Bosphorus Queen*, n 14, para 108.

[22] See the data and studies referred to at the beginning of this chapter and in ch 1.

one specific ship and the 'damage' it inflicts on the coastal State's inhabitants is, in the view of this author, too weak to support such use of article 220(6).[23] This conclusion – that there is a need for a causal link between a specific violation and concrete visible damage – is supported by the statement made by Advocate General Wahl in his Advisory Opinion in the *Bosphorus Queen.*[24]

The applicability of article 220(6) can also be discussed concerning violations of discharge limits for wash water from open loop scrubber systems, including national limits adopted in accordance with article 211(6), which, under article 220(8), applies to article 220(6). Although such a violation would lead to the discharge of water with a heightened pH-value and different pollutants in lesser amounts, it cannot be presumed that this would constitute *major damage* (or the threat thereof) to the coastal State's coastline or interests. As previously noted, though, if the ship calls into a port or at an off-shore terminal in a coastal State after its violation in the EEZ (or territorial sea), the wider jurisdiction pursuant to article 220(1) applies, enabling the coastal State to take action against violations of sulphur limits as well as wash water discharge limits.

Thought should also be given to situations where a foreign ship has violated air pollution regulations, such as MARPOL Annex VI, during its transit of the EEZ – and the coastal State cannot stop this passage under article 220(5)–(6) – but the same ship at a later time, perhaps weeks or months afterwards, calls into a port in the coastal State. Could the coastal State take measures against the earlier EEZ violation at this point because of the article 220(1) jurisdiction it has? This author believes this is possible, provided the coastal State has solid evidence of the violation and any national or international limitation periods, for instance under article 228(2) (see chapter 12), have not been breached. One might call this *delayed enforcement*. Article 220(6) does not present a bar to such delayed enforcement against ships that later voluntarily call into a port in the coastal State. The violation is covered by article 220(1).

Following this train of thought, a ship's violation of MARPOL Annex VI during its innocent passage or transit passage in an international strait in the territorial sea could also be subject to enforcement measures at a later time, if the ship were voluntarily to call into a port in the coastal State, pursuant to article 220(1). This enforcement would not have any ties to the provisions of parts II and III of UNCLOS governing these forms of passage, described in chapter 3, including article 220(2) and article 233 of part XII, as these legal regimes merely focus on the coastal State's not hindering innocent or transit passage by stopping and inspecting a ship. Delayed enforcement in response to a violation of the sulphur

[23] An exception to this conclusion might be if a ship, in violation of IMO regulations, were to release an extremely harmful (toxic) pollutant gas into the atmosphere, which could travel from the EEZ (minimum 12 nm from shore, ie outside territorial waters) in over land and directly harm (poison) humans. This would create the link needed to invoke art 220(5)–(6).

[24] Advocate General Wahl, Advisory Opinion in the *Bosphorus Queen*, n 14, paras 106–108, delivered on 28 February 2018.

limits of MARPOL Annex VI, or other regulations on air pollution (GHGs, etc), would merely reflect that the ship had violated article 220(1). As mentioned, such enforcement is subject to the ship's voluntarily calling into a port or at an off-shore terminal in the coastal State, to the existence of clear evidence of the previous violation and to the fact that any limitation period, such as the three-year limit pursuant to article 228(2), has not been exceeded.

G. Detention and the Posting of Financial Security: Article 220(7)

Article 220(7) specifies that any detention of a foreign ship by a coastal State in accordance with article 220(6) must be ended if 'bonding or other appropriate financial security' has been assured in accordance with international procedures. This approach coincides with the procedural guarantees set out in article 226(1)(b), described in chapter 4.

The opposite conclusion must be that the coastal State is permitted to detain the vessel until such appropriate financial security is posted by the shipowner.

It must be presumed that if the detention of the ship in the EEZ poses a risk of further pollution of the marine environment or a risk to navigational safety in the area, the coastal State can *force* the ship to sail to a port or off-shore terminal in the coastal State. Also, if a shipowner cannot provide the required economic security within a reasonable time (days or weeks), the wellbeing of the crew or ship could dictate that the vessel should be forced to call into a port in the coastal State.

If a ship calls *voluntarily* into a port in the coastal State after the infringement in the EEZ (or in the territorial sea), the State can institute proceedings in accordance with article 220(1), which includes detaining the vessel pursuant to article 226(1)(b) as article 226(1)(a) directly refers to article 220.

The legality of such detention, including the size of the required financial security following article 226(1)(b), can be assessed by the competent judicial organ under part XV on dispute settlement (see article 292).

H. Enforcing National Regulations under Article 211(6): Article 220(8)

As already mentioned several times in this chapter – and in chapter 7 – article 220(8) refers to the specific national rules for protection of the marine environment in the EEZ adopted by the coastal State in accordance with article 211(6). Article 220(8) stipulates that article 220(3)–(7) apply to violations of such nationally adopted EEZ regulations.

The reason why there is no reference to article 220(1)–(2) in article 220(8) is of course due to the geographical scope of those provisions, as article 220(2) deals with infractions occurring in the territorial sea and article 220(1) deals with all violations within waters under a coastal State's jurisdiction, provided the ship afterwards calls upon a port or at an off-shore terminal in the coastal State.

It should be noted that even though article 220(8) does not refer to article 220(1), the wording of the latter paragraph indicates that violations of nationally adopted laws may be the subject of enforcement measures taken by the coastal State, including in the EEZ. The precise wording of article 220(1) is that the State may

> institute proceedings in respect of any violation of its laws and regulations adopted in accordance with this Convention or applicable international rules and standards for the prevention, reduction and control of pollution from vessels when the violation has occurred within the territorial sea or the exclusive economic zone of that State.

The reference to 'its laws and regulations adopted in accordance with this Convention' must, inter alia, be a reference to national laws adopted in accordance with article 211(6).

II. Article 220 Used in Conjunction with Article 111

Under article 111(1) of UNCLOS, coastal States with waters adjacent to the high seas have the *right of hot pursuit* following a foreign ship's suspected violation of coastal State rules committed in internal or territorial waters or in the contiguous zone, or, under article 111(2), in the EEZ or on the continental shelf. This was briefly discussed in chapter 3.[25] This right allows the coastal State to pursue a foreign ship *in continenti* (uninterrupted) from its own waters and out onto the high seas, where it can arrest the ship and escort it back to the State (see article 111(7)).

Article 111 does not specify what kind of regulations are enforceable in accordance with the article, so the applicable rules must be those that are enforceable within the different areas. This means that rules protecting the marine environment, including the sulphur regulations of MARPOL Annex VI, are encompassed by article 111, pertaining to (suspected) violations committed in internal and territorial waters and in the EEZ.

Article 111(1) provides that if a foreign ship is within the contiguous zone, as defined in article 33, pursuit may only be undertaken if it is for a violation of a right protected in that 24 nm zone, which does not include environmental legislation. Nonetheless, it should be remembered that the 200 nm EEZ, in accordance

[25] The provisions of UNCLOS regulating the contiguous zone (art 33) and the continental shelf (pt VI) are also discussed in ch 3.

with article 57, stretches from the baseline, allowing the coastal State to enforce violations of environmental law in the contiguous zone area from the 12th to the 24th nm from the baseline in accordance with article 111(2).

The same problem arises regarding coastal State jurisdiction on the continental shelf, given the reference in article 111(2) to the right of hot pursuit's applying *mutatis mutandis* there. The same reasoning and conclusion as stated above can be applied because of the overlapping geographical 200 nm area of the EEZ and the continental shelf.[26]

Consequently, if a ship violates the sulphur limits in regulation 14 of Annex VI in *internal or territorial waters*, a coastal State may, in principle, follow the ship *in continenti* out onto the high seas and exercise jurisdiction over it in accordance with article 220(1)–(2) of UNCLOS.

If a ship violates regulation 14 in the *EEZ*, the restrictions under article 220(5)–(6) apply regarding stopping and investigating the vessel and prosecuting any such violation. This presumably bars the coastal State from conducting hot pursuit for such a violation, recalling the previous discussion and conclusion in this chapter (sections I.D and I.E) that sulphur emissions (discharges) cannot meet the requirements of causing (or threatening to cause) significant – and tangible – pollution, nor major damage.

Before exercising such hot pursuit (for suspected Annex VI violations in internal and territorial waters) at the discretion of the coastal State, a couple of essential principles should be kept in mind. First, the requirements of article 111 for conducting hot pursuit – set out in chapter 3 – must be observed. For example, ensuring that the pursuit ends as soon as the ship enters the territorial sea of its own State or of another State (article 111(3)),[27] that it has been established that the believed violation took place within the areas specified in article 111(1)–(2) and that the foreign ship has ignored any signal or radio transmission to stop (article 111(4)), and that the pursuit is carried out by a ship or aircraft clearly marked and identifiable as being in government service (article 1111(5)).

Second, one should be mindful of the fact that article 111 does not grant a coastal State an independent extraterritorial right to prescribe and enforce legislation outside its territory, as it concerns infringements that took place within areas under the coastal State's jurisdiction.

Yet even though these restrictions and limitations apply, article 111 can prove to be a powerful practical and legal tool for a coastal State when it comes

[26] It is not relevant in this context to discuss coastal State jurisdiction on a prolonged 350 nm continental shelf, as the added 150 nm do not result in any wider jurisdiction for the coastal State to take action against general violations of environmental legislation.

[27] LM Paul, 'Using the Protective Principle to Unilaterally Enforce Transitional Marine Pollution Standards', Paper presented at the Second International Conference on Marine Debris, 2–7 April 1989, Honolulu, Hawaii, 1051, available at http://webcache.googleusercontent.com/search?q=cache:iNC0b6JIg1cJ:swfsc.noaa.gov/publications/TM/SWFSC/NOAA-TM-NMFS-SWFSC-154_P1045.PDF+&cd=1&hl=da&ct=clnk&gl=no.

to taking action against infringements of air pollution regulations, such as MARPOL Annex VI or future GHG rules, that take place in its internal or territorial waters, and over which it has jurisdiction pursuant to article 220 and to article 222.

As noted in chapter 3, the description of pursuit by aircraft under article 111(6), including the reference to the application of the *mutatis mutandis* principle, has led this author to conclude that hot pursuit can be carried out by a *drone* transmitting visual information direct to coastal State authorities. Such a drone could also have a *sniffer* attached, which could measure the levels of SOx, NOx, CO_2, etc released from a ship. A drone, with a sniffer attached, could therefore detect a violation by a foreign ship in internal or territorial waters of the coastal State and inform the authorities of it, which could then hail (contact) the ship. If the ship did not answer and kept on sailing, the drone could pursue the ship out of internal and territorial waters and out onto the high seas, keeping and relaying visual contact at all times, until a vessel or aircraft from the coastal State was able to intercept, stop and arrest the ship. Under article 111(3), this is possible until the ship enters the territorial sea of another coastal State or of the flag State. A coastal State will nevertheless have to exercise this means of enforcement in a responsible and dutiful manner, as article 111(8) stipulates that if a ship is stopped or arrested in circumstances that do not justify the exercise of hot pursuit then it (the shipowner) must be compensated by the coastal State for any economic loss or damage sustained.

III. Article 220 Read in Conjunction with Articles 223, 230 and 231

Although the procedural safeguards of section 7 of part XII of UNCLOS are discussed in chapter 12 of this book, three of the provisions within that section should briefly be examined here in conjunction with the application of article 220.

A. Coastal State Obligations According to Article 223

It was noted in chapter 8 that flag States, in accordance with article 223, are required to allow representatives from the IMO and an injured coastal State to attended legal proceedings for violations of rules for the protection of the environment, and to let these representatives enjoy international (diplomatic) rights.

The same obligation falls upon coastal States when initiating legal proceedings pursuant to article 220, widening the circle of representatives to include those from the flag State (see article 223).

B. Coastal State Obligations under Article 230

Article 230(1)–(2) set a procedural bar on coastal States' imposition of penalties other than fines for *all* violations occurring in the EEZ and for *non-wilful* or *minor* violations occurring in the territorial sea. This means that only infringements that take place in internal waters, or deliberate infringements resulting in serious pollution taking place in the territorial sea, can be subject to other penalties such as imprisonment. And then only provided the basic *recognised rights of the accused* are observed by the coastal State in accordance with article 230(3).

All violations of environmental regulation, including MARPOL Annex VI and other rules on air pollution, occurring in the EEZ, which are acted on by a coastal State in accordance with article 220(1) (where the ship calls into a port), can only be met with monetary penalties (fines) pursuant to article 230(1).[28]

Violations of MARPOL Annex VI occurring in the *territorial sea*, which are enforced in accordance with article 220(1)–(2) pursuant to article 222, can also only be met with fines according to article 230(2), as it is very unlikely that such infringements of air pollution regulations – given the argument in sections I.E and I.F regarding the applicability (or lack of applicability) of article 220(5)–(6) – will result in tangible *serious* pollution.

Article 230 does not, in principle, present a bar to coastal States' penalising violations of MARPOL Annex VI in *internal waters* with non-monetary penalties, including imprisonment. It must be considered very unlikely, however, that this possibility would be applied, particularly due to the principle of proportionality, as it often would be the shipowner that would be considered the main culprit in such violations, as the owner is the entity that stands to gain from the violations, not the master or crew, who are those who will be in the custody of the coastal State. Proven complicity by the master and/or other crew members could, of course, influence this judgement, especially if it relates to falsifications of documents or giving false statements to Port State Control Officers, as discussed in chapter 4.

C. Coastal State Obligations under Article 231

Article 231 of UNCLOS requires all States to promptly notify the flag State and other States concerned, for example coastal States affected by a polluting incident, of any enforcement measures taken in accordance with the provisions of section 6 of part XII of the Convention. This includes enforcement measures taken by coastal States in accordance with article 220 (and article 222).

[28] The same applies to penalties imposed for violations covered by art 220(5)–(6), eg for violations of MARPOL Annex I or II.

The reference to 'any enforcement measures', as a general rule, includes any questioning, stopping, investigation, detection and legal proceedings carried out by a coastal State pursuant to article 220(1)–(8). An exception to this is found in the second sentence of article 231, which explicitly reduces the coastal State's reporting duties, if a violation was committed in the *territorial sea*, to notification of any measures taken in legal proceedings.

The lack of any reference in article 231 to infringements occurring in *internal waters* may be interpreted in one of two ways. Either the coastal State is under no obligation to inform the flag State of any measures taken, or the requirement in article 231 (second sentence) to inform about measures taken in legal proceedings also implicitly applies to measures taken against violations occurring in internal waters. As a coastal State has full sovereignty and jurisdiction in internal waters pursuant to article 2(1) and no exceptions apply (opposite ended conclusion of article 2(3)), it must be assumed that it is the first-mentioned possibility that prevail. Thus, coastal States are *not* legally bound to inform a flag State of any enforcement measure taken in accordance with article 220(1) for violations that occur within in its internal waters.

A coastal State should nonetheless always endeavour to voluntarily inform flag States of any legal proceedings commenced with regard to violations in its internal waters. Especially as such violations can be met with non-monetary penalties, for example imprisonment, following article 230. This practice is recommended not only from a *cooperative viewpoint*, for maintaining good relations with other (flag) States, but also from a *procedural point of view*, in case a flag State decides to bring proceedings before an international court in accordance with article 287 of part XV of UNCLOS (see chapter 12) to question the legality of the enforcement by the coastal State. Any information provided to the flag State in a timely manner would help the coastal State in such proceedings.

Article 231 specifies that the information to the flag State should be provided to *consular officers* and, if possible, to the *maritime authority* of the flag State.

IV. Other Provisions of Part XII Bestowing Rights on Coastal States

Several provisions of UNCLOS besides articles 211, 212 and 222 – as mentioned in chapter 7 – provide a special legal basis for coastal States to adopt and implement laws and regulations for the protection of the marine environment. These include article 21(1)(f), with regard to innocent passage through the territorial sea, article 42(1)(b), pertaining to transit passage through an international strait, and article 210 on dumping.

Laws adopted under article 21(1)(f) are, as already concluded in section I.B, enforced in accordance with article 220(2).

Enforcement measures for any breach of article 42(1)(b) are taken in accordance with article 233 (see section IV.C); and enforcement measures for any infringement of article 210 are undertaken by the coastal State in accordance with article 216, which is briefly examined in section IV.A following.

Article 221 is also examined to determine whether its extraterritorial jurisdiction provides coastal States with the possibility of enforcing violations of MARPOL Annex VI on the high seas (see section IV.B).

A. Enforcing Dumping Violations under Article 210: Article 216

As mentioned in chapter 7, article 210 of UNCLOS provides that all States, which includes coastal States, must adopt laws preventing pollution of the marine environment by *dumping*. These laws must, at a minimum, be as effective as international regulations on dumping (ie the London Convention[29]), in accordance with article 210(6). These laws and regulations can be enforced by the coastal State in accordance with article 216(1)(a) within its territorial sea or EEZ, or on its continental shelf.

As the reference to the *continental shelf* is without any further specified restrictions, this should allow coastal States that – pursuant to article 76(4)–(8) of UNCLOS (see chapter 3) – have extended their continental shelf to 350 nm from the baseline, to exercise jurisdiction over dumping violations in that area in accordance with article 216(1)(a). Normally, the continental shelfs rights in part VI of UNCLOS do not give coastal States any extended legal basis for enforcing rules and regulations for the protection of the marine environment. But because of the clear and unambiguous wording of the provision, and the fact that dumping potentially can damage the seabed and the resources in the subsoil, article 216(1)(a) must be a *lex specialis* right for coastal States to enforce environmentally founded anti-dumping laws on the continental shelf, as well as on a potentially 350 nm recognised prolonged continental shelf.

Following the definition of the term 'dumping' found in article 1(1)(5) of UNCLOS – and in article III(1)(a) of the London Convention – article 216 does *not* encompass discharges (including the release of emissions[30]), as 'dumping' is there defined as the deliberate disposal of waste and man-made structures, such as ships and platforms, at sea. This is underlined by the definition of the term 'discharge' in article 2(3) of the MARPOL Convention, as article 2(3)(b)(i) clearly defines 'discharge' as not including dumping.

[29] Convention for the Prevention of Marine Pollution by Dumping of Wastes and Other Matter (adopted 13 November 1972, entered into force 30 August 1975) 1046 UNTS 120 ('the London Convention').

[30] The term 'discharge' in art 218 can encompass emissions. See ch 10.

Article 216 therefore does not grant any jurisdiction for coastal States to enforce violations of MARPOL Annex VI or (future) GHG regulations.[31]

B. Jurisdiction to Prevent Pollution Resulting from a Maritime Casualty: Article 221

Article 221 of UNCLOS allows coastal State to

> take and enforce measures beyond the territorial sea proportionate to the actual or threatened damage to protect their coastline or related interests, including fishing, from pollution or threat of pollution following upon a maritime casualty or acts relating to such a casualty, which may reasonably be expected to result in major harmful consequences.

Article 221 provides a coastal State with an extraterritorial jurisdictional basis for enforcing laws and regulation for the protection of its coastlines and interests beyond the territorial sea. As the article, unlike the other provisions of part XII (including articles 216, 220 and 222), does not limit the jurisdiction to territorial waters, the EEZ or the continental shelf, the phrase 'beyond the territorial sea' must encompass the coastal State's exercise of extraterritorial jurisdiction on the high seas.

The wide-ranging geographical scope of the article is curtailed by the reference to pollution (or the threat thereof) following 'a maritime casualty', which in article 221(2) is defined as a 'collision of vessels, stranding or other incident of navigation, or other occurrence on board a vessel or external to it resulting in material damage or imminent threat of material damage to a vessel or cargo'.

The jurisdictional basis of article 221 is therefore closely linked to the Intervention Convention and the legal basis this Convention provides for coastal States to protect themselves from the pollution that may occur when ships collide, also on the high seas.

Because of the requirement for the pollution to stem from a collision at sea, or an incident on board resulting in material damage, article 221 contains *no* legal basis of for taking action against violations of air pollution regulations, including Annex VI of MARPOL, on the high seas.

Even if a situation could be conceived of where non-foreseeable damage on board (threatening the safety of the ship) results in a violation of MARPOL Annex VI, it should be remembered that, as described in chapter 2, these violations are exempt from prosecution pursuant to article 3 of Annex VI.

Article 221 therefore does not provide coastal States with any extraterritorial jurisdictional basis for enforcing violations of regulation 14 Annex VI on the

[31] Art 216 can, however, provide a jurisdictional basis for taking measures against violations of reg 8.1 of MARPOL Annex I, as described in ch 14.

high seas. And even if applied, the situation would most likely be covered by the scope of article 3 of Annex VI, being a force majeure-like circumstance precluding wrongfulness.

C. Taking Measures against Violations by Ships Making a Transit Passage through a Strait: Article 233

Article 233 of UNCLOS[32] stipulates, as a general rule, that the right of a foreign ship to make a transit passage through an international strait is not prejudiced by the provisions of part XII, including article 220. The exception is the right of coastal States to enforce international rules pertaining to the discharge of oil and noxious substances, implemented in accordance with article 42(1)(b) of UNCLOS. This could, inter alia, include MARPOL Annexes I and II.

If a ship violates such rules while conducting a transit passage through an international strait that borders a coastal State, and this violation results in 'causing or threatening major damage to the marine environment of the straits' then the coastal State 'may take appropriate enforcement measures', provided it respects the safeguards provisions of section 7 of part XII of UNCLOS.

This means that if a foreign ship violates MARPOL Annex I or Annex II while making a transit passage through an international strait, this violation can be acted upon by the coastal State in accordance with article 233. 'Appropriate enforcement measures' must refer to the proportionate approach of article 220(3)–(7), first requiring information from the ship, then stopping and investigating it, and then – if the evidence so warrants – detaining and prosecuting it.

A foreign ship's violation of MARPOL Annex VI, or any other international air emission regulations such as the rules on CO_2/GHG, is not covered by article 42(1)(b) and therefore cannot be the subject of action by a coastal State pursuant to article 233.

V. Conclusion on Coastal State Enforcement

Coastal States are able to take measures against violations of nationally adopted rules and internationally implemented regulations for the protection of the marine environment in accordance with several different provisions of part XII of UNCLOS, depending on the regulatory nature of the legislation. For example, article 216 provides a legal basis for acting against dumping violations, including infringements of the London Convention, and article 221 gives an extraterritorial

[32] Art 233 is in principle a procedural safeguard provision found in section 7 of pt XII of UNCLOS, but its direct reference to bordering coastal States' taking 'appropriate enforcement measures' makes it eligible to be studied in this chapter and in general in ch 12.

basis for countering pollution originating in collisions at sea (with principles following the Intervention Convention). Also, article 233 offers an opportunity for enforcing regulations prohibiting the discharge of oil and other noxious substances into an international strait.

When it comes to coastal State enforcement of legislation pertaining to air pollution from ships, such as MARPOL Annex VI, the legal basis for this is found in article 220 and article 222 of UNCLOS. The geographical scope of the latter provision is limited to the *internal* and *territorial waters* of the State, and article 222 contains no detail on how such enforcement is to be conducted in these areas; thus article 220(1) and (2) provide a supplementary legal basis for such enforcement.

Article 220(1) allows for the taking of measures against all violations in internal and territorial waters and in the EEZ, if the ship afterwards voluntarily calls into a port or at an off-shore terminal in the coastal State.

Article 220(2) provides a special legal basis for enforcing national and international environmental legislation, including on air pollution, adopted in accordance with article 21(1)(f), applying to foreign ships making an *innocent passage* of the territorial sea.

All infringements of sulphur and GHG limits that occur in the *EEZ* can therefore only be acted against by the coastal State pursuant to article 220(1). This is conditional on the ship's afterwards voluntarily calling into a port or at an off-shore terminal in the coastal State. Violations that take place in the EEZ and where the ship is not calling into a port in the coastal State, are subject to the provisions of article 220(3)–(8).

All violations, including of MARPOL Annex VI, can result in the coastal State's requiring the ship to provide certain information if it has *clear grounds* for presuming that a violation has occurred (see article 220(3)).

The more stringent requirements of article 220(5) (significant pollution of the marine environment) and (6) (major damage to the coastline or related interests) make it difficult, if not impossible, to establish the necessary causal link between a specific infringement and specific damage when discussing the application of these provisions on the violation of air pollution rules such as MARPOL Annex VI. The scope of article 220(6) has been somewhat clarified by the ECJ and Advocate General Wahl in the 2018 *Bosphorus Queen* case.[33]

Article 220(7) deals with the lifting of detention in situations governed by article 220(6), provided appropriate economic security has been posted. Article 220(8) clarifies that any national EEZ laws, adopted in accordance with article 211(6), can be enforced by the coastal State in accordance with article 220(3)–(7).

Article 111 of UNCLOS allows a coastal State to conduct a hot pursuit, for example by using a drone with a sniffer attached and a video uplink, pursuant to the *mutatis mutandis* reference to pursuit effected by aircraft in article 111(6).

[33] The *Bosphorus Queen* has been reviewed in ch 5, section V.B.

Article 223 requires coastal States to let representatives from the IMO, other injured (coastal) States and the flag State to attend any legal proceedings initiated in accordance with article 220.

Article 230 dictates that violations of MARPOL Annex VI, and of other environmental legislation, that take place in the EEZ or in the territorial sea can only be met with monetary penalties (fines) imposed by the coastal State. Violations in internal waters can be met with non-monetary penalties, such as imprisonment, but this is often reserved for more serious environmental violations than breaches of regulation 14 of MARPOL Annex VI.

Article 231 requires a coastal State to promptly notify the flag State about any enforcement measure taken against a ship flying its flag. However, if an infringement occurs in the territorial sea then the coastal State is only obligated to inform about any measures taken in legal proceedings. It is under no obligation to notify regarding enforcement measures taken for violations in internal waters, although it is recommended that it does so.

10

The Special Jurisdiction for Port States: Article 218

As already noted in previous chapters, article 92 of UNCLOS[1] not only reaffirms the flag State principle, but it also stipulates that there are exceptions to this principle codified in UNCLOS.

Article 218(1) of part XII of UNCLOS must be considered one of these exceptions, as it provides *port States* with an extraterritorial jurisdictional basis for taking action against violations of international rules for the protection of the marine environment outside areas under their jurisdiction, for example on the high seas.

The port State legal entity found in article 218 was formally introduced with the adoption of UNCLOS in 1982. This new entity joined the flag States and coastal States. It may be described as a State with one or more ports and/or off-shore terminals into or at which foreign ships voluntarily call.

A *coastal State*, with jurisdiction in the different areas and zones as described in chapter 3 and with the right to enforce in them pursuant to article 220, as described in the previous chapter, can therefore also be described as a *port State*.[2] Especially considering that article 220(1), as well as article 218(1), refers to ships voluntarily calling into a port or at an off-shore terminal in the State. The difference lies in the geographical scope of where the infringement has been committed (consummated) by a foreign ship:

(a) If a ship, voluntarily calling upon a port, has committed a violation in the *EEZ*, or in *internal* or *territorial waters* of the State, that State exercises it jurisdiction as a *coastal State* and enforces the regulations pursuant to article 220(1), along with article 222 of UNCLOS regarding violations of air pollution regulations.

(b) If a ship, voluntarily calling into a port, has committed a violation outside those areas, for example on the high seas[3] – or in the waters of another

[1] United Nations Convention on the Law of the Sea (adopted 10 December 1982, entered into force 16 November 1994) 1833 UNTS 3.

[2] A State – besides being a coastal State and port State – can also be deemed a flag State if vessels are flying its flag and are listed in the State's ship registry.

[3] See LM Paul, 'Using the Protective Principle to Unilaterally Enforce Transitional Marine Pollution Standards, 1053. Paper following the Second International Conference on Marine Debris, 2–7 April 1989,

State[4] – then the State has jurisdiction over this violation by virtue of being a *port State* in accordance with article 218.[5]

Consequently, the jurisdictions under article 220 and article 218 do not overlap but merely complement each other, as article 218 extends the State's jurisdiction beyond the EEZ.

The use of a different term for that legal entity (port State) is therefore justified, as the effects of the violation do not necessarily have any impact on the enforcing State, nor does the infringement take place in an area under its jurisdiction. Thus the grounds for enforcement pursuant to article 218(1) are very different from those applying to enforcement pursuant to article 220 – and to articles 221 and 222 for that matter.

It should be noted that article 218 contains the only references to *port States* in UNCLOS, not only within part XII, but throughout the whole Convention. The term 'port State' is only used explicitly in article 218(4), but given the heading of article 218, 'Enforcement by port States', and that the possibility of using article 218 rests on the ship's sailing into port (or to an off-shore terminal), this book will use the term 'port State' when referring to States exercising jurisdiction in accordance with article 218(1)–(3).

This is also supported by how the term 'port State' is used in article 218(4), as it clearly refers back to the State entities mentioned in article 218(1)–(3). For instance, the first line of article 218(4) refers to '[t]he records of the investigation carried out by a port State pursuant to this article'. And then it states that 'Any proceedings instituted by the port State on the basis of such an investigation may, subject to section 7 …'.

It is therefore safe to conclude that when article 218(1) refers to *States* undertaking investigations and – if the evidence so warrants – initiating proceedings against foreign ships, it is a reference to *port States*.

Even though no article in UNCLOS, besides article 218, mentions *port States*, article 226, relating to PSC of foreign ships, contains a direct reference to article 218 and thereby implicitly to port States. The same applies to article 217(4) pertaining to flag States' having to respect port State jurisdiction, given the explicit reference in that paragraph to article 218.

Honolulu, Hawaii, 1051, available at http://webcache.googleusercontent.com/search?q=cache:iNC0b6JIg1cJ:swfsc.noaa.gov/publications/TM/SWFSC/NOAA-TM-NMFS-SWFSC-154_P1045.PDF+&cd=1&hl=da&ct=clnk&gl=no.

[4] A Pozdnakova, *Criminal Jurisdiction over Perpetrators of Ship-Source Pollution* (Martinus Nijhoff, 2012) 153.

[5] If a ship, voluntarily calling into a port, is subject to an inspection in port, this is regulated in accordance with arts 224–227 of UNCLOS, as described in ch 4. The State entity that carries out such inspections in port is often referred to as a *port State*, leading to the term Port State Control (PSC). Yet UNCLOS does not use the term 'port State' for describing such inspections, referring instead to States in general – see art 226. As UNCLOS only uses the term 'port State' in art 218 to describe a State's jurisdiction beyond its normal boundaries, ie on the high seas, this book will also use the term to describe this legal entity with far-reaching extraterritorial jurisdiction. The book also uses the term, as applied in ch 4, to describe States that inspect foreign ships at berth in their ports.

I. Article 218 in General

Article 218 consists of four paragraphs, with a distinction between article 218(1) and article 218(2)–(4):

- Article 218(1) provides the important legal basis for port States to exercise their extraterritorial jurisdiction for taking measures against discharge violations that are committed outside the State's territory, meaning on the high seas or in the waters of another State.
- Article 218(2)–(4) clarify the overlapping jurisdiction that may occur if a discharge violation takes place inside another State's territory, as both that State and the port State will have jurisdiction over the same infringement. As article 218(2) refers to violations committed in another State's internal water, territorial sea or EEZ, or to the damage (or threat thereof) to another State, this State, and its overlapping jurisdiction, must be deemed a *coastal State*. This book will therefore refer to article 218(2)–(4) as resolving the matter of overlapping (competing) jurisdictions between a port State pursuant to article 218(1) and a coastal State pursuant to article 220.[6] This is also consistent with how article 218(4) explicitly applies the term 'coastal State', clearly referring to the same State entity mentioned in article 218(2)–(3). This is the same conclusion as was reached regarding the use of the term 'port State' in article 218(4), referring to the same State entity in article 218(1)–(3).

Although it may resolve the question of overlapping jurisdiction between port and coastal States, article 218 does not provide any clarification of the overlapping jurisdiction between a flag State according to article 217 and a port State pursuant to article 218(1), pertaining to discharge violations on the high seas. This overlapping jurisdiction is, however, resolved though article 228(1), as it determines which State can assert primary jurisdiction over such a violation.

This is underlined by the aforementioned article 217(4), referring not only to article 218, but also to article 220 and article 228, which provides that flag States must respect port and coastal State jurisdictions under article 218 and 220, and how these are resolved in accordance with article 228.

The analysis of article 218 in this chapter will primarily focus on article 218(1), as that paragraph provides the essential basis for establishing that a non-flag State (a port State) can take measures against infringements of certain international (International Maritime Organization (IMO)) rules for the protection of the marine environment on the high seas, which is the subject of this book.

The content of article 218(2)–(4) is studied in section VI of this chapter.

[6] Cf art 222 regarding violations of air pollution regulations in internal and territorial waters.

II. Requirements for Exercising High Seas Jurisdiction: Article 218(1)

The precise wording of article 218(1) is reproduced here, as it will be a continual point of reference throughout the entire book from this point on. Article 218(1) reads:

> When a vessel is voluntarily within a port or at an off-shore terminal of a State, that State may undertake investigations and, where the evidence so warrants, institute proceedings in respect of any discharge from that vessel outside the internal waters, territorial sea or exclusive economic zone of that State in violation of applicable international rules and standards established through the competent international organization or general diplomatic conference.

Thus article 218(1) sets out numerous requirements that port States must meet to exercise the extraterritorial jurisdiction under the provision. These are:

- that the vessel (foreign ship) must voluntarily call into a port or at an off-shore terminal of the port State;
- that the port State may only institute legal proceedings where the evidence so warrants;
- that the violation must have taken place outside the port State's own waters;
- that it must be a violation of an international accepted rule or standard; and finally, and most importantly when discussing enforcement of air pollution regulations such as MARPOL Annex VI[7] or future greenhouse gas (GHG) measures,
- that it must be a *discharge* violation – *any* form of discharge.

These requirements will be examined in the following subsections, with close and enquiring focus on the discharge reference in article 218(1), this requirement therefore being being analysed in its own separate section (section III).

A. The Ship Must Voluntarily Call into a Port or at an Off-shore Terminal

It is a requirement for the application of article 218(1) that the ship must voluntarily call into a port or at an off-shore terminal in the port State, meaning that the

[7] International Convention for the Prevention of Pollution from Ships (adopted on 11 February 1973, as modified by the Protocol of 17 February 1978, entered into force 2 October 1983) 1340 UNTS 61 (MARPOL), Annex VI, IMO Publication: IMO-520E.

ship cannot be forced to enter the port.[8] If a State were to force a ship into port, for instance by invoking article 220(6)–(7)[9] or following a hot pursuit according to article 111, that (coastal) State could not also exercise jurisdiction in accordance with article 218(1) over any previously detected discharge violations outside the EEZ.

Circumstances amounting to force majeure forcing a ship into port, such as weather conditions, damage to on-board equipment, etc, should also be regarded as circumstances that could preclude the application of article 218(1), provided these extraordinary circumstances can be proved. It should be recalled that many IMO rules have built-in exemptions for force majeure situations, which bar a (port)State from imposing sanctions. See, for example, regulation 3 of MARPOL Annex VI.

B. Legal Proceedings May be Instituted Only Where the Evidence So Warrants

The requirement that legal proceedings may be instituted only where *the evidence so warrants* must be a demand for the port State to have sufficient proof of a violation before prosecuting the ship(owner). The on-board (PSC) investigations envisaged in article 218(1) ('may undertake investigations') could provide such adequate evidence to 'institute proceedings'. Also, if the violation concerns the discharge of oil or other visible substances, aerial photos, recordings, etc linking the discharge to a specific ship should be sufficient to establish that a violation has occurred.

Proving the discharge of invisible airborne pollutants, such as sulphur emissions, can be much more difficult. As addressed in previous chapters, however, the technical advancements in this field are positive, and it is therefore not inconceivable that some of the technologies described, such as mandatory use of Continuous Emission Monitoring Systems (CEMSs) or detection via the use of *sniffers* attached to drones or planes, used in conjunction with *fuel sampling* and *fuel calculation*, will allow port States to detect and sufficiently prove violations of the 0.5% sulphur limit on the high seas. This would fulfil the onus under article 218(1) to have evidence that *warrants* the institution of legal proceedings.

[8] For the purpose of clarity, the reference to 'off-shore terminals' will not be repeated throughout the chapter from this point on, but should be considered implied whenever reference is made to a foreign ship's calling into a 'port'.

[9] It should be recalled that a coastal State detention in the EEZ pursuant to art 220(6) can result in the coastal State's forcing the ship into port for reasons to do with environmental or navigational safety, or if the required economic security (bail) cannot be provided by the shipowner in a timely manner, or if concerns for the wellbeing of the crew or ship dictate that it should go into port.

C. The Violation Must have Taken Place Outside Port State Territory

The geographical scope of article 218(1) is unique, as it provides port States with an extraterritorial jurisdictional basis for enforcing measures against discharge violations outside their own territory, which encompasses violations taking place on the *high seas*. This conclusion is also reached by Advocate General Kokott in her Opinion in the *Intertanko case*[10] before the European Court of Justice (ECJ), discussed in chapter 5. Advocate General Kokott emphasises, in paragraph 61 of the Opinion[11] – when referring to article 218 of UNCLOS – that 'Such proceedings presuppose that the State concerned is entitled to impose penalties for such discharges on the high seas.'

The reason this author describes this jurisdictional basis as being 'unique' is that it conflicts with a basic principle within international law that was described in chapter 6 – a principle that stipulates that a State's omnipotent right to *prescribe* rules (including implementing IMO regulations) applying outside its territory is curtailed by its limited jurisdictional right to *enforce* and *adjudicate on* violations of those rules. This basic principle can nevertheless be set aside by an international convention pursuant to the *first Lotus principle*, established in the *Lotus case*, also described in chapter 6.[12]

Article 218(1) of UNCLOS represents such an exception, codified in a convention bestowing extraterritorial enforcement rights on a (port) State, as the 167 State parties to UNCLOS have accepted this by adopting and ratifying the Convention, which includes the contents of part XII and the extraterritorial port State jurisdiction of article 218.

It should also be noted that article 218(1) has a focus on enforcing *international* rules and not nationally adopted rules (see section II.D), which eases any concerns relating to allowing extraterritorial enforcement.

D. It Must be a Violation of an International Accepted Rule or Standard

Article 218(1) refers to the 'violation of applicable international rules and standards established through the competent international organization or general diplomatic conference'. This means that the extraterritorial jurisdiction of article 218(1) applies only to international rules implemented by the port State and not to national rules adopted in accordance with articles 211 or 212.

[10] Case C-308/06 *Intertanko, Intercargo, Greek Shipping Co-operation Committee, Lloyd's Register, International Salvage Union v Secretary of State for Transport*, ECLI:EU:C:2008:312.

[11] Advocate General Juliane Kokott's Advisory Opinion of 20 November 2007 in the *Intertanko case*.

[12] *SS Lotus (France v Turkey)* PCIJ Rep Ser A No 10.

The reference to these international rules' being established through the 'competent international organization' is, as concluded in previous chapters, a direct reference to the IMO.[13] *Discharge* violations therefore must mean IMO regulations on the hindrance of pollution of the marine environment through discharges from ships.

Without anticipating the detailed discussion of the next section relating to the term 'discharge' encompassing emissions from ships, it should briefly be noted that the MARPOL Convention,[14] including all of its Annexes, and the BWM Convention[15] represent IMO rules and standards that fall within the scope of article 218(1) as they focus on prohibiting the discharge of polluting substances into the sea or air. Also, future IMO regulations on GHG would be covered by article 218(1).[16]

The London Convention on dumping is not subject to the jurisdiction of article 218, as UNCLOS, the London Convention and the MARPOL Convention all define 'dumping' in a manner that does not include 'discharge'.[17] Violations of dumping rules can be acted on by coastal, port and flag States in accordance with article 216 of UNCLOS.

The Hong Kong Convention on Ship Recycling[18] and the Nairobi Convention on Wreck Removal[19] are also not covered by the scope of article 218, as they do not concern themselves with prohibiting any discharges from ships.

Potential future IMO regulations on Black Carbon and Grey Water, and updated standards on open loop scrubber discharge limits could be covered by the reference in article 218(1) to international rules and standards.

III. Discharge Violations

The scope of article 218 of UNCLOS is somewhat narrow compared to that of article 217 and article 220(1), as the provision applies exclusively to *discharge violations*. It is therefore essential to establish exactly what forms of violations are covered by the term 'discharge'. The initial step towards determining this is

[13] See *Implications of the United Nations Convention on the Law of the Sea for the International Maritime Organization*, LEG/MISC.8 (2014) 56.

[14] See Y Tanaka, *The International Law of the Sea*, 2nd edn (Cambridge University Press, 2015) 296.

[15] The International Convention for the control and management of ship's ballast water and sediments, 2004 (adopted 13 February 2004, entered into force 8 September 2017) ('the BWM Convention').

[16] The enforcement of the current regulations – MARPOL Annexes I–VI (but not reg 14) and the BWM Convention – and future GHG regulations through art 218 is analysed in Part III of this book.

[17] See art 1(1)(5) of UNCLOS; art III (1)(a) of the Convention for the Prevention of Marine Pollution by Dumping of Wastes and Other Matter (adopted 13 November 1972, entered into force 30 August 1975) 1046 UNTS 120 ('the London Convention'); and art 2(3)(b)(i) of the MARPOL Convention.

[18] The Hong Kong International Convention for the Safe and Environmentally Sound Recycling of Ships, 2009 (adopted 15 May 2009, not yet entered into force) ('the Hong Kong Convention').

[19] The Nairobi International Convention on the Removal of Wrecks (adopted 18 May 2007, entered into force 14 April 2015) ('the Nairobi Convention').

to examine whether UNCLOS itself defines the term: unlike other terms, such as 'pollution of the marine environment' and 'dumping', the Convention does not define 'discharge'.

If the extent and applicability of a term or word in a convention needs defining, international law lays down basic principles of interpretation and clarification. These principles are embodied in section 3 (articles 31–33) of the Vienna Convention on the Law of Treaties (VCLT)[20] pertaining to 'Interpretation of treaties'. The conclusion resulting from this interpretation can then be compared to definitions of the term (here 'discharge') in other relevant international legislation, in this instance such as article 2(3) of the MARPOL Convention, which also applies to its Annexes pursuant to article 1(2), and how the term is already *used* in regulation 12 of MARPOL Annex VI.

First, though, the context of how the term is applied in article 218(1) must be established, as well as a comparison of its use in other provisions of UNCLOS.

A. Any Discharge is Covered by Article 218(1)

The term 'discharge' is, as established in chapter 9, applied in article 220(5)–(6) to describe how a discharge from a ship can cause (or threaten to cause) significant pollution, which may result in major damage to the coastline and the related interests of the coastal State. The term 'discharge' is thus used with certain restrictions regarding what forms of discharge are covered by these paragraphs: the discharge must result in tangible pollution that has caused or may cause visible damage.

Article 42(b) of UNCLOS also uses the term 'discharge' in a very specified context, referring to regulations prohibiting 'the discharge of oil, oily wastes and other noxious substances in the strait', against which measures can be enforced in accordance with article 233. Again, the scope of the term is limited by its application in the provision to 'only' cover the discharge of oil, oily wastes and other noxious substances.

When studying article 218(1) it becomes clear that no such limitations are set on the application of the term 'discharge'. On the contrary, the wording of the paragraph unambiguously gives the term the widest possible use and interpretation by referring to '*any* discharge'.[21]

No other provision in UNCLOS that uses the term 'discharge' includes such reference to *any* form of discharge. It must therefore be the intention that the term in article 218(1) is to cover all possible applications. This means that the term can be interpreted in the broadest conceivable way when using the provisions of section 3 VCLT.

[20] Vienna Convention on the Law of Treaties 1969 (adopted 22 May 1969, entered into force 27 January 1980) 1155 UNTS 331 (VCLT).

[21] Emphasis added.

That such wording – 'any discharge' – is to be regarded in its widest possible sense, allowing for the broadest possible interpretation, was confirmed by the ECJ in the *Bosphorus Queen*,[22] discussed in chapter 5. The Court specified in paragraph 81, when determining the extent of the term 'any resources' in article 220(6) of UNCLOS:

> It must be observed that Article 220(6) of the Montego Bay Convention [UNCLOS] refers to damage caused or the threat of damage to 'any' resources of the territorial sea or the EEZ of the coastal State. Therefore, that provision must be interpreted broadly in that regard, which is clear from its wording, and must not be understood as excluding certain resources from the scope of that provision.

This confirms that the use of the word 'any' allows for application of the broadest possible meaning of the word following, be it 'discharge' or 'resources'. The term 'discharge' in article 218(1) can therefore be interpreted in the broadest possible sense of the word.

B. Interpretation of the Term 'Discharge' in Accordance with Article 31 VCLT

The VCLT codifies widely recognised principles of international treaty law, many of which represent principles of customary law. The VCLT has been ratified by 115 States. Forty-five other States have signed the treaty, obliging them not to act contrary to the principles of the Convention according to article 18(a) VCLT.

All States must, besides the 115 ratifying States and the 45 signatory States, abide by the principles of interpretation under articles 31 and 32 VCLT, as these represent international customary law. This is affirmed in a report by the International Law Commission (ILC) to UN General Assembly, in which the Commission said, 'The VCLT rules on treaty interpretation – articles 31 and 32 – are recognized as customary law and widely applied in the WTO system.'[23]

Articles 31 and 32 VCLT determine how a word or a term in a treaty or convention should be interpreted when establishing the meaning and extent of this. This includes interpreting words and terms in UNCLOS, as expressed in another ILC report from 2017:

> The rules of international law relating to the protection of the atmosphere and other relevant rules of international law, including inter alia the rules of international trade and investment law, of *the law of the sea* and of international human rights law, should, to the extent possible, be identified, interpreted and applied in order to give

[22] Case C-15/17 *Bosphorus Queen Shipping Ltd Corp v Rajavartiolaitos*, ECLI:EU:C:2018:557.

[23] ILC, *Fragmentation of international law: difficulties arising from the diversification and expansion of international law*, prepared by Martti Koskenniemi, Report A/CN.4/L.682 (presented at the 58th session in Geneva, 1 May–9 June and 3 July–11 August 2006) 89, para 168.

> rise to a single set of compatible obligations, in line with the principles of harmonization and systemic integration, and with a view to avoiding conflicts. This should be done in accordance with the relevant rules set forth in the Vienna Convention on the Law of Treaties of 1969, including articles 30 and 31, para 3 (c), and the principles and rules of customary international law.[24]

This conclusion is supported by the ECJ in the *Bosphorus Queen case*, where the Court in paragraph 67 directly stated that '[i]n order to interpret the provisions of the Montego Bay Convention [UNCLOS] it is necessary to refer to the rules of customary international law reflected by Article 31 of the Vienna Convention'.

The 2017 ILC Report and the statement of the ECJ in the *Bosphorus Queen* refer to article 31 VCLT and not article 32. The reason for this is found in the wording of article 32, which requires article 31 to be used as the primary interpretational source, article 32 being applied as a supplement to article 31 when necessary.

Article 31(1) VCLT provides that 'A treaty shall be interpreted in good faith in accordance with the ordinary meaning to be given to the terms of the treaty in their context and in the light of its object and purpose.' The term 'discharge' in article 218(1) of UNCLOS is therefore to be interpreted in accordance with article 31 VCLT.

The reference to the interpretation's being done in good faith and 'in accordance with the ordinary meaning to be given to the terms of the treaty' denotes that the word or term being interpreted in accordance with article 31 VCLT should be understood in accordance with its *ordinary meaning*. The ordinary meaning of a word or term is best found by using a dictionary and/or a thesaurus. To determine the ordinary meaning of the word 'discharge'[25] pursuant to article 31(1) VCLT – thereby determining the extent in article 218(1) in UNCLOS – recognised English dictionaries and thesauruses must be used. For example:

- 'Allow (a liquid, gas, or other substance) to flow out from where it has been confined.'[26]
- To 'emit', 'let out', 'give off', 'exude', 'ooze', 'jet'.[27]
- '[T]o send out a substance, especially waste liquid or gas' and 'Emitting and ejecting' (with further definitions of 'emissions', 'Carbon Emissions', and 'Emanate from/through').[28]

[24] ILC, *United Nations – Annual Report of the International Law Commission, Sixty-ninth session*, Report A/72/10 (1 May–2 June and 3 July–4 August 2017) ch VI, 'Protection of the atmosphere', 155 (emphasis added).

[25] The English version of UNCLOS (ie the word 'discharge') is used for determining the scope of art 218(1), as the English version, along with the French, Russian, Spanish, Chinese and Arabic versions, is one of the confirmed authentic original texts under art 320 of UNCLOS, which are those that should form the basis of any interpretation in accordance with art 33(2) VCLT.

[26] See at https://en.oxforddictionaries.com/definition/discharge (definition no 2).

[27] See at https://en.oxforddictionaries.com/thesaurus/discharge.

[28] See at https://dictionary.cambridge.org/dictionary/english/discharge?q=discharge.

- '[T]o give outlet or vent to: emit'; 'vehicles discharging exhaust fumes'.[29]
- To 'discharge gas' and send 'substances into the air' (eg originating from a factory).[30]

To show that these definitions of the term 'discharge' are not new, *Longman: Dictionary of Contemporary English* (1987)[31] defines 'discharge' as meaning, inter alia, 'to send, pour, or let out (gas, liquid etc): The chimney discharges smoke'.

This author concludes, based on these definitions, that an interpretation of the term 'discharge' pursuant to article 31(1) VCLT leads to the term in article 218(1) of UNCLOS, inter alia, covering *emissions* from ships, such as sulphur and CO_2 (GHG) emissions.[32] This is based on the references to 'discharge' encompassing the release of *gases*, the release of *smoke* from chimneys and *exhaust fumes* from vehicles; and especially references to 'emitting', which includes the release of *carbon emissions*.

This means that smoke, gas, exhaust fumes and emissions, including carbon (and sulphur) emissions, emanating from the 'chimney' (funnel) of a ship are covered by the term 'discharge' in article 218(1) of UNCLOS. A ship's emissions of SOx, NOx, CO_2 (GHG), ozone-depleting substances (ODSs) and other regulated air pollutant substances are therefore covered by article 218(1).

This conclusion must also be in accordance with the *object* and *purpose* of part XII of UNCLOS, as expressed in article 31(1) (in fine) VCLT, as the provisions of part XII seek to protect the marine environment from pollution, which also encompasses pollution that can damage human health, in accordance with the definition in article 1(1)4) of UNCLOS.

Port States can therefore exercise extraterritorial jurisdiction in accordance with article 218(1) over violations of IMO regulations on air pollution, for instance when taking action against violations of the 0.5% sulphur limit according to regulation 14.1.3 of MARPOL Annex VI on the high seas.

C. The Definition of 'Discharge' in the MARPOL Convention

The conclusion – set out in the previous section – that 'discharge' in article 218(1) covers emissions from ships, is supported by the way in which 'discharge' is defined in article 2(3) of the MARPOL Convention. (Unlike UNCLOS, the MARPOL Convention does contain a definition of the term.) This definition also applies to the Annexes to the MARPOL Convention (see article 1(2)). This includes Annex VI, which does not set out its own definition of the word.

[29] See at https://www.merriam-webster.com/dictionary/discharge (definition 2.d).
[30] See at https://dictionary.cambridge.org/dictionary/english/discharge?q=discharge.
[31] *Longman: Dictionary of Contemporary English*, 2nd edn (Longman, 1987).
[32] As art 31 VCLT provides a clear conclusion on the definition and extent of the term 'discharge', it is not necessary to resort to art 32.

The definition of 'discharge' in the MARPOL Convention is of great interpretational value, as MARPOL is a highly recognised legal instrument within international law for the protection of the marine environment, with 158 States parties representing over 99% of the world's merchant fleet tonnage. It would therefore be preferable for there to be consistency between the general (formal) jurisdictional basis provided in article 218 of UNCLOS and the specific (material) environmental regulation of Annex VI, regarding how the term 'discharge' is applied and defined, as the rules in Annex VI are enforced within the framework of UNCLOS.[33]

This need for a consistent approach to enforcement between the two conventions is emphasised by the cross-references in article 9(2) of the MARPOL Convention and article 237 of UNCLOS, each imposing obligations on the States parties to respect the principles outlined in the other, as described in chapter 3.

The MARPOL Convention defines 'discharge of harmful substances' in article 2(3)(a)–(b), with article 2(3)(a) making a 'positive demarcation' of the term by stating what it covers, while article 2(3)(b)*bis*(i)–(iii) make a 'negative demarcation' by asserting what the term does not cover, such as dumping, exploration of the seabed and scientific research. The negative listing in article 2(3)(b) will not be addressed further, as it contains no relevant references or examples pertaining to whether the term 'discharge' can cover emissions. This is in contrast to article 2(3)(a), which sets out useful examples for determining the extent of the expression.

Article 2(3)(a) of the MARPOL Convention provides:

> (a) Discharge, in relation to harmful substances or effluents containing such substances, means any release howsoever caused from a ship and includes any escape, disposal, spilling, leaking, pumping, emitting or emptying …

First, it must be established that the reference to 'harmful substances' covers air pollutants such as sulphur and GHG, because of their adverse impacts on both the environment and human health.[34]

The reference to 'any release howsoever caused' encompasses unintended discharges.[35] The word 'any' – as was the case with 'any discharge' in article 218 of UNCLOS – clearly indicates, in this context too, that the broadest possible application of the term 'release' is intended.[36] Operational discharges of harmful

[33] A Mihneva-Natova, *The Relationship between United Nations Convention on the Law of the Sea and the IMO Conventions* (UN and The Nippon Foundation of Japan Fellowship, 2005) 16, where she states 'Article 2(2) and (3) of MARPOL includes a definition of "harmful substances" which is entirely compatible with the definition of "pollution of the marine environment" included in Article 1(4) of the Law of the Sea Convention.'

[34] See the description of the harmful effects from sulphur pollution in ch 1, and from GHG pollution in Part IV of this book.

[35] A Kanehara, 'Environmental Protection of Ocean and Flag-State Jurisdiction', Paper presented at the 8th Conference of SCA Qingdao, China 27–30 May 2008, 5.

[36] See H Ringbom, *The International Legal Framework for Monitoring and Enforcing Compliance with the Sulphur in Fuel Requirements of MARPOL Annex VI*, a CompMon Report, 4th version (2017) 32, fn 104.

substances resulting from a combustion process on board are therefore covered by this. For example, the release of SOx, NOx and CO_2 emissions following combustion of marine fuel to create energy and propulsion.

An initial analysis of article 2(3) establishes that nothing in the provision bars or limits the definition of 'discharge' from applying to emissions being discharged from ships as part of a combustion process. Still, the positive list of defining examples in article 2(3)(a) will be scrutinised next, to determine whether the MARPOL Convention, and therefore Annex VI, defines the term 'discharge' to cover emissions from ships.[37]

The examples of 'discharge' provided in article 2(3)(a) are: 'escape', 'disposal', 'spilling', 'leaking', 'pumping', 'emitting' or 'emptying'. Initially it is worth noticing that the provision makes no reference to, or gives any indication of, any of these terms having a higher interpretational value than the others for determining the scope of the term 'discharge', so all these definitions are equal in their capacity to clarify the meaning of 'discharge' pursuant to a reverse conclusion from article 31(4) VCLT.

The references to 'disposal, spilling, leaking, pumping … [and] emptying' relate to the 'classical' use of the term 'discharge' in connection with the release of a substance (liquid, etc) into the sea. This encompasses the discharge of oil, chemicals, sewage, garbage and ballast water, etc covered by MARPOL Annexes I–V and the BWM Convention.

The reference to 'emitting' is, however, extremely relevant when determining whether the term 'discharge' can cover emissions, as 'emitting' is the present participle of the verb 'to emit'. The noun 'emission' – which also derives from the verb – means, inter alia, 'a substance which is emitted'.[38] This is supported by how the term 'emitting' is defined and exemplified in various English dictionaries and thesauruses, once again applying the interpretational principles of article 31(1) VCLT.[39] 'Emitting' ('emit') is defined as meaning:

- 'Produce and discharge (something, especially gas or radiation).' The same dictionary gives the following use of the word in a sentence: 'even the best cars emit carbon dioxide'.[40]
- '[D]ischarge', 'release', 'give off', 'effuse', 'vent', 'give vent to'.[41]

[37] Professor Henrik Ringbom also notes that 'article 2(3)(a)contains a very broad definition of the term "discharge", covering "any release howsoever caused from a ship"': H Ringbom, 'Enforcement of the Sulphur in Fuel Requirements: Same, Same But Different', available at https://webcache.googleusercontent.com/search?q=cache:EBk8-0aReA0J:https://www.duo.uio.no/bitstream/handle/10852/61600/SO-Artikel-Ringbom.pdf%3Fsequence%3D4%26isAllowed%3Dy+&cd=1&hl=da&ct=clnk&gl=no, 14, fn 54.

[38] *Concise Oxford English Dictionary*, 11th edn rev'd (Oxford University Press, 2006).

[39] The English version of the MARPOL Convention is used, and thus the word 'emitting', as the English version is one of the officially confirmed authentic texts under art 20 of the MARPOL Convention, and therefore to be used when interpreting terms under art 33(2) VCLT.

[40] See at https://en.oxforddictionaries.com/definition/emit.

[41] See at https://en.oxforddictionaries.com/thesaurus/emit.

- '[T]o send out a beam, noise, smell, or gas'.[42]
- '[T]o send forth (liquid, light, heat, sound, particles, etc); discharge'.[43]

Webster's Electronic Dictionary also gives the following example of use of the term 'emitting': 'chimneys emitting thick, black smoke'.[44]

Accordingly, it can be concluded that the definition of 'discharge' in article 2(3)(a) of the MARPOL Convention, by referring to 'emitting', allows for the term to cover emissions, including sulphur and CO_2 emissions being released from ships as a result of the combustion of marine fuels.[45]

D. MARPOL Annex VI Uses the Term 'Discharge' to Cover Emissions

That this conclusion, in accordance with article 1(2) of the MARPOL Convention, also applies to its Annexes is fortuitous, as MARPOL Annex VI already uses the term 'discharge' to describe the release of emissions.

Regulation 12 of MARPOL Annex VI regulates ships' emissions of ODSs. Regulation 12.7 provides that a ship must carry an 'Ozone-depleting Substance record book' where, inter alia, information on any accidental release of ODSs into the atmosphere is to be logged by the ship.

Regulation 12.7.3 specifies that this record book shall include information on the 'discharge of ozone-depleting substances to the atmosphere'. This indisputably establishes that the release of emissions regulated in MARPOL Annex VI, including sulphur emissions pursuant to regulation 14, can be described as the *discharge of emissions* into the atmosphere.

E. Conclusion: The Term 'Discharge' in Article 218 as Encompassing Emissions

The result of interpreting the term 'discharge' in article 218(1) of UNCLOS in accordance with article 31(1) VCLT confirms that 'discharge' can cover the release of *emissions*. This is in full alignment with how the term is defined in article 2(3)(a) of the MARPOL Convention, and with how it is already applied in regulation 12.7.3 of MARPOL Annex VI.

Given this result, this author concludes that the term 'discharge' in article 218(1) of UNCLOS can cover the release of emissions from ships, including sulphur emissions. This allows port States to assert the extraterritorial jurisdictional basis

[42] See at https://dictionary.cambridge.org/dictionary/english/emit?q=emitting.
[43] See at http://www.dictionary.com/browse/emitting.
[44] See at https://www.merriam-webster.com/dictionary/emitting.
[45] Ringbom, n 37, 14, fn 54.

of the provision for taking measures against violations of the 0.5% sulphur limit in regulation 14.1.3 of MARPOL Annex VI, and also against infringements that take place on the high seas.

This conclusion also applies to the enforcement of other applicable IMO regulations, such as the other Annexes of the MARPOL Convention and the BWM Convention, and will apply to any future regulations on CO_2 (GHG) emissions. This is described in Part III (chapters 14 and 15) of this book.

IV. Port State Enforcement of MARPOL Annex VI on the High Seas

A port State can enforce the 0.5% sulphur limit on the high seas pursuant to article 218(1) of UNCLOS, provided the evidence of a violation so warrants and provided the ship afterwards voluntarily calls into a port in the State.

This means that port States can impose fines that remove (confiscate) all savings achieved by violating regulation 14.1.3 of Annex VI, including on the high seas. Such fines can also include a dissuasive punitive element, which increases in the event of aggravating circumstances, for instance in cases of recidivism. The fines prescribed by port States should therefore met the same criteria that an effective flag State fine must meet, as described in chapter 8, modelled on the principles found in article 18 of the Sulphur Directive.[46]

It should be noted that the jurisdiction for port States to take action against violations in accordance with article 218(1) does not require the enforcing port State to be the next port of call for the ship after the infringement, if any previous port States into which the ship had called have not taken measures to address the violation. If a ship calls into a port after committing a violation on the high seas, but manages to leave the port without that port State's taking action in response to the violation, perhaps due to a short stay in port or the port State's being unable to take enforcement measures, then other port States into which the ship subsequently calls can investigate – and if the evidence so warrants – initiate legal proceedings for the violation that took place during a previous 'leg' (part) of the vessel's journey.

Once a port State has effectively taken measures against the violation, other port States are barred from doing so in respect of the same violation due to the principle of *ne bis in idem*.[47] The same applies if the flag State has taken effective measures as regards the violation. The principle of *ne bis in idem* is codified in article 228(2) of UNCLOS, as discussed in chapter 12.

As the foreign ship, pursuant to article 218(1), has voluntarily called into a port in the port State, the principle of *no more favourable treatment* (NMFT) applies,

[46] Directive 2016/802 of the European Parliament and of the Council of 11 May 2016 relating to a reduction in the sulphur content of certain liquid fuels [2016] OJ L132/58.

[47] A Henriksen, *International Law* (Oxford University Press, 2017) 96.

enabling the port State to investigate and prosecute the foreign ship for violations of MARPOL Annex VI even though it is flying a flag from a (flag) State that is not party to the Annex.[48] (See regulation 10.3 of Annex VI and article 5(4) of the MARPOL Convention.[49])

Such a violation of regulation 14.1.3 of Annex VI can be the subject of enforcement measures by the port State as part of a PSC inspection,[50] as regulation 11.6 of MARPOL Annex VI stipulates:

> The international law concerning the prevention, reduction, and control of pollution of the marine environment from ships, including that law relating to enforcement and safeguards, in force at the time of application or interpretation of this Annex, applies, mutatis mutandis, to the rules and standards set forth in this Annex.

Regulation 11 deals with the detection of and enforcement against violations of MARPOL Annex VI through PSC, and article 218(1) of UNCLOS must fall within the category of 'international law concerning the prevention, reduction, and control of pollution of the marine environment from ships, including that law relating to enforcement', especially as article 218 is found in section 6 of part XII of UNCLOS relating to 'enforcement'.

The interpretation of the term 'discharge' does therefore apply *mutatis mutandis* to Annex VI, although no necessary changes are required as this UNCLOS interpretation coincides with the existing definition of the term in the MARPOL Convention and how it is already applied in Annex VI (regulation 12.7.3).[51]

The reference in regulation 11.6 of MARPOL Annex VI to 'safeguards' must, inter alia, relate to the safeguards set out in section 7 of part XII of UNCLOS, as

[48] JA Zerk, *Extraterritorial Jurisdiction: Lessons for the Business and Human Rights Sphere from Six Regulatory Areas* (Harvard Corporate Social Responsibility Initiative, 2010) 194.

[49] See ch 4 for further information on the NMFT principle.

[50] Coordinated inspection campaigns between the different Memorandums of Understanding (MoUs), eg Paris MoU, Tokyo MoU and Indian Ocean MoU, as described in ch 4, could prove to be valuable tools of enforcement (and deterrence) in response to violations of the 0.5% sulphur limit on the high seas, as the participating port States can take measures against these violations pursuant to art 218(1) of UNCLOS and by applying the NMFT principle.

[51] Dr Erik J Molenaar and Professor Henrik Ringbom have argued that reg 11.6 of Annex VI, by itself, constitutes a legal basis for interpreting art 218 of UNCLOS to encompass emissions. E Molenaar, *Coastal State Jurisdiction over Vessel-Source Pollution* (Kluwer Law International, 1998) 508–09; and Ringbom, n 37, 14. This author respectfully disagrees with this stance as, inter alia, the *lex superior* principle prevents MARPOL Annex VI – a IMO regulation with 96 parties – from interpreting the scope of art 218 of UNCLOS – a UN framework convention (constitution) for the law of the sea with 167 parties. Using the VCLT to interpret UNCLOS is conversely a recognised approach within international law for determining the scope of a provision such as art 218. Also, the 'principle of legality' (*nullum crimen sine lege, nulla poena sine lege*) dictates that criminal liability should only be based upon infringements of legislation that is expressed with adequate precision and clarity. (See, eg, paras 70–71 of the *Intertanko case* (n 10) for the application of this principle.) Basing a port State's extraterritorial high seas jurisdiction on a *mutatis mutandis* reference found in a lower-ranking legislative text can hardly meet this criterion. Essentially, reg 11.6 of Annex VI merely allows for art 218 of UNCLOS to be interpreted so as to encompass emissions by way of art 31 VCLT. Such a conclusion is also supported by Advocate General Juliane Kokott's Advisory Opinion in the *Intertanko case*. The statement made by the Advocate General not only reflects her own opinion, but also refers to official statements made by different EU Member States and EU institutions: 'However, as Denmark, France, the Council

they also represent international law. Article 228(1), which determines the overlapping high seas jurisdiction between port and flag States, is such a safeguard provision as is found in section 7. This provision is analysed in the next chapter. Other procedural safeguards, including articles 223, 230 and 231 of section 7 of part XII of UNCLOS, are, however, also relevant when discussing port State enforcement of MARPOL Annex VI.

A. Port State Obligations under Article 223

Article 223 of UNCLOS requires port States – as was the case for flag States and coastal States (see chapters 8 and 9) – to allow representatives from the IMO, the flag State and an injured coastal State to attend any legal proceedings instigated in accordance with article 218(1). These representatives must enjoy international (diplomatic) rights.

B. Port State Obligations under Article 230

Port States can only sanction a violation that is acted on in accordance with article 218 with a *monetary penalty* (fine) pursuant to article 230(1), as the violation by definition will take place outside the prosecuting (port) State's territorial sea – a requirement for invoking article 218(1). This restriction applies irrespective of whether it was a deliberate infringement or not.

As noted previously, the extraterritorial jurisdictional basis of article 218(1) allows port States to impose fines that strip all savings made from the violation, and which include a punitive element. For example, for violations of the 0.5% sulphur limit in Annex VI on the high seas.

C. Port State Obligations under Article 231

Article 231 of UNCLOS requires all States, including port States, to promptly notify the flag State and other States concerned of any enforcement measure taken in accordance with the provisions of section 6 of part XII, such as article 218.

A port State that investigates or institutes legal proceedings against a foreign ship for a violation of regulation 14.1.3 of Annex VI on the high seas must immediately

and the Commission in particular correctly point out, basing oneself strictly on the wording of Marpol 73/78 would lead to absurd results. Even discharges resulting from intentional damage to a ship or its equipment would be permitted as long as neither the master nor the owner acted with intent or recklessly. It is thus necessary not only to interpret Marpol 73/78 in isolation on the basis of its wording, but also to take account of its objectives and its function within the framework of the Convention on the Law of the Sea' (Advocate General Juliane Kokott's, Advisory Opinion in the *Intertanko case*, n 11, paras 86–87). Reg 11.6 cannot, by itself, lift art 218 of UNCLOS to cover emissions, but can facilitate the application of the VCLT interpretation.

inform the flag State thereof, meaning *consular officers* and, if possible, *the maritime authority* of the flag State.

This information, provided by the port State to the flag State in accordance with article 231, also plays an extremely vital role when determining whether a flag State will lose its jurisdiction over such a high seas violation pursuant to article 228(1). This would occur, for instance, if the flag State did not respond to an infringement effectively within the six-month deadline set out in the provision – a deadline that begins with the article 231 notification (see chapter 11).

V. Article 218(1) Represents Customary International Law

All States that are parties to UNCLOS can apply article 218(1) to assume extraterritorial jurisdiction over MARPOL Annex VI violations on the high seas, regardless of whether these States also are parties to the VCLT, as the interpretation provisions of VCLT (articles 31 and 32) represent customary international law, recalling the previously reproduced statement of the ILC,[52] and can therefore be applied by all States.

Some States, including the US, are not parties to UNCLOS, but if article 218(1) of the Convention – like articles 31 and 32 VCLT – can be seen as representing customary international law, it would allow these non-party States to apply the same high seas enforcement principles pertaining to discharge violations.

This author believes that article 218(1) codifies basic principles of customary international law for the protection of the marine environment (including human health) from pollution, including air pollution.[53] This belief is founded upon the assumption that the principles of article 218(1) represent *consistent repetition of a particular behaviour* (the objective criterion),[54] as States, for *a considerable period of time*[55] – dating back to before UNCLOS, with the adoption of the Intervention Convention in 1969[56] – have exercised extraterritorial enforcement outside their own waters to protect themselves from high seas pollution.

[52] See extract in the text at n 24 above.

[53] Paul, n 3, 1054.

[54] Henriksen, n 47, 25.

[55] The fact that the principles of art 218(1) have been known since the adoption of UNCLOS in 1982 may in itself be enough to amount to *consistent repetition of a particular behaviour* based on the ICJ's reference in para 74 of the judgment in the *North Sea Continental Shelf cases* (*North Sea Continental Shelf cases, Federal Republic of Germany/Denmark and Federal Republic of Germany/Netherlands*, Judgment, 20 February 1969, [1969] ICJ Rep 3), where the Court emphasised that although a rule has only been applied for a 'short period of time', that does not automatically bar it from becoming a customary rule. See this example in Henriksen, n 47, 25.

[56] The Convention Relating to Intervention on the High Seas in Cases of Oil Pollution Casualties (adopted 29 November 1969, entered into force 6 May 1975) 970 UNTS 211 ('the Intervention Convention').

This also builds on the assumption that the principles of article 218(1) represent generally accepted and applied principles of international law (*opinio juris sive necessitatis*,[57] ie the subjective criterion) as the 167 parties to UNCLOS have accepted the legal basis of article 218(1).[58] Discussions on meeting the subjective and objective criteria are also found in chapter 4 regarding the principle of NMFT.

States that are not parties to UNCLOS, for example the US, can therefore also apply the principles of article 218(1) of UNCLOS, as it represents customary international (maritime) law.

Conversely, those non-party States should also respect the principles of article 218(2)–(4) and article 228(1) pertaining to overlapping jurisdiction with coastal and flag States, if they exercise such extraterritorial jurisdiction in accordance with *customary law*, as non-party States should not be able to achieve a more favourable inter partes position than those (coastal and flag) States that are party to UNCLOS.

VI. Limitation of Port State Jurisdiction: Article 218(2)–(4)

As the geographical scope of article 218(1) enables a port State to exercise jurisdiction over discharge violations taking place outside its own territory, it not only covers violations on the high seas, but also, in principle, infringements occurring within the territory (internal or territorial waters, or in the EEZ) of another (coastal) State that could also have a jurisdictional claim over these violations pursuant to article 220 of UNCLOS.

Article 218(2)–(4) seek to clarify the potential overlapping jurisdiction between a port State and coastal State relating to a violation that occurred within the territory of the latter. It should be remembered that in these situations, the ship that commits the violation is *foreign* to both the coastal State and the port State, by flying the flag of a third (flag) State.

Article 218(2) provides the essential legal basis for determining such overlapping jurisdiction, while article 218(3) merely requires the port State to carry out investigations when requested. Article 218(4) imposes certain procedural obligations upon the port State to deliver records, evidence and any economic security posted to the coastal State.

Article 218(2)–(4) contain several references to the *flag State*, but it should be noted that all these references 'merely' require the port State to investigate, institute proceedings and inform when so requested by the flag State. These provisions

[57] See Henriksen, n 47, 27.
[58] GG Schram, 'Havretten og de nordiske lande' (1984) 1 *De nordiske juristmøter* 401, 419.

do entitle a flag State to require a port State to stop any ex officio proceedings instituted. As stated on numerous previous occasions in this book, this question – of overlapping jurisdiction between a port State (article 218(1)) and a flag State (article 217) over violations occurring on the high seas – is resolved through article 228(1).

A. Respecting Coastal State Jurisdiction: Article 218(2)

Article 218(2) curtails the broad geographical port State jurisdiction of article 218(1) by providing that if a discharge violation takes place in another (coastal) State's internal or territorial waters or in the EEZ, or the offence damaged or threatened a (coastal) State, then the port State is barred from instituting proceedings unless the affected coastal State – or the flag State – so requests.[59] Or to rephrase: the extraterritorial jurisdiction of the port State according to article 218(1) is subordinated to the coastal State's territorial jurisdiction pursuant to articles 220 and 222.

An interesting question could arise regarding violations of air pollution regulations, such as regulation 14 of MARPOL Annex VI, that take place in a coastal State's EEZ. A port State would have jurisdiction over this violation in accordance with article 218(1) given the geographical scope of the provision. Coastal States also have jurisdiction over violations in their EEZ pursuant to article 220(1) and (5)–(6). But if the foreign ship does not call into a port in the coastal State then the right of the coastal State to take measures against all violations under article 220(1) becomes irrelevant.

The coastal State's legal basis for acting in response to such violations is therefore article 220(5)–(6), which confer the rights to stop and detain a vessel, and to initiate proceedings. Yet, recalling the discussion in chapter 9, it could be difficult for a coastal State to invoke these rights if the infringement concerns violations of air pollution regulations, as it could be hard to establish the necessary *causal link* proving that a specific violation has caused specific significant pollution that could threaten to damage the coastline or the interests of the coastal State. Also, article 222 is not applicable, as the geographical scope of that provision does not encompass the EEZ.

One could therefore argue that a port State can take enforcement measures against such violations of air pollution regulations, for example of the 0.1% or 0.5% sulphur limits in MARPOL Annex VI, committed in the EEZ, despite any coastal State objections.

This conclusion might not present a problem if attention is given to the fact that article 220(5)–(6) 'merely' concern the coastal State's right to stop, detain

[59] See Tanaka, n 14, 296.

and prosecute a foreign ship that is sailing through the EEZ, thereby respecting a foreign ship's right to transit an EEZ in accordance with article 58 of UNCLOS. The same procedural safeguards are not relevant in article 218(2) situations, as the foreign ship has voluntarily called into a port in the port State pursuant to article 218(1), thereby subjecting itself to PSC, NMFT and article 218 jurisdiction.

Had the ship called into a port in a coastal State after the EEZ violation, that coastal State's right to take measures against such air pollution infringements pursuant to article 220(1) would not be in question. As affirmed in chapter 9, coastal States can – where the evidence so warrants – prosecute proven air pollution violations in the EEZ if a ship later (perhaps weeks or months after the violation) calls into a port in the coastal State, provided any limitation periods have not been breached, for example the three-year cut-off period found in article 228(2). A port State must therefore respect article 218(2) and article 220(1) pertaining to the coastal State's right to enforce all rules and regulation, including MARPOL Annex VI (or GHG regulations), in the coastal State's EEZ, if the ship, possibly at a later time, calls into port in the coastal State. Consequently, the port State can only proceed against these violations if so requested by the coastal State. Any legal proceedings already instituted by the port State must be suspended, and all evidence, case records and any economic security posted must be transferred to the coastal State pursuant to article 218(4).

B. Duty to Seek to Comply with Requests from other States: Article 218(3)

Article 218(3) requires port States – *as far as practicable* – to accommodate requests from another State to investigate possible discharge violations pursuant to article 218(1). The provision refers to accommodating requests from (coastal) States regarding violations taking place within, or inflicting damage to, their internal or territorial waters or EEZ.

It also explicitly demands that port States comply with requests from flag States to investigate such possible violations, 'irrespective of where the violation occurred'. Even though these last few words of the provision have an extremely broad geographical scope, the *investigation* mentioned would often involve conducting a PSC and submitting any other evidence (sniffer or CEMS measurements, etc) the port State can obtain and provide, *as far as practicable*.

The requirement under article 218(3) for port States to comply with a request to investigate by a flag State should be read in conjunction with article 217(5) (see chapter 8) pertaining to the requirement for all States to endeavour to respond to a flag State's request for assistance in clarifying the circumstances of a specific case. Those two provisions supplement each other, although the applicability of article 218(3) is limited to discharge violations.

C. Transferring all Evidence, Records and Economic Security: Article 218(4)

Article 218(4) (first sentence) requires a port State to provide a coastal State or flag State with the records of an investigation for determining a ship's violation occurring within the waters of, or potentially causing damage to, another (coastal) State.

The second sentence of article 218(4) specifies that the port State must suspend any legal proceedings instituted by it, if the coastal State where the violation occurred requests this. The port State must, pursuant to the third sentence of the paragraph, subsequently transfer any evidence, relevant information, case records and any financial security posted by the shipowner to the coastal State.

Finally, the fourth sentence of article 218(4) provides that when a port State has suspended its proceedings and has transmitted all records, evidence, etc, it is precluded from continuing those proceedings.

Article 218(4) (first sentence) sets itself apart from the rest of the sentence in the paragraph by referring to *flag States*, alongside coastal States, as parties entitled to receive any of the findings and evidence provided by the investigations carried out in accordance with article 218(2)–(3).

Article 218(4) (second to fourth sentence) grants exclusive rights for coastal States to suspend any legal proceedings initiated by port States and have all information *and economic security* transferred to them, thereby bringing the port State's enforcement of the case to an end. The flag State is not given such rights in article 218 to demand that a port State suspend its proceedings; that is found in article 228(1). This could be an expression of fact that the coastal State's *territorial jurisdiction*, pursuant to article 220(1) is stronger than the *extraterritorial jurisdiction* of port States (article 218) and of flag States (article 217).[60]

Article 228(1), which resolves overlapping flag and port State jurisdiction over (discharge) violations on the high seas, takes a more considered approach to which State has the final primary jurisdiction, compared to the no-nonsense style of article 218(4) (second and third sentences), which grants the coastal State full discretionary rights to decide whether the port State's initiated proceedings should be suspended or not.

VII. Article 218 Provides Grounds for Complete Enforcement of the Sulphur Limits

Port States have, in accordance with article 218(1), an exterritorial jurisdictional basis for taking action against violations of the emissions regulations

[60] These basic forms of jurisdiction are described in ch 6.

(MARPOL Annex VI) outside their own territory. Article 218(2)–(4) nevertheless set *explicit limits* on this enforcement in other (coastal) States' waters without consent.

On the other hand, article 218(2)–(3) also provide – in the view of this author – *implicit possibilities* for ensuring complete port State enforcement against all discharge violations, irrespective of where these take place, if consent is given by coastal States. This is of the upmost importance when addressing the effective enforcement of the sulphur limits under regulation 14 of MARPOL Annex VI.

As already affirmed in section III.E and section IV, the extraterritorial jurisdiction of article 218(1) allows port States to impose fines that remove (confiscate) all savings achieved by not complying with regulation 14.1.3 on the high seas. It also allows for any fine to include a dissuasive punitive element, which increases in the event of aggravating circumstances. If a foreign ship has violated regulation 14 of Annex VI by using non-compliant fuel throughout its *entire journey* before coming into port in the port State, it may have breached the 0.1% or 0.5% limit in Annex VI while sailing in the waters (EEZs, etc) of another (coastal) State, *and* the 0.5% limit whilst sailing on the high seas. The coastal State in question could, in accordance with the principles set out in article 218(2)–(3), give its consent to the port State's taking enforcement measures against those violations too.

The fines imposed by the port State could therefore involve confiscating not only the savings made from violating regulation 14.1.3 on the high seas, but also those savings achieved by violating the limits in another (coastal) State's waters, including the waters of the State where the ship last called into port if the ship was already non-compliant by the time it left the port. This conclusion applies irrespective of whether a 0.5% limit applies in the waters of the coastal State pursuant to regulation 14.1.3 of Annex VI, or whether it is a Sulphur Emission Control Area (SECA) zone with a 0.1% limit in accordance with regulation 14.4.

The required coastal State consent can be given on an *ad hoc* basis, but it will also be possible for all States – for example through the internationally coordinated PSC regimes of the different MoUs described in chapter 4 – to make *ex ante* multilateral agreements that include tacit consent from all participating States to allow all other participating States to enforce such violations on their behalf and vice versa. A port State that is party to such an agreement and/or MoU would automatically be able to institute legal proceedings for violations taking place in another party's (coastal State's) waters pursuant to article 218(1)–(3) without explicit consent.[61] The port State could, of course, choose to inform any affected

[61] If the ship has violated reg 14 while sailing in the waters of its flag State, that State may choose not to give its consent for port State enforcement for this part of the violation, given its dual role as a flag State and coastal State (see art 218(2)–(3). But that State cannot – in its sole capacity as flag State – bar the port State from initiating proceedings for violations taking place on the high seas (see art 218(1)) as this falls outside the legal basis of art 218(2)–(3). This question of competing jurisdiction between a port State and flag State for violations on the high seas must be resolved in accordance with art 228(1).

coastal State of its pending enforcement measures, as this would simultaneously fulfil the requirement in article 231 (see section IV.C).

If the ship continued to violate regulation 14 of Annex VI whilst sailing into the port of the port State, the port State can of course also take action against this part of the violation by virtue of its *coastal State* jurisdiction pursuant to article 220(1). It may be recalled from chapter 1 that one State can be viewed as a *port State* and as a *coastal State*, depending on where a violation has taken place.[62] A State's use of article 218(1) to bring measures against violations on the high seas does not bar it from also exercising jurisdiction over the same ship in accordance with article 220(1) for violations of Annex VI taking place within its territory, for example if the ship continues its infringement through the EEZ, territorial sea and in internal waters coming into port. This applies irrespective of that State's having a 0.5% sulphur limit or a 0.1% sulphur limit in its waters.

A. The Taking of End-to-end Enforcement Measures by Port States in Response to MARPOL Annex VI Violations

These conclusions allow for the taking of 'end-to-end' enforcement measures by port States in the face of violations of MARPOL Annex VI, regardless of where these violations occur. The jurisdictional legal basis for this is found in article 218(1) for the part of the violation taking place on the high seas, article 218(2)–(3) (pursuant to article 218(1)) for the part of the violation taking place within the waters of another (coastal) State(s) and article 220(1) regarding any part of the violation that took place in the (port) State's own waters as the ship voluntarily sailed into port. This provides the jurisdictional basis for port States to impose fines that remove (confiscate) all savings achieved, and which include a dissuasive punitive element, for the *entire* violation.

To use the example given in chapter 1, say that a large container vessel sails from China to Europe, enabling it (the shipowner) to save $750,000 by being continuously non-compliant with regulation 14 of MARPOL Annex VI. It is to be taken that this constant non-compliance can be proved in the European port State into which the ship next calls, for example by measurements made by an on-board CEMS or a sniffer-attached drone used in conjunction with PSC fuel sampling, etc.

The European port State can then impose a fine removing (confiscating) all the savings achieved over the entire trip if China, the coastal State where the violation first occurred (when the ship left the Chinese port and sailed out), and any other affected coastal States (States whose waters the ship may have passed through on its way in to the European port State) give their consent to this enforcement on their behalf pursuant to article 218(2)–(3).

[62] It can, of course, also be a *flag State*.

The European port State has jurisdiction over the part of the violation that takes place on the high seas in accordance with article 218(1); and if the ship continued its violation whilst sailing into port, the port State also has jurisdiction over that final part of the voyage in accordance with article 220(1) given its own rights as a coastal State. The European State can therefore impose a fine that removes all savings and which entails a dissuasive punitive element. A fine of a minimum $1 million would not be considered out of place in this example, where three-quarters of it ($750,000) represent confiscation of the savings achieved by the violation and the final one-quarter ($250,000) represents the dissuasive punitive element.

In the event of aggravating circumstances being present, for example repeated violations involving such high levels that they yield an economic gain (saving), the punitive element should increase accordingly to be *dissuasive*. Fines in the range of tens of millions of dollars would not be disproportionate in these cases.

Granted, the procedural logistics of such enforcement can seem tortuous for a port State if consent from all coastal States affected needs to be obtained on an *ad hoc* basis. But if *ex ante* agreements can be made between States giving tacit consent to such enforcement, presumably under the auspices of the different MoUs, it would be easier for a port State to expedite the enforcement procedure.[63] The MoUs could also make such agreements inter partes, for instance between the Tokyo, Paris and Indian MoUs, perhaps also involving the US PSC. Such a unified approach to enforcement would also align itself with the basic principle of *ne bis in idem* codified in article 228(2) (*in fine*), which requires other (coastal or port) States to respect proceedings already instituted.[64]

It is important to keep in mind, when reading this conclusion, that it presupposes that the port State, in accordance with article 228(1), has primary jurisdiction over the part of the violation that took place on the high seas, as this jurisdiction overlaps with the flag State's high seas jurisdiction pursuant to article 217. Article 228(1) is analysed in chapter 11.

VIII. Conclusion on Port State Enforcement under Article 218

To sum up, article 218(1) of UNCLOS confers an extraterritorial right for port States to investigate and – if the evidence so warrants – bring measures of

[63] One could of course argue that even if such *ex ante* consent was not provided by all relevant States, and the port State had to obtain *ad hoc* consent in each case, the benefits that may be gained from effective enforcement of reg 14 of MARPOL Annex VI on a global scale, ie preventing 137,000 early deaths and 7.6 million children from developing asthma, and also ensuring a level playing field within the shipping industry, would warrant the extra work it would require by the port State authorities to obtain such consent.

[64] Art 228(2) and (3) are discussed in ch 12.

enforcement against violations of international (IMO) discharge regulations occurring outside their own territory, for example on the high seas, provided the infringing ship voluntarily calls into a port in the port State.

The term 'discharge' in article 218(1) ('any discharge') is, by applying article 31(1) VCLT, interpreted to encompass *emissions* from ships, including emissions of SOx, NOx and CO_2. This conclusion is supported by how 'discharge' is defined in article 2(3)(a) of the MARPOL Convention, applicable to all of the Annexes pursuant to article 1(2), and how it is already used in regulation 12.7.3 of MARPOL Annex VI, referring to the discharge of ODSs into the atmosphere.

States that are not party to UNCLOS can presumably also invoke the extraterritorial principles of article 218(1) as the provision represents customary international law. These non-party States must, however, also respect the principles of article 218(2)–(4) and article 228(1) pertaining to a port State's overlapping jurisdiction with other (coastal and flag) States.

Port States can – by virtue of the NMFT principle codified in regulation 10.3 of MARPOL Annex VI referring to article 5(4) of the MARPOL Convention – take measures against such violations committed by all ships, regardless of their flag.

If, after a violation, a ship calls into a port in a port State that does not proceed against this violation, other port States (into which the ship subsequently calls) can institute measures against such previous violations, provided the evidence so warrants. The principle of *ne bis in idem* bars a port State from proceeding against a violation once measures have been effectively taken by a coastal State, flag State or another port State.

Regulation 11.6 of MARPOL Annex VI allows port States to take enforcement measures pursuant to article 218(1) of UNCLOS during a PSC, provided all the procedural safeguards in section 7 of part XII of the Convention are met. This includes ensuring that IMO and flag State representatives can oversee any legal proceedings (article 223) and that only monetary penalties (fines) are imposed for violations on the high seas (article 230(1)). Also, the flag State must promptly be informed by the port State of any investigation or proceedings initiated in accordance with article 218(1) (article 231).

Article 218(2)–(4) explicitly obliges the port State to respect the territorial jurisdiction of a coastal State (article 220) pertaining to violations taking place within its waters. However, article 218(2)–(3) also implicitly provides a legal basis for (coastal) States to give consent, perhaps tacit *ex ante* consent within the different MoUs, allowing other (port) States to act in response a violation on their behalf. These port States can also take action against violations in their own waters, for example during the last part of the voyage before coming into port, as a (coastal) State, in accordance with article 220(1).

Article 218, applied in conjunction with article 220(1), provides port States with the jurisdiction to take action against a violation of regulation 14 of MARPOL Annex VI from beginning to end, irrespective of where that violation may occur. This extraterritorial jurisdiction under article 218(1) also provides port States with a legal basis for imposing a fine that removes (confiscates) all savings gained from the entire trip, as well as including a dissuasive punitive element, which increases in the event of aggravating circumstances.

Port State enforcement on the high seas pursuant to article 218(1) must be done in accordance with article 228(1) to determine which State has primary jurisdiction, as the port State's jurisdiction overlaps with the flag State's jurisdiction (see article 217).

The same principles could apply to the enforcement of future IMO regulations on GHG, depending on which legislative measures are adopted for achieving the necessary GHG reductions. This could involve the use of zero-carbon or fossil-free fuels, speed reductions, etc, as the term 'discharge' in article 218(1) does not apply to the enforcement of all proposed GHG legislative measures.[65]

[65] See ch 15.

11

Resolving Overlapping Jurisdiction: Article 228(1)

As noted many times in previous chapters, article 228(1) of UNCLOS[1] tackles the problem of overlapping[2] *extraterritorial jurisdiction* between a flag State, according to article 217, and a port State, according to article 218(1), for discharge violations on the high seas, which includes violations of regulation 14.1.3 of MARPOL Annex VI.[3] It also resolves the matter of overlapping jurisdiction between a flag State (article 217) and a coastal State pursuant to article 220 for violations occurring within the territory of the latter.[4] The question of coastal State jurisdiction overlapping with port State jurisdiction is, as discussed in chapter 10, resolved in accordance with article 218(2)–(4).[5]

Although article 228 is a provision found within section 7 of part XII of UNCLOS on procedural *safeguards* – while articles 217, 218 and 220 appear in section 6 on *enforcement* – article 228(1) calls for thorough analysis in a separate chapter, as this provision can, alongside article 218(1), be regarded as one of the few codified exceptions to the flag State principle referred to in article 92 of the Convention. This is also emphasised by the explicit reference in article 217(4) to articles 218, 220 and 228 pertaining to flag States' having to respect port State and

[1] United Nations Convention on the Law of the Sea (adopted 10 December 1982, entered into force 16 November 1994) 1833 UNTS 3.

[2] While some use the terms *competing jurisdictions* or *concurrent jurisdictions*, this book refers to *overlapping jurisdictions*.

[3] International Convention for the Prevention of Pollution from Ships (adopted on 11 February 1973, as modified by the Protocol of 17 February 1978, entered into force 2 October 1983) 1340 UNTS 61 (MARPOL), Annex VI, IMO Publication: IMO-520E. It may be recalled from chs 7 and 8 that any reference in this book to flag State enforcement of MARPOL Annex VI according to art 217 entails an implicit reference to art 222.

[4] It may be recalled from chs 7 and 9 that any reference in this book to coastal State enforcement of MARPOL Annex VI in accordance with art 220(1) entails an implicit reference to art 222 pertaining to enforcement in internal and territorial waters. Coastal State enforcement of Annex VI, and other air pollution regulations, in the EEZ is solely based on art 220, as the geographical scope of art 222 does not encompass the EEZ.

[5] If a violation occurs within the waters of the flag State then this State has sole jurisdiction over the violation.

coastal State jurisdictions, and to respect how article 228 resolves the flag State's overlapping jurisdiction with these.

It could be argued that article 218(1) provides port States with extraterritorial jurisdiction to take proceedings and other measures against discharge violations on the high seas (*enforce*), and that article 228(1) curtails that jurisdiction by limiting the right to complete proceedings (*adjucate*) in certain circumstances.[6] The right for port States to enforce applicable international rules and standards includes the right to investigate, detain, require financial security and *institute* legal proceedings. However, the right for port States to *complete* (execute) the legal proceedings and claim the posted financial security to cover any fine (irrespective of any protests made by the flag State), is found in article 228(1). A port State cannot execute its enforcement measures unless the requirements in article 228(1) are fulfilled.

Article 228(2), which sets out a three-year limitation period, and article 228(3), pertaining to the flag State's right to always enforce violations, are studied in chapter 12 with the other procedural safeguards in section 7 of UNCLOS.

I. The Wording and Overall Content of Article 228(1)

Article 228(1) is a very long and detailed paragraph. To aid comprehension, the four sentences in the paragraph are quoted separately below, as follows:

- 'Proceedings to impose penalties in respect of any violation of applicable laws and regulations or international rules and standards relating to the prevention, reduction and control of pollution from vessels committed by a foreign vessel beyond the territorial sea of the State instituting proceedings shall be suspended upon the taking of proceedings to impose penalties in respect of corresponding charges by the flag State within six months of the date on which proceedings were first instituted, unless those proceedings relate to a case of major damage to the coastal State or the flag State in question has repeatedly disregarded its obligation to enforce effectively the applicable international rules and standards in respect of violations committed by its vessels.'
- 'The flag State shall in due course make available to the State previously instituting proceedings a full dossier of the case and the records of the proceedings, whenever the flag State has requested the suspension of proceedings in accordance with this article.'
- 'When proceedings instituted by the flag State have been brought to a conclusion, the suspended proceedings shall be terminated.'

[6] The right for a State to enforce and adjudicate on (and prescribe) legislation is discussed in ch 6.

- 'Upon payment of costs incurred in respect of such proceedings, any bond posted or other financial security provided in connection with the suspended proceedings shall be released by the coastal State.'

The first sentence stipulates the *main rule* and the *two exceptions* regarding how the overlapping jurisdiction between a flag State and either a coastal State or a port State should be resolved.

The second sentence relates to the application of the *main rule* by setting criteria that must be met by the flag State.

The third and fourth sentences lay down procedural requirements that must be fulfilled if the flag State invokes the main rule.

Sections II and III of this chapter will focus on the scope of article 228(1) and the main rule in the first sentence, along with the criteria for applying that rule (in the second sentence) and the procedural follow-up requirements (in the third and fourth sentences). Section IV will focus on the two exceptions provided for in the first sentence, as from the onset they – if applicable – will bar the flag State from invoking the main rule. Special attention will be given to the second exception in the first sentence as it specifically relates to the overlapping jurisdiction on the high seas between a flag State and port State, also covering enforcement of MARPOL Annex VI and other International Maritime Organization (IMO) air pollution regulations.

II. The Scope of Article 228(1)

The first part of the first sentence of article 228(1) sets the parameters for the scope and application of the provision, referring to:

> Proceedings to impose penalties in respect of any violation of applicable laws and regulations or international rules and standards relating to the prevention, reduction and control of pollution from vessels committed by a foreign vessel beyond the territorial sea of the State instituting proceedings …

The initial reference to imposing penalties narrows the extent of the regulation to cover the aforementioned execution of enforcement by a port State or coastal State, as it does not hinder initial enforcement by these States by investigation, detention,[7] the requiring of financial security and the bringing of proceedings pursuant to articles 218 or 220.

The term 'any violation' covers all forms of violations, including discharge.

The 'applicable laws and regulations or international rules and standards relating to the prevention, reduction and control of pollution from vessels'

[7] The legality of a detention, including the proportionality of requested financial security, can be brought before the International Court of Justice (ICJ), International Tribunal of the Law of the Sea (ITLOS), etc in accordance with the expedited procedure described in art 292. See ch 12.

encompass all national and international regulations for the protection of the marine environment. The broad reference to 'international rules' covers all environmental IMO regulations and conventions, including, but not limited to, the London Convention,[8] the Anti-fouling Convention,[9] the Hong Kong Convention,[10] the Nairobi Convention[11] and – of course – the MARPOL Convention and its Annexes I–VI. It obviously also encompasses future IMO regulations on greenhouse gases (GHGs).

National laws could be coastal State laws adopted in accordance with articles 211, 212 or 234 of UNCLOS. No State can enforce national laws that regulate foreign ships on the high seas, as the high seas are considered *res communis*, as described in chapter 3.[12]

The words 'committed by a foreign vessel beyond the territorial sea of the State instituting proceedings' sets the geographical scope of the provision by limiting its application to violations taking place outside a (coastal) State's territorial sea, for instance in an EEZ or on the high seas. Article 228(1) can therefore be applied to determine which State has primary jurisdiction when a flag State's jurisdiction overlaps with a port State's high seas jurisdiction or with a coastal State's EEZ jurisdiction.

The reverse conclusion from this provides the first important result of the analysis of article 228(1): If a foreign vessel commits a violation in a coastal State's *internal waters* or in the *territorial sea*, that coastal State has *primary jurisdiction* over this violation in accordance with article 220(1)–(2) (and article 222 regarding air pollution), as compared with the flag State's jurisdiction pursuant to article 217, as it falls outside the scope of article 228(1). This conclusion is in complete alignment with article 2(1) of UNCLOS, which provides that coastal States enjoy sovereignty over these areas. This means that a coastal State always has full jurisdiction over violations of any national or international sulphur rules that take place within internal or territorial waters, with the exceptions mentioned in chapter 3 pertaining to innocent and transit passage.

[8] Convention for the Prevention of Marine Pollution by Dumping of Wastes and Other Matter (adopted 13 November 1972, entered into force 30 August 1975) 1046 UNTS 120 ('the London Convention').

[9] International Convention on the Control of Harmful Anti-fouling Systems on Ships (adopted 5 October 2001, entered into force 17 September 2008) ('the Anti-fouling Convention' or 'AFS Convention').

[10] The Hong Kong International Convention for the Safe and Environmentally Sound Recycling of Ships, 2009 (adopted 15 May 2009, not yet entered into force) ('the Hong Kong Convention').

[11] The Nairobi International Convention on the Removal of Wrecks (adopted 18 May 2007, entered into force 14 April 2015) ('the Nairobi Convention').

[12] Flag State jurisdiction to adopt national laws that apply to ships flying its flag should not be regarded as an exception to the *res communis* principle, as flag State jurisdiction fully governs ships under the State's flag sailing on the high seas and does not confer a right for the flag State to adopt and enforce national legislation over certain areas of the high seas and the foreign ships sailing thereon.

III. The Main Rule in Article 228(1)

The main rule in article 228(1) is found in the next part of the first sentence of the paragraph, which establishes that any proceedings initiated by a coastal or port State

> shall be suspended upon the taking of proceedings to impose penalties in respect of corresponding charges by the flag State within six months of the date on which proceedings were first instituted …

In other words, the main rule in article 228(1) provides that such proceedings instituted by a coastal State for an EEZ violation or by a port State for a high seas (discharge) violation shall be suspended if the flag State makes such a request, provided it fulfils the two initial requirements for doing so:

(a) the flag state must make such a request for suspension within *six months* of the instituting of the earlier proceedings brought by another State; and
(b) the flag State must brings *corresponding charges* against the ship(owner).[13]

A. The Flag State Must Act within Six Months

The six-month deadline in article 228(1) should – to some extent – be read in conjunction with article 231, which, it will be recalled (see chapters 9 and 10), requires coastal States and port States to promptly inform the flag State of any investigations instituted or procedural steps taken. Yet there is not complete overlap between the applicability of the two articles. The requirement in article 231 to notify the flag State applies from the time *any measure* was taken by a coastal State or port State, whereas the six-month deadline in article 228(1) begins from when the earlier *proceedings* were instituted.[14]

This could, in principle, mean that two separate notifications should be sent to the flag State from the enforcing coastal State or port State: one in accordance with article 231, when an enforcement measure is initiated; and one when the proceedings are instituted, to start the six-month deadline under article 228(1). However, the term 'proceedings' in article 228(1) should, in the view of this author, be understood to cover more than just proceedings in a court or the proceedings linked to imposing an administrative fine, as the six-month deadline would seem strangely long, considering how quickly such court proceedings can be concluded in most countries. 'Proceedings' must therefore be presumed to cover more than 'just' the

[13] Y Tanaka, *The International Law of the Sea*, 2nd edn (Cambridge University Press, 2015) 295.

[14] The exception in art 231 (second sentence), that notification is not required until legal proceedings are instituted pertaining to violations occurring in internal or territorial waters, falls outside the geographical scope of art 228(1).

days in court when the case is presented and decided; it must also encompass the preceding investigation, including the collection of evidence and establishing the case.

The case, including the investigation, is often handled by the maritime (Port State Control (PSC)) or environmental authorities before being transferred to the police and/or public prosecutor, unless the authorities themselves have the right to impose administrative fines. As such investigative measures often begin at the same time that other enforcement *measures* are taken, such as detention and demands for economic security (bail), etc during PSCs, the notification from the coastal or port State to the flag State under article 231 would encompass all such procedural enforcement actions, which would also mark the beginning of the six-month deadline for the flag State to act pursuant to article 228(1).

Such an article 231 notification should explicitly refer to the possibility of legal proceedings being instituted and the penalties that may result, given the safeguarding nature of articles 228 and 231. This is particularly the case with regard to a port State's enforcement on the high seas in accordance with article 218(1), as this is an exception to the established flag State principle.[15]

Articles 228(1) and article 231 might also be read in combination with article 223, another section 7 safeguarding article. Article 223 requires coastal and flag States to 'facilitate the attendance at such proceedings [instituted under Part XII] of official representatives of the competent international organization [and] the flag State'. One could argue that in order to facilitate the attendance of representatives from the flag State (and the IMO), information on the proceedings must be provided.

The coastal State or port State should thus always strive to provide the flag State with clear and explicit notification of any measures instituted, so that it can specifically identify and *prove* when the six-month period began, allowing the notifying State to conclude the proceedings, such as by imposing a fine and claiming any financial security posted, after this deadline has expired.

i. The Six-month Deadline Used in the Context of Enforcement of MARPOL Annex VI

The reverse conclusion to be drawn from article 228(1) is that if a flag State *does not* require a coastal or port State to suspend its proceedings within the six months after these began, the flag State automatically forfeits its right to assume primary jurisdiction over the violation. If a flag State's request for

[15] It should be noted that these provisions of UNCLOS have a fixed focus on the interactions between States. Separate notifications of any enforcement measures commenced or legal proceedings instituted by a coastal or port State would inevitably also be forwarded to the shipowner, who may subsequently contact its flag State directly to obtain its assistance. Such 'second-hand' flag State notification cannot by itself replace an official article 231 notification.

suspension, or other protest, is submitted *after* the six-month deadline has expired, the coastal State or port State can ignore such request/protest, complete the proceedings and execute enforcement.

This could become a focal point for ensuring effective enforcement of the 0.5% global sulphur limit pursuant to regulation 14.1.3 of Annex VI, as *open registry flag States* can be divided into two categories in this regard: those flag States that *actively* protect their flag and the ships flying it; and those that *passively* offer an open registry by essentially not doing anything.

It must be presumed that an *active* open registry flag State will respond to an article 231 notification within six months. *A passive* open registry flag *State* may, however, be unable (due to a limited maritime administration, etc) or unwilling (for political, economic or other reasons) to react to such notifications within six months of when proceedings were instituted.

This could prove to be the first important piece of the puzzle that falls into place to ensure that coastal States and port States can undertake enforcement of regulation 14.1.3 of Annex VI in EEZs and on the high seas. For example, if a (passive) flag State, for reasons unknown, is *unable or unwilling*[16] to respond to a port State's notification of any proceedings instituted, it would allow the port State to take action against the violation of the 0.5% limit, resulting in *end-to-end enforcement*, as described in chapter 10.[17]

B. Corresponding Charges Must be Brought by the Flag State

The second requirement that flag States must meet before being able to invoke the main rule of article 228(1) and suspend proceedings instituted by a coastal State or port State, is to take over the proceedings with the intent to 'impose penalties in respect of corresponding charges by the flag State'.

[16] This phrase, used in this context for describing flag State involvement, is borrowed from the '*unable or unwilling doctrine*' relating to the discussions within international law for justifying a State's military actions – see art 51 of the UN Charter (Charter of the United Nations and Statute of the International Court of Justice (adopted 26 June 1945, entered into force 24 October 1945) 1 UNTS XVI) on self-defence – against non-State actors (terrorist organisations) located within a foreign State where the government is *unwilling or unable* to prevent the use of its territory for such non-lawful activities. See JD Olin, 'The Unwilling and Unable Doctrine Comes to Life', quoting a letter by Samantha Powers to the UN, available at http://opiniojuris.org/2014/09/23/unwilling-unable-doctrine-comes-life/. This author does not intend in any way to belittle or dilute this important concept and how it is normally applied.

[17] The term 'end-to-end enforcement' covers a port State's measures of enforcement against the violation of reg 14 of Annex VI on the high seas and in foreign and own waters pursuant to art 218(1) and (2)–(3) and art 220(1) (and art 222). The fine imposed by the port State can confiscate all savings and include a punitive element.

This precondition establishes that the flag State does not have carte blanche to suspend proceedings within six months; rather, the flag State must prove to the coastal State or port State that it will effectively proceed against the violation committed by a ship flying its flag.

Given that the scope and purpose of article 228(1) is to ensure enforcement of the IMO's environmental legislation, the term 'corresponding charges' must be interpreted in the narrowest of ways, leaving little or no leeway for flag States to avoid enforcing a violation effectively. This would mean, for example, that if a port State has instituted proceedings against a ship for violating regulation 14.1.3 of MARPOL Annex VI on the high seas, the flag State must bring an identical charge covering the same infringement

One might even go so far as to claim that if the port State, meaning, for example, the maritime authority or public prosecutor, intends to impose a fine of a certain amount and this amount is based on recognised principles for calculating such fines, the duty to 'impose penalties in respect of corresponding charges' would require the flag State to impose a fine of at least the same amount. It must be emphasised that such a demand would not mean that all States must wholly align their sanctions for penalising violations of MARPOL Annex VI; it merely suggests that such sanctions must meet some minimum principles, such as those set out in article 18 of the EU Sulphur Directive,[18] to be considered *effective* and to avoid circumvention of the regulations.[19]

It will be recalled from chapter 8 that this corresponds with the general obligation under article 217(1) for flag States to enforce *effectively*, and that a flag State must ensure that its national laws will result in such effective enforcement and dissuasive penalties in accordance with article 217(8). The obligation for flag States to bring corresponding charges that ensure effective enforcement following a violation is therefore to be found in article 228(1) *and* in article 217(1) (and article 222 pertaining to Annex VI violations).

As the requirement for corresponding charges in article 228(1) is a precondition to the suspension of a coastal State's or port State's proceedings, it must imply the right for those States to deny a flag State's request for suspension if it does not assure (guarantee) them of its intention to bring such corresponding charges, which includes effective sanctions. Whether the flag State actually lives up to such assurances can subsequently be confirmed by the coastal State or port State in accordance with article 228(1) (second sentence).

[18] Directive (EU) 2016/802 of the European Parliament and of the Council of 11 May 2016 relating to a reduction in the sulphur content of certain liquid fuels [2016] OJ L132/58.

[19] As noted in previous chapters, fines for infringements of the sulphur limit could be calculated based on the parameters set out in art 18 of the EU Sulphur Directive, requiring that all savings are removed (confiscated) and a dissuasive punitive element, which increases in the event of aggravating circumstances, imposed.

C. The Procedural Requirements after Invoking the Main Rule

If a flag State can meet the conditions for invoking the main rule in article 228(1), that is, acting within six months and instituting its own proceedings on corresponding charges, the coastal State or port State in question is required to suspend its own proceedings. The operative word in this context is 'suspend', implying that the coastal State or port State proceedings are merely 'put on hold' when a flag State invokes the main rule. This assumption is confirmed by the words of the *third sentence* of article 228(1): 'When proceedings instituted by the flag State have been brought to a conclusion, the suspended proceedings shall be terminated.'

Before the coastal State or port State is required to terminate its proceedings – and thereby also required to release any financial security posted pursuant to the *fourth sentence* of the paragraph – the flag State must meet the explicit and implicit demands of the *second sentence* of article 228(1):

> The flag State shall in due course make available to the State previously instituting proceedings a full dossier of the case and the records of the proceedings, whenever the flag State has requested the suspension of proceedings in accordance with this article.

This second sentence thereby extends the list of requirements to which the flag State must adhere if it wishes to assert primary jurisdiction over a violation committed in an EEZ or on the high seas, by establishing an *explicit* post-suspension obligation to provide the coastal or port State with a full dossier and the records of the proceedings.

This author sees this obligation as including an *implicit* right for the coastal or port State to review these case records and dossiers, and to evaluate whether the flag State has actually fulfilled the requirement to bring *corresponding charges* pursuant to article 228(1) that must result in *effective enforcement* in accordance with article 217(1). If the coastal State or port State concludes that the flag State did not bring corresponding charges resulting in effective enforcement, that coastal or port State – following the reverse conclusion drawn from the third sentence of article 228(1) – is entitled to *resume* its legal proceedings, as these were merely *suspended* until this point. This includes claiming any financial security posted by the shipowner, as this has not yet been released, following the reverse conclusion drawn from the fourth sentence.

If a flag State does not fulfil the requirement in the second sentence of article 228(1) by providing a full dossier of the case and the records of the proceedings', it must be taken that corresponding charges have not been brought, which also allows the coastal State or port State to recommence its suspended proceedings. Such a failure to act by the flag State would in many ways equate to a flag State's lack of compliance with article 217(6) and (7) pertaining to providing information to an informing coastal State or port State (and the IMO and all States).

Unlike the reporting obligation in article 217(7), the reporting obligation in article 228(1) does not require the flag State to inform the IMO or other States of its effective and corresponding enforcement. The flag State can therefore forward the information directly to the port State without using the GISIS module.[20] However, the *port State* could log any non-compliance by the flag State in GISIS, for example if the flag State does not notify its enforcement or if it has brought non-corresponding charges, as this is relevant to the port State's subsequently resuming the, up until that point, suspended proceedings.

A flag State cannot avoid providing such mandatory information on corresponding charges by referring to any national law barring this (see article 27 of the Vienna Convention on the Law of Treaties (VCLT)[21]).

D. Conclusion on Flag States' Utililisation of the Main Rule

A flag State can, pursuant to article 228(1), assume primary jurisdiction over a violation of laws or regulations for the protection of the marine environment committed by a ship under its flag while it was sailing in a foreign coastal State's EEZ or on the high seas. A coastal State's jurisdiction over violations occurring in internal or territorial waters is not affected, following a reverse conclusion drawn from article 228(1).

A coastal State or port State has jurisdiction to use its powers of enforcement, such as by investigating, detaining and indicting a foreign ship(owner) pursuant to article 220 and article 218 of UNCLOS, provided it informs the flag State thereof in accordance with article 231.

Article 228(1) allow flag States to stop any resulting proceedings by a coastal State or port State to impose penalties, before a fine is imposed and any financial security posted is claimed.

This assertion of primary jurisdiction requires the flag State to fulfil three obligations (two initial and one subsequent). The two initial obligations require the flag State:

(a) to request the coastal State or port State to suspend its proceedings within six months of their first being instituted; and
(b) to institute its own legal proceedings in respect of *corresponding charges*, which include effective enforcement (fines) (first sentence).

[20] GISIS (the Global Integrated Shipping Information System) is an IMO-administered online reporting database, where the Member States can report different information relating to IMO regulations, including MARPOL Annex VI. See further ch 4.

[21] Vienna Convention on the Law of Treaties 1969 (adopted 22 May 1969, entered into force 27 January 1980) 1155 UNTS 331 (VCLT).

The third and subsequent obligation of the flag State is:

(c) to present a full dossier and records from the proceedings to the coastal State or port State, thereby allowing that State to evaluate whether the flag State did institute proceedings in respect of corresponding charges (second sentence).

If the flag State fulfils these three obligations then the coastal State or port State is required to fully *terminate* the, up until this point, suspended proceedings and release any financial security posted by the shipowner (third and fourth sentences). On the other hand, if either of the two initial obligations are not met by the flag State, the coastal State or port State has the right to *refuse* the flag State's request for suspension.

If the flag State does not meet the third obligation, proving that effective corresponding charges were brought, the coastal or port State can *resume* its suspended legal proceedings, imposing a fine and claiming (invoke) any security posted.

It is important to note that a coastal State or port State also can *refuse* a request for suspension from the onset even though a flag State fulfils all three requirements for invoking the main rule, provided one of the *two exceptions in* article 228(1) applies (see further section IV of this chapter).

i. Conclusion on Utilising the Main Rule for Enforcement of Annex VI

In terms of enforcing the sulphur rules in MARPOL Annex VI, this means that some open registry flag States that have a somewhat passive approach to their duties, being either unable or unwilling to fulfil them, would lose their primary jurisdiction by not acting within six months.

Even if such a timely protest is issued by the flag State, it must bring corresponding charges that ensure the imposition of a penalty that meets the criteria set out in article 18 of the EU Sulphur Directive, by removing (confiscating) all savings achieved by the infringement and also imposing a dissuasive punitive element. This would also meet the requirement in article 217(1) pertaining to effective enforcement.

If a flag State does not assure the coastal State or port State of its intention to bring corresponding charges, the request for suspension can be denied. If such assurance is provided, the result of the enforcement must be notified to the coastal State or port State, which can then satisfy itself that corresponding charges were brought and that they resulted in effective enforcement.

If such information is not provided, or it shows ineffective enforcement, the coastal State or port State can lift the suspension and resume its proceedings.

A port State can, if any of these circumstances (missing the six-month deadline, failing to bring corresponding charges, etc) leads to the flag State's not being able to invoke the main rule, assert full extraterritorial jurisdiction over the entire violation of MARPOL Annex VI from end to end, pursuant to article 218(1) and (2)–(3) and article 220(1), as described in chapter 10.

The port State can on these grounds impose a fine that strips all savings from the whole violation, and which includes a dissuasive punitive element that can increase if, for example, a shipowner has violated regulation 14 of Annex VI on other occasions.

Even if a flag State meets all requirements for invoking the main rule, the port State can deny such a request for primary jurisdiction if the *second exception* codified in article 228(1) applies (see section IV.B).

IV. The Two Exceptions in Article 228(1)

While the *first clause* of the first sentence of article 228(1) confirms that a flag State is permitted to invoke the main rule – provided all the requirements for doing so are met – the *second clause* of the first sentence limits the main rule by stating that the flag State is able to utilise the main rule, and thereby suspend any proceedings instituted by a coastal or port State,

> unless those proceedings relate to a case of major damage to the coastal State or the flag State in question has repeatedly disregarded its obligation to enforce effectively the applicable international rules and standards in respect of violations committed by its vessels.

Thus, article 228(1) contains two *exceptions* to the main rule:

(a) The *first exception* relates to the violations' having caused 'major damage' to a coastal State.
(b) The *second exception* pertains to a flag State's repeatedly having 'disregarded its obligation to enforce effectively'.

Before taking a closer look at these exceptions, it should be noted that, for the purposes of saving space and simplification, this book will refer to the first exception as 'the first exception in article 228(1)' or simply 'the first exception'. The second exception will be referred to as 'the second exception in article 228(1)' or simply 'the second exception'.

A. Major Damage to Coastal States: The First Exception in Article 228(1)

The first exception in article 228(1) is quite simple and straightforward in its wording and application, as it covers cases of 'major damage' to a coastal State. The geographical scope of the exception covers violations occurring in a coastal State's EEZ, recalling that violations taking place in internal and territorial waters invariably fall under the primary jurisdiction of the coastal State.

Unlike the second exception (see section IV.B), this first exception does not refer to its solely covering international rules and regulations; instead, the reference in the first clause of article 28(1) is to the 'applicable laws and regulations or international rules and standards'. This widens the regulatory scope of the first exception to encompass violations of international rules and of nationally adopted EEZ laws pursuant to article 211(6) or article 234, provided these violations result in *major damage*.

Such national and international rules for environmental protection in the EEZ, including air pollution regulations, are enforceable in accordance with article 220(1), provided the ship voluntarily goes into port in the coastal State. They are also enforceable pursuant to article 220(5)–(6) pertaining to non-air pollution discharge regulations.

If such an EEZ violation results in major damage to the coastal State, the first exception allows the coastal State to deny a request from a flag State for suspension of the coastal State's proceedings even though the requirements for applying the main rule are met by the flag State.[22]

Special extraterritorial rights for coastal States to protect themselves from major damage in the EEZ are also found in article 220(6). Yet the first exception in article 228(1) sets itself apart from article 220(6), as it is not limited to discharges and does not include violations that merely *threaten* to cause major damage. The first exception only covers violations that have actually resulted in major damage to the coastal State.

It is at the discretion of the coastal State whether damage sustained by a violation can be deemed major damage that will allow the coastal State to refuse a flag State's request for suspension. Exercising this right to determine if this first exception applies, is not without limitations for the coastal State. Such decisions are taken in light of the 'liability of States' in article 232 of UNCLOS, under which the coastal State becomes liable for any injuries to or loss for a shipowner attributable to its enforcement. The coastal State may also incur State liability towards the flag State pursuant to article 235 after refusing suspension on unlawful grounds.

However, the right to deny a flag State's request for suspension, by claiming that the first exception applies, is therefore not without its limitations and consequences. Especially, as a flag State can bring a case before ITLOS or the ICJ, etc,[23] claiming that the coastal State's assessment of the violation as causing major damage is incorrect.

Regarding enforcement of MARPOL Annex VI, and air pollution regulations in general, the discussion in chapter 9 also applies here. A mere *causal link* between a violation of air pollution regulations in an EEZ and the 'damage' it inflicts on the coastal State is, in the view of this author, too weak to establish that the violation

[22] Tanaka, n 13, 295.

[23] See ch 12 for proceedings being brought before ITLOS or the ICJ, etc in accordance with pt XV of UNCLOS.

of regulation 14 of MARPOL Annex VI has caused *major damage* to the coastal State.[24]

A coastal State therefore cannot refuse a request by a flag State for suspension by invoking the first exception in article 228(1) for a violation of the Annex VI sulphur limits in an EEZ. The same applies to enforcement of other air pollution regulations, including GHG regulations, in the EEZ.[25]

This could seem like a loophole for ensuring global compliance in EEZs with the 0.1% and 0.5% sulphur limits in MARPOL Annex VI. But, before so concluding, certain other points should be kept in mind:

(a) The flag State's request for suspension must meet all the preconditions for invoking the main rule in article 228(1), including bringing corresponding charges within six months and subsequently presenting documentation for this.
(b) Even though the *first exception* in article 228(1) might not allow a coastal State to deny a flag State's request for suspension if it concerns non-major damage, the *second exception* in the provision might, as it applies to coastal State proceedings for EEZ violations and port State proceedings for high seas (discharge) violations.

The second exception in article 228(1) is therefore deemed the more interesting – and potentially more important – when trying to establish how regulation 14 of MARPOL Annex VI is best enforced on a global scale, especially when trying to minimise the impact of *active open registry States'* offering *flags of convenience.*

B. Repeatedly Disregarding Obligations to Enforce Effectively: The Second Exception in Article 228(1)

It ought to be underlined initially that the word 'or' between the first and second exceptions in article 228(1) clearly indicates that these are two separate exceptions. While the first exception refers to major damage to a coastal State, the focus of the second is on countering a flag State's lack of enforcement in general by concentrating on past performance. This opens up the geographical scope of the second exception to encompass all legal proceedings covered by article 228(1), including proceedings instituted by coastal States according to article 220(1) for EEZ

[24] The same argument can be made for violations of nationally adopted discharge limits from open loop scrubbers in the EEZ under art 211(6) or art 234, as a violation thereof does not result in *major damage* to a coastal State.

[25] As was also stated in ch 9. An exception to this conclusion might be if a ship, in violation of IMO rules, were to release an extremely toxic pollutant (gas, etc) into the atmosphere, which can travel from the EEZ over land and directly harm (poison) humans. This would create the necessary link for invoking the first exception in art 228(1).

violations and those instituted by port States pursuant to article 218(1) for high seas discharge violations.[26]

The second exception in article 228(1) allows a coastal State or port State to refuse a flag State's request for suspension, even though all the preconditions for invoking the main rule have been met, if

> the flag State in question has repeatedly disregarded its obligation to enforce effectively the applicable international rules and standards in respect of violations committed by its vessels.

The wording of the second exception in article 228(1) thus establishes *three conditions* that must be fulfilled before a coastal State or port State can apply this exception:

(a) The flag State must repeatedly have disregarded its obligations.
(b) These obligations must relate to effective enforcement.
(c) The disregarded obligations must be found within international rules and standards.

To analyse these requirements in the most logical way, the order in which they are listed above will be reversed.

i. Obligations Found within International Rules and Standards

The second exception, unlike the first, only concerns itself with a flag State's disregard of *international rules and standards*, and therefore does not cover nationally adopted rules pursuant to articles 211, 212 or 234.

The international rules that *are* covered by the second exception are the 'usual suspects' in the regulatory library of the IMO, such as the BWM Convention,[27] the London Convention, the Anti-Fouling Convention and – of course – the MARPOL Convention, including its Annexes.

The reduced regulatory scope of the second exception, compared to that of the first exception, which covers national adopted rules, probably relates to the wider geographical scope of the second exception, as it also covers violations on the high seas.

ii. Flag State Obligations Relating to Effective Enforcement

The reference in the second exception to the flag State's having 'disregarded its obligation to enforce effectively' must be studied more closely to determine the obligations to which this refers.

[26] Tanaka, n 13, 297.

[27] The International Convention for the control and management of ship's ballast water and sediments 2004 (adopted 13 February 2004, entered into force 8 September 2017) ('the BWM Convention').

The reference to 'its obligation to enforce' is, in the view of this author, not a reference to one specific rule (article) in UNCLOS, although written in the singular form, but a reference to all part XII provisions that confer an obligation on flag States to enforce the applicable rules and standards.[28] The reference to 'its obligation' should therefore be recognised as a reference to all the provisions in part XII that create obligations for flag States, such as articles 216, 217, 222 and 223.

The specific mentioning of 'its obligation' is therefore linked to the repeated failure of the flag State in question to meet the obligation for effective enforcement under a particular article, for example where a flag State repeatedly fails to effectively enforce international rules on dumping (the London Convention) in accordance with article 216 of UNCLOS.

The requirement under article 223 for all States – including flag States – to allow representatives from an injured coastal State and the IMO to attend any legal proceedings brought for part XII violations is also an obligation encompassed by the second exception in article 228(1).

As this book focuses on enforcement on international rules on air pollution, such as MARPOL Annex VI and future GHG regulations, the flag State obligations in article 222 and article 217 require further examination.

Initially, it should be recalled (see chapters 7 and 8) that although article 222 provides a *lex specialis* legal obligation for flag States to take enforcement measures against violations of air pollution regulations, this by itself provides no specific definition of the extent of the enforcement obligation or what is required to fulfil it. Article 222 should therefore be used in conjunction with article 217 covering the general requirements with which flag States must always comply when enforcing all forms of IMO regulations for the protection of the marine environment, including regulation 14 of MARPOL Annex VI.

The specific details of article 217(1)–(8) are set out in greater depth in chapter 8, but to target and determine how article 217 relates to the application of the second exception in article 228(1), the relevant paragraphs of article 217, and their relevance to enforcing MARPOL Annex VI, will be examined below.

a. Flag State Obligations in Article 217 That Do Not Relate to the Second Exception in Article 228(1)

The first step is to eliminate those paragraphs of article 217 that are without greater relevance for establishing whether a flag State has repeatedly disregarded its obligation to effectively enforce regulation 14 of MARPOL Annex VI, pursuant to the second exception in article 228(1).

[28] Otherwise the exception ought to have included a direct reference to which specific articles were covered by it, just as art 220(8) refers to art 211(6), or how art 222 relates to art 212, or art 233 to art 42.

As concluded in chapter 8, article 217(5) does not burden the flag State with an obligation but bestows a right for the flag State to request assistance from other States to determine whether a violation has occurred. This paragraph is therefore without relevance to invoking the second exception in article 228(1).

The same applies with regard to article 217(2) and (4), which require a flag State to detain a ship under its flag when the vessel does not meet the required international standards and rules (article 217(2)), and to conduct investigations into possible violations of which the flag State itself (ex officio) becomes aware (article 217(4)). Although these regulations impose obligations on flag States, the common denominator of these requirements is that it will be difficult – if not impossible – for other States or the IMO to determine and prove that a particular flag State has failed to fulfil them.

The reasoning behind this conclusion is that a flag State, with regard to article 217(2), only has jurisdiction to detain a vessel flying its flag whilst it is at berth in a port in the flag State or sailing in waters under its sovereignty.[29] This makes it difficult to prove that the flag State had knowledge of the ship's non-compliance while it was sailing or at berth at that time, particularly as regards the question of which marine fuel was being used.[30] The same problem arises in connection with proving the *mens rea* (guilty mind) of a flag State when determining (under article 217(4)) whether it should have initiated an *ex officio* investigation of a possible violation of which it had itself become aware, meaning without official information from another State or the IMO.

The statement made by this author in chapter 8 (section IV) might be recalled, pertaining to article 217(4)'s covering *unofficial* information provided by other maritime actors, such as pilots, other ships, etc. It would be difficult to prove that a flag State has disregarded its article 217(4) obligation by not responding to such information, as it left to the flag State to validate and assess the usefulness of such information, which is sometimes provided anonymously.

This author believes that ironclad evidence is needed to establish that a flag State has disregarded its article 217 obligation, so that breaches of article 217(2) and (4) are unlikely to be deemed a *disregard* in the context of the second exception in article 228(1).

[29] As described in ch 8, some flag States conduct *flag State inspections* of ships flying their flag while the ship is in port in a foreign State. Such a flag State inspection on foreign territory is based on an agreement with the foreign port State and often does not allow the flag State to unilaterally detain the ship without assistance from the foreign port State authorities. As the flag State inspection is not a mandatory obligation for port States but based on a voluntary agreement, it is without relevance to the applicability of the second exception in art 228(1).

[30] International sniffer-attached drones will often not be permitted to fly in the ports and internal waters of a flag State. Its should be recalled that a State has full sovereignty in the air space over these areas; see art 2(2) of UNCLOS, which allows the State to refuse to allow such overflights and measurements.

b. Flag State Obligations in Article 217 Relating to the Second Exception in Article 228(1)

The main obligation to which flag States must adhere is found in article 217(1), which requires a flag State to ensure that international rules for the protection of the marine environment are effectively enforced, irrespective of where a violation occurs.

When focusing on *effective enforcement*, the criteria for imposing effective penalties (fines) have been described numerous times in previous chapters of the book, thus only a short restatement of them is necessary at this point.

For a flag State to fulfil the requirement in article 217(1) (and article 4(4) of the MARPOL Convention) for effective enforcement of the MARPOL Annex VI sulphur rules in the face of violations, a fine must, as a minimum, always meet the conditions set out in article 18 of the EU Sulphur Directive, as these are universally applicable. An effective flag State fine must thus:

(a) deprive the ship(owner) of all savings achieved (including confiscating all profits made from the violation); and
(b) contain a proportionate, yet dissuasive, punitive element that increases in the event of aggravating circumstances such a deliberate or gross violations, or – most importantly – where the shipowner has continually violated the same regulations.

A flag State can also impose sanctions other than a fine, including imprisonment, as the safeguarding principle in article 230, which as a main rule restricts sanctions to monetary penalties, does not apply to flag States. In fact, such sanctions can always be imposed by a flag State, even after a fine has been levied by another (coastal or port) State (see article 228(3), discussed in chapter 12).

Yet when discussing effective enforcement in the context of the second exception in article 228(1), the absolute minimum level of acceptable enforcement must represent the bar that must be reached by the flag State. A monetary penalty (fine) is therefore used as the appropriate sanction for determining effective flag State enforcement.

Consequently, if a flag State's enforcement in response to a MARPOL Annex VI violation does not, as a minimum, meet these above-mentioned conditions (confiscating savings, being dissuasive), the flag State will have disregarded its obligation to enforce effectively in accordance with article 217, and has thereby also disregarded its obligation to enforce effectively within the scope of the second exception in article 228(1).

It is the responsibility of the flag State – in accordance with article 217(3) – to ensure that all ships flying its flag have the required international certificates on board proving compliance with IMO rules and regulations for the protection of the marine environment. This includes the International Air Pollution Prevention Certificate (IAAP Certificate) and the International Energy Efficiency Certificate (EEI Certificate), as stipulated in Annex VI. Article 217(3) further requires a flag

State to ensure that ships under its flag are subject to periodic surveys to verify that the certificates and the actual circumstances on board match.

The issuing of certificates or the completion of periodic surveys may not seem to amount to effective enforcement as covered by the second exception. Nonetheless, it must be remembered that these formal requirements of article 217(3) ensure that a ship complies with the material requirements of the different IMO regulations, including the sulphur limits in regulation 14 of MARPOL Annex VI.

A flag State's failure to issue certificates or conduct a survey is a violation of article 217(3), and would thus be a disregard of its obligation to enforce according to the second exception in article 228(1). Such violations can be established (proved) during PSCs, as the initial document control of a PSC would involve checking that the ship carries all such required and up-to-date certificates, the flag State must ensure are present.[31]

It should be recalled that a flag State's private, inter partes agreement with a *classification society* for conducting such surveys and the issuing of certificates on behalf of the flag State does not relieve the flag State of its duties and accountability under article 217(3). A failure to comply by a classification society will externally – towards the rest of the UNCLOS community – constitute a flag State's disregard of article 217, and will therefore amount to a disregard covered by the second exception in article 228(1).

The obligation for a flag State to investigate violations and – where the evidence so warrants – institute proceedings against a ship flying its flag pursuant to article 217(6) constitutes an obligation the flag State must fulfil.[32] If a flag State receives an official written request from another (coastal or port) State to investigate an alleged violation or to institute proceedings (if such evidence can be provided), it obliges the flag State to act accordingly. An example of evidence provided by another State that obliges a flag State to institute proceedings could, as described in chapter 8, be measurements made by a Continuous Emission Monitoring System (CEMS) or a sniffer attached to a drone, aircraft, etc. It is a disregard of the flag State's article 217 obligation, and thereby a second exception disregard in the context of article 228(1), if the flag State does not comply with the requirement in article 217(6) to investigate and – where the evidence so warrants – institute proceedings for violations, for example, of MARPOL Annex VI.

Article 217(7) requires flag States to *promptly* notify an informing coastal State or port State and the IMO of any steps taken by the flag State and the outcome of any legal proceedings (under article 217(6)), which includes disclosing the size of any fine imposed. The flag State must at a later stage also make the same information available to all States.[33]

[31] Increased digitalisation of international certificates could help prove a violation of the art 217(3) obligation, thereby allowing it to be more easily seen as a *disregard* in the 'second exception' context.

[32] This obligation is also found in art 4(1) (2nd sentence) of the MARPOL Convention.

[33] It is recalled that art 27 VCLT bars a flag State from excusing itself from providing such information by making reference to its national laws, eg laws on public access.

It must be presumed that the IMO Legal Committee would be able to offer some non-binding, uniform interpretational guidance on how the word 'promptly' should be interpreted, as flag States are under the same obligation to provide information pursuant to article 4(3) of the MARPOL Convention and regulation 11.4 of Annex VI for enforcing, inter alia, regulation 14 of MARPOL Annex VI.

If a flag State does not present the mandatory information within the timeframe set out in article 217(7) it will constitute a violation that is a disregard in the context of article 228(1).

Article 217(8) requires flag States to ensure that their national laws allow for penalties that are 'adequate in severity to discourage violations wherever they occur' for sanctioning ships under their flag if these violate international rules for the protection of the marine environment. If a flag State is unable to impose such discouraging penalties for violations of regulation 14 of Annex VI, for instance due to the existence of an *administrative fining* system that employs a maximum on how high a fine can be, it cannot meet the obligation of effective enforcement pursuant to article 217(1). And if a flag State fails to effectively enforce by imposing a *discouraging* fine because of its national legislation, it also fails to fulfil the requirement under article 217(8) (and article 217(1)), which is a disregard covered by the second exception in article 228(1).

To this end it should be recalled that open registry flag States were previously separated into two categories in section III.A.i of this chapter: *active flag States*, which protect ships under their flag by responding to coastal or port State information notified under article 231 and protest legal proceedings initiated by these; and *passive flag States*, which offers sanctuary to ships under their flag by not engaging in any type of enforcement or responding to a coastal State or port State. A *passive* flag State would presumably often repeatedly disregard its article 217(7) obligation to provide any information to other States and the IMO after receiving an official request in accordance with article 217(6). An *active* flag State, in comparison, would often provide information, but it would violate its obligation to enforce effectively under article 217(1), perhaps by referring to the fact that its national legislation does not permit higher fines, which is covered by article 217(8). The second exception in article 228(1) 'catches' both active and passive open registry flag States offering flags of convenience, as article 217(1) and (6)–(8) are covered by this exception.

Finally, after dissecting the different paragraphs of article 217, it should be noted that if a flag State fails to meet the third requirement of the second sentence of article 228(1) for invoking the main rule, that is, subsequently informing the coastal State or port State of its effective enforcement after bringing corresponding charges, this would, as previously described in this chapter, allow the coastal State or port State to resume the suspended proceedings. But this lack of information should *also* be counted as a disregard in terms of applying the second exception in article 228(1), and should therefore be included in the continuous assessment of how many disregards a flag State has made before this can be deemed as being done 'repeatedly'.

iii. A Flag State's Repeated Disregard of its Obligations

If a flag State fails to meet any of the relevant article 217 obligations, this constitutes a disregard of an obligation covered by the second exception in article 228(1). The next part of the second exception analysis therefore requires us to establish how many times a flag State can disregard obligations before it is deemed to have done so repeatedly, thereby allowing all coastal States and port States to deny all requests from that flag State to invoke the main rule and suspend any proceedings instituted.

Before addressing this question, however, the breadth of the scope of article 228(1) must be determined, pertaining to whether a flag State's disregard of its different obligations codified in part XII can be accumulated.

The applicability of the second exception in article 228(1) should – in the view of this author – be interpreted in the most restrictive way, as it sets aside the fundamental *flag State principle* by allowing port States to exercise primary extraterritorial jurisdiction over discharge violations on the high seas pursuant to article 218(1). The reference to 'its obligation' – in the singular form – is therefore interpreted to relate to recurring disregards of the same part XII obligation, for example article 216, or article 217 or article 223. Thus, disregards of these obligations do not accumulate.

The subsequent reference to 'the applicable international rules and standards' – in the plural form – indicates that all IMO environmental regulations are covered by the second exception. This means that even though a flag State has disregarded its obligation under article 216 of UNCLOS to take action against dumping violations of the London Convention, and has disregarded its obligation to enforce other IMO Conventions (for instance MARPOL Annexes I–VI in accordance with article 217), and even if it fails to allow representatives from the IMO and other States to attend proceedings in accordance with article 223, *these disregards of articles 216, 217 and 223 cannot be accumulated.*

If, however, a flag State on several separate occasions disregards its obligation to enforce IMO rules that are covered by the same part XII obligation, for example disregarding its obligation to enforce MARPOL Annex I and Annex VI according to article 217, *these separate disregards of article 217 can be accumulated* as these are disregards of the same obligation.

The next part of the equation involves determining what the term 'repeatedly' covers. It undoubtedly covers more than *one* disregard, but exactly how many disregards are allowed can be difficult to set in stone, as several factors come into play. This author can therefore merely offer some suggestions as to how this term can be defined and applied.

One obvious solution might, of course, be to define 'repeatedly' as a specific number, for example 5, 10 or 25 disregards. This solution might create some unwanted results, though, as 10 disregards in enforcing the Anti-Fouling Convention might not result in as much damage to the marine environment or human health as 10 disregards in enforcing MARPOL Annex II or Annex VI, or the BWM

Convention, but it still results in a flag State's loss of its primary jurisdiction due to activation of the second exception in article 228(1) – a result that might be deemed disproportionate in comparison to the disregarded obligation.

A more nuanced solution could therefore be to define the term 'repeatedly' according to each individual IMO convention for the protection of the marine environment. As just stated, the reference in the second exception to 'applicable international rules and standards' in the plural does not bar the accumulation of a flag State's disregards in enforcing different IMO conventions pursuant to article 217. But even though this is possible, it does not necessarily mean that it is the best solution, as the differences in scope and content of the different IMO regulations may make it necessary to differentiate.

It could therefore be envisioned that the IMO would offer some non-binding, uniform interpretation as to how many times a flag State might disregard its obligations according to article 217 (or article 216 pertaining to dumping) when enforcing a specific convention, before the flag State loses its jurisdiction over enforcement of this environmental IMO convention. The IMO could, for example, deem that the potential vast and damaging pollution following a *single, specific violation* of MARPOL Annexes I or II could justify only very few article 217 disregards from a flag State before it would be deemed to have *repeatedly* failed in its duty, making the second exception in article 228(1) applicable. Violations of Annex VI could also be deemed as having such an *overall* negative impact on the global environment that only few article 217 disregards would be allowed.

It should be recalled that the non-specific obligation in article 222 for ensuring flag State compliance with air pollution regulations is supported (absorbed) by the more detailed regulations under article 217. It is therefore article 217 that is used as the relevant 'minimum yardstick' to assess flag State enforcement of IMO air pollution regulations, including MARPOL Annex VI and future GHG regulations.

The IMO could supplement this approach by adopting a *sliding enforcement scale*, which would allow a more lenient approach towards flag State enforcement of new conventions that have recently come into force. The number of flag State disregards permitted in enforcing this new legislation would gradually be reduced. This approach acknowledges that flag State authorities might need a transitional period within which to adapt their investigations under and, in particular, their enforcement of such new regulations. This approach would also allow the IMO to set a timeframe for when a disregard from a flag State becomes outdated. The number of disregards relating to a certain IMO regulation could, for instance, be assessed over a constantly rolling two-year period.

Another approach for determining whether a flag State *repeatedly* has disregarded its obligation to enforce, for example in accordance with article 217, would be to focus on the subjective performance of each flag State individually, rather than identifying a specific objective number for each convention. This would allow for other relevant factors to be considered, such as the total number of ships flying the flag of the State, what types of ship are predominantly being registered under the flag, and how many violations the flag State has dealt with effectively compared to those violations that might have 'slipped through the net'.

The reason for applying this approach would be to ensure that flag States that have large ship registries are not put at a disadvantage or subject to disproportionately harsh criteria. For example, a large flag State may have taken effective enforcement measures against 990 violations out of 1,000 committed by vessels under its flag, but if the specific cut-off limit is '10 disregards' then that flag State would be found to have *repeatedly* disregarded its obligations as covered by the second exception in article 228(1), despite having a 99% effective enforcement record. In comparison, a small flag State with few ships and nine disregards would not exceed the cut-off limit of 10 disregards, although the nine disregards in theory could represent a mere 10% effective enforcement record. These are of course extreme examples, but they underline the problem with applying specific number of disregards to define the term 'repeatedly'.

A flag State's performance should therefore be calculated on an individual, subjective basis, relating to each flag State's distinct circumstances. This approach could perhaps apply to some of the same calculating models and algorithms that are used to designate a flag State as being a so-called *black-*, *gray-* or *white-* listed flag State, as described in chapter 4. This would help to ensure that flag States are not penalised disproportionately by other States invoking the second exception of article 228(1).

iv. Consequences of Invoking the Second Exception in Article 228(1)

A coastal State or port State can invoke the second exception in article 228(1) if the required conditions are met, meaning that a flag State repeatedly has disregarded a part XII obligation, for instance article 217.

A flag State that fulfils this criterion is put on the proverbial *black list*, as all coastal States and port States can immediately sanction vessels coming from that flag State for violations of the relevant IMO environmental regulations, without having to await the flag State's response within the six-month deadline to the information provided under article 231.

This means that port States would automatically and immediately be able to conduct the previously described *end-to-end enforcement*[34] in response to violations of regulation 14 of MARPOL Annex VI over all ships coming from such a (black-listed) flag State, which repeatedly has disregarded its obligations

[34] The legal basis for end-to-end enforcement is, as described in ch 10, found in art 218(1) and (2)–(4) and art 220.

to enforce pursuant to article 217, thereby making the second exception in article 228(1) applicable.

Or, to rephrase, the flag State loses its right to assert primary extraterritorial jurisdiction over all violations in response to which it is required to act in accordance with article 217. As a result, the flag State is unable to enforce all IMO regulations covered by that part XII provision, for example all violations of MARPOL Annex VI committed by ships flying its flag. All relevant safeguarding measures should thus be followed when establishing that the second exception is applicable, as this is a very damaging sanction (and label) for a flag State – that it cannot enforce MARPOL Annex VI.

The loss of jurisdiction for a flag State pursuant to the second exception in article 228(1) should consequently be *temporary*, being limited to a specific period, such as 12 or 24 months. This *loss-of-jurisdiction period* could be followed a *probationary period*, during which the flag State's jurisdiction is restored, allowing it to assume primary jurisdiction, for example according to article 217 over all MARPOL Annex VI violations.

During this probationary period, which also could be a 12- or 24-month period or perhaps longer, the flag State would be under observation to ensure that it enforces effectively pursuant to article 217, making a lower, tolerated level of acceptable disregards over that period. Hence the reference to 'probation'.

Ensuring that the these strengthened enforcement obligations are not set aside during the probationary period should be a task performed under the auspices of the IMO or collectively by States, for example through the Memorandums of Understanding (MoUs).

On a final note, it could be argued that even though the consequences of applying the second exception in article 228(1) might seem very severe for a flag State, that State has violated basic principles within UNCLOS and international law, failing to fulfil obligations it has voluntarily undertaken by becoming a party to UNCLOS and/or MARPOL Annex VI. This violates the basic principle *pacta sunt servanda*, codified in article 26 VCLT, which also is embodied in article 235(1) of UNCLOS on State responsibility. The harshness of the second exception is therefore justified, as the flag State has violated the *pacta sunt servanda* principle on a recurring basis.

V. If Flag States are Not Party to UNCLOS or Annex VI

It should be noted that the starting point for the discussions in this chapter has been that the flag State in question is party to UNCLOS and to the violated IMO environmental regulation, for example MARPOL Annex VI. However, the legal consequences if a flag State is *not* party to either UNCLOS and/or the violated IMO regulation should, for reasons of clarity, also be explored.

A. The Flag State is Not Party to UNCLOS but is Party to an IMO Regulation

If the flag State is *not* party to UNCLOS but *is* party to the violated IMO regulation, the principles of article 217 and article 228 of the Convention should apply, as they – as was the case with article 218(1) – codify international customary law by representing 'consistent repetition of a particular behaviour', and also represent generally accepted and applied principles of international law (*opinio juris sive necessitatis*).

This means that the flag State must accept losing its natural primary jurisdiction[35] to other (coastal or port) States – which are party to UNCLOS – as they enforce in accordance with articles 218, 220 and article 228. And the non-party flag State is required to comply with the enforcement *obligations* pursuant to article 217 despite the flag State's not being a party to the Convention.

Conversely, the coastal and port States must bestow on the non-party flag State the same safeguarding *rights* under part XII, such as informing it of any investigations and proceedings instituted in accordance with the provisions of article 231, allowing flag State representatives to attend the proceedings in accordance with article 223 and permitting the non-party flag State to assume primary jurisdiction by fulfilling the three requirements for applying the main rule in article 228(1).

The flag State should not be held accountable under State liability as it is not party to UNCLOS and therefore not bound by the principle of *pacta sunt servanda* in article 26 VCLT and article 235(1) of UNCLOS.

Finally, it should be recalled that such a flag State that is party to MARPOL Annex VI but not to UNCLOS, is required to take enforcement measures against violations of Annex VI in accordance with the regulations in the Annex, including regulation 10 and regulation 11, as well as the principles of the MARPOL Convention for effective enforcement and informing other parties and the IMO thereof (see article 4), as these also apply to the Annexes.

B. The Flag States are Party to UNCLOS but Not Party to an IMO Regulation

Where the flag State *is* party to UNCLOS but *not* party to an IMO regulation, for example where the flag State is not a party to MARPOL Annex VI, the flag State is not obliged to enforce these environmental rules in accordance with UNCLOS.

The principle of *no more favourable treatment* (NMFT) will nevertheless, as described in chapter 4, ensure that ships hailing from that flag State are subject

[35] The flag State principle – as derived from the *Lotus case* (*SS Lotus (France v Turkey)* PCIJ Rep Ser A No 10) – applies to all flag States, including flag States that are not party to UNCLOS.

to the same requirements of the IMO regulations, including the sulphur limits in regulation 14 of MARPOL Annex VI, pursuant to regulation 10.3 of Annex VI and article 5(4) of the MARPOL Convention.

C. The Flag State is Party to Neither UNCLOS Nor an IMO Regulation

The previous conclusion still applies, even if a flag State is party to neither UNCLOS nor the IMO's environmental regulations, as ships under its flag must adhere to the IMO rules when voluntarily calling into the ports of another (port) State that is party to the IMO regulations, due to the principle of NMFT.[36]

VI. Conclusion on Article 228(1)

Article 228(1) establishes how any overlapping (competing) jurisdiction between a flag State and a coastal or port State is resolved.

The geographical scope of the provision covers violations that occur in a coastal State's EEZ or on the high seas. This leads to the reverse conclusion that a coastal State always has primary jurisdiction over violations taking place in its internal and territorial waters.[37]

The main rule in article 228(1) provides that the flag State has primary jurisdiction as long as it can meet three requirements – two initial preconditions and one subsequent requirement. The conditions for invoking the main rule are:

(a) that the flag State must within *six months* request the coastal State or port State to suspend its proceedings;
(b) that the flag State acknowledges its intent to institute proceedings on corresponding charges for the same violation.

If the flag State cannot meet these two preconditions, the coastal State or port State can refuse the request for suspension.

If it fulfils the preconditions then the coastal State or port State proceedings are suspended, but the flag State is subsequently obliged to inform the coastal State or

[36] The consequences of a State's threatening to withdraw from UNCLOS or the IMO regulations, such as MARPOL Annex VI, must therefore be considered limited as far as enforcement is concerned. It should be also recalled that such non-party States – whether a coastal State or a port State – cannot distort the effective enforcement of such IMO regulations, even if the violating ship first calls into a port in the non-party State after the infringement. If a non-party State does not take measures against a violation then another coastal State or port State can, at a later time, carry out the enforcement in response to the specific infringement, if the ship calls into a port in that State.

[37] The applicable exceptions described in ch 3 pertaining to sailing in the territorial sea and referenced in art 2(3), still apply.

port State of the outcome of its proceedings instituted on corresponding charges, including the size of any fine imposed. If the flag State does not provide such information, or if the information reveals ineffective enforcement, the coastal State or port State may *resume* its proceedings, including imposing a fine and claiming any financial security posted.

If the flag State fulfils all three conditions for invoking the main rule, the coastal State or port State must terminate its proceedings and release any financial security posted, provided one of the two exceptions in article 228(1) does not apply. If one the two exceptions *does* apply, it allows the coastal State or port State to refuse any initial request from the flag State for suspension, irrespective of whether the flag State fulfils all the requirements for invoking the main rule.

The *first exception* in article 228(1) refers to major damage to a coastal State following a violation in the EEZ. It is at the discretion of a coastal State whether pollution originating in a violation in the EEZ has resulted in *major damage*; the mere threat of major damage – unlike in article 220(6) and article 233 – is insufficient. A coastal State is subject to strict liability if it invokes the first exception, as article 232 makes it responsible for any damage a shipowner might suffer from unlawful enforcement. It might also incur State liability in accordance with article 235.

The *second exception* in article 228(1) covers flag States that have repeatedly disregarded a part XII obligation to enforce IMO regulations for the protection of the marine environment. The flag State must have repeatedly disregarded the same enforcement obligation, for example article 217, which encompasses enforcement of MARPOL Annex VI.

The term 'repeatedly' is open-ended, and several solutions for defining it could be applied, for example setting a specific cut-off number of allowed disregards for each IMO regulation. Another solution might be to focus on a flag State's individual ability to enforce compared to the size of its ship registry, etc, for instance by applying some of the principles used for calculating and determining whether a flag State is deemed to be on the MoUs' black, grey or white lists.

A flag State's loss of primary jurisdiction over certain violations due to application of the second exception should be limited to a specific period, perhaps one or two years, followed by a probationary period.

12

Safeguards, Liability and the Settlement of Disputes

This chapter will review some of the safeguard and liability provisions of sections 7 (articles 223–233) and 9 (article 235) of part XII of UNCLOS.[1] However, as many of these provisions have been discussed in some depth in previous chapters (eg articles 224–227 in chapter 4 and article 228(1) in chapter 11), they will not be revisited in detail here.

The chapter will also attempt to provide insight into the regulations in part XV of UNCLOS covering the settlement of disputes, including the possibility of referring a dispute on MARPOL Annex VI[2] enforcement to the International Court of Justice (ICJ), the International Tribunal on the Law of the Sea (ITLOS) or resolving it through arbitration.

I. Safeguards: Section 7 of Part XII of UNCLOS

The safeguard provisions in section 7 of part XII of UNCLOS provide a certain framework, including limitations, for flag States, coastal States and port States when exercising their jurisdiction in accordance with the enforcement regulations in section 6. The previously examined article 228(1) is an example, as it sets the limits on how coastal and port States may exercise their 'section 6 jurisdiction' (pursuant to articles 218 and 220) on the high seas and in an EEZ. Article 228(1) also places certain obligations on flag States, which they must fulfil in order to invoke this safeguard, provided neither of the two exceptions to it (examined in chapter 11) applies. Article 228 (2)–(3) set different restrictions on the application of the first paragraph of that provision, which is discussed in section I.C of this chapter. This, and the other safeguarding provisions of section 7, tries to strike a fair balance between the interests of the different States.[3]

[1] United Nations Convention on the Law of the Sea (adopted 10 December 1982, entered into force 16 November 1994) 1833 UNTS 3.

[2] International Convention for the Prevention of Pollution from Ships (adopted on 11 February 1973, as modified by the Protocol of 17 February 1978, entered into force 2 October 1983) 1340 UNTS 61 (MARPOL), Annex VI, IMO Publication: IMO-520E.

[3] This expression is borrowed from the European Court of Justice (ECJ), which used it to describe the overall purpose of UNCLOS in Case C-308/06 *Intertanko, Intercargo, Greek Shipping*

A. Measures to Facilitate Proceedings: Article 223

Article 223 provides that all States must 'facilitate the attendance at … proceedings [instituted pursuant to Part XII] of official representatives of the competent international organization, the flag State and any State affected by pollution arising out of any violation'. This means that when a flag State initiates proceedings against a ship under its flag following a report from another State in accordance with article 217(6), or if a violation has resulted in damage to a (coastal) State, it must allow official representatives from those States and the International Maritime Organization (IMO) to attend these proceedings.

This obligation also covers the situation where a flag State has instituted proceedings by bringing *corresponding charges* in accordance with article 228(1), leading to the suspension of a coastal State's or port State's proceedings. Representatives from the coastal State or port State in question, and from the IMO, should be allowed to attend these proceedings.

Coastal States and port States are also required, by this provision, to allow representatives from the flag State and the IMO to participate in any legal proceedings initiated. For example, when a coastal State initiates proceedings in accordance with article 220(1), or a port State in accordance with article 218(1).

Some (flag, coastal or port) States may have criminal codes, etc that do not allow for such participation, specifically when dealing with criminal cases against private individuals. This should, in principle, not be an acceptable excuse (see article 27 of the Vienna Convention on the Law of Treaties (VCLT).[4] Criminal proceedings against private individuals (such as the master of the ship or a crew member) could perhaps be allowed to be carried out without representatives being present if the court were to decide to 'close the doors' of the proceedings, as this would be in alignment with United Nations (UN) principles of a fair trial.[5] Also, the other States and the IMO could, as a discretionary courtesy, refrain from invoking the right to participate in proceedings against private individuals. On the other hand, representatives should always be allowed to participate in criminal proceedings against a legal entity, often the shipowner.

Article 223 also stresses that 'The official representatives attending such proceedings shall have such rights and duties as may be provided under national laws and regulations or international law.' The reference to their enjoying rights under international law could perhaps be seen as a reference to the international rights enjoyed by diplomats and consuls in accordance with the Vienna Convention

Co-operation Committee, Lloyd's Register, International Salvage Union v Secretary of State for Transport, ECLI:EU:C:2008:312, para 58. See ch 5.

[4] Vienna Convention on the Law of Treaties 1969 (adopted 22 May 1969, entered into force 27 January 1980) 1155 UNTS 331 (VCLT).

[5] See art 10 of the Universal Declaration of Human Rights (UDHR), available at https://www.un.org/en/universal-declaration-human-rights/.

on Diplomatic Relations[6] and the Vienna Convention on Consular Relations,[7] as these officials are normally described as being *representatives* of the sending State. This would mean, for example, that any documents, etc belonging to these official representatives would be covered by the diplomatic privilege set out in article 24 of the Vienna Convention on Diplomatic Relations.

B. Exercise of Power and Investigating Foreign Vessels: Articles 224–227

The applicability of articles 224–227 is closely examined in chapter 4 relating to Port State Control (PSC) enforcement by port States, and is therefore only briefly described in this section:

- Article 224 requires all States to ensure that enforcement is exercised by government officials, or by using clearly marked and identifiable warships, ships and military aircraft.
- Article 225 stipulates that all States have a duty to ensure that their enforcement does not result in adverse consequences to the safety of navigation or the protection of the marine environment.
- Article 226 concerns the (PSC) inspection of foreign vessels, distinguishing between an initial inspection of documents and a detailed inspection of the ship, the latter requiring clear grounds.
- Article 227 bars all States from discriminating against foreign vessels, a requirement also implicitly stipulated by the 'no more favourable treatment' (NMFT) principle.[8]

C. Restrictions on the Instituting of Proceedings: Article 228(2)–(3)

Article 228(1) resolves the question of the overlapping jurisdiction between flag States and coastal or port States. This is described in further detail in chapter 11.

[6] Vienna Convention on Diplomatic Relations 1961 (done at Vienna 18 April 1961, entered into force 24 April 1964) 500 UNTS 95.

[7] Vienna Convention on Consular Relations 1961 (done at Vienna 24 April 1963, entered into force 19 March 1967) 596 UNTS 261.

[8] This non-discrimination obligation in art 227 and the NMFT principle must be seen in the light of art 217(2)–(3), which require all (flag) States to ensure that ships under their flags follow the same rules as all foreign vessels, thereby ensuring that all ships in a port adhere to the same rules, irrespective of which flag they fly.

Article 228(2) contains two obligations that must be observed, setting a *three-year limitation period* and basically codifying the principle of *ne bis in idem*, which stops a person (or company) from being tried for the same offence twice.

It must initially be highlighted that the safeguards in article 228(2) only apply to enforcement by coastal and port States, as the provision refers to *foreign vessels*. This means that flag States are not covered by the provision, so they can prosecute ships under their flag beyond the three-year limit and despite another State's already having penalised a ship for its violation. The special prerogative for flag States to take measures against violations is described in article 228(3).

A coastal State or port State must respect the three-year limitation period laid down in the first half of article 228(2) when proceeding against violations committed by foreign vessels, for example when a port State takes action in response to a violation of MARPOL Annex VI on the high seas pursuant to article 218(1).[9] The limitation period is calculated from the time when the violation occurred.

Should the national laws of a coastal or port State provide the accused (shipowner) with a more advantageous procedural safeguard, for example by offering a two-year limitation period, this should be observed.[10] This interpretation also aligns itself with the requirement found in article 230(3) of UNCLOS, which stipulates that 'the recognized rights of the accused shall be observed'.

The limitation period should be suspended, or be considered not to apply, if a shipowner deliberately seeks to prolong a legal investigation or proceedings, for example by avoiding responding to notifications from the authorities, or by not having a clear point of contact (address, phone or email) that authorities from other States can easily use to get in touch with the owner.

Further, if a flag State suspends the legal proceedings of a coastal or port State by invoking the main rule under article 228(1), it automatically also suspends the limitation period in accordance with article 228(2).

This suspension of the limitation period will overlap with the suspension of the legal proceedings. A flag State could otherwise circumvent the safeguard protection of article 228(1) by suspending the proceedings of a coastal or port State and waiting over three years before imposing an ineffective penalty or not bringing corresponding charges, for instance by completely dismissing the case.

The second half of article 228(2) stipulates that a coastal State or port State shall not institute legal proceedings against a foreign vessel if such proceedings have been commenced in accordance with article 228(1). This codifies the basic principle of *ne bis in idem*.

It should be noted that article 228(2) explicitly refers to proceedings under article 228(1). This means that article 228(2) provides that coastal States or

[9] And in the waters of other (coastal) States according to the *end-to-end enforcement* described in ch10.

[10] Conversely, the wording of art 228(2) sets the three-year limitation as a maximum that must always be observed. So even if the national laws of a coastal State or port State offer a four-year limitation period, the three-year period in art 228(2) supersedes this.

port States must respect any legal proceedings instituted by the flag State that have resulted in effective enforcement. They must also respect any proceedings commenced by another (coastal or port) State regarding the same violation. This situation could arise, for example, if a violating ship has passed through different EEZs, causing major damage to several coastal States, giving all these States primary jurisdiction (and extinguishing the flag State's jurisdiction) pursuant to the *first exception* in article 228(1). Or, if several port States wish to sanction a foreign vessel's discharge violation on the high seas and the vessel flies the flag of a flag State that repeatedly has disregarded its obligation to enforce effectively, pursuant to the *second exception* in article 228(1), which, in principle, gives all port States the right to assert primary jurisdiction over the violation according to article 218(1).

Article 228(3) clarifies the position of the flag State in this context by stipulating that neither article 228(1) nor article 228(2) prohibits that flag State from taking 'any measures, including proceedings to impose penalties, according to its laws irrespective of prior proceedings by another State'. This means that it is the prerogative of the flag State to commence proceedings for a violation committed by a ship under its flag, regardless of whether this violation has been acted on by another State and whether the flag State wishes to sanction a violation that occurred more than three years ago. This regulation aligns itself with the flag State principle, as it enables a flag State to take measures against all violations committed by a vessel under its flag.

It should be noted that article 228(3) does not influence the right of a coastal State or port State to assume primary jurisdiction in accordance with article 228(1). It 'merely' grants the flag State the sovereign right always to proceed against violations, thereby allowing the flag State to supplement any sanction imposed by a coastal State or port State, for example by imposing non-monetary penalties, such as imprisonment, for violations that took place on the high seas or in an EEZ, as coastal States and port States do not have the jurisdiction to do so (see article 230).

D. Instituting Civil Proceedings: Article 229

Article 229 clarifies that none of the provisions of UNCLOS, including part XII, hinders the right of any State to institute *civil proceedings* for any loss or damage it may have been caused following a ship's pollution of the marine environment.

For example, pursuant to article 220, a coastal State can impose a fine for violations that take place inside its waters, but if any cost directly related to the violation is not covered by the fine then article 229 can be invoked. This could perhaps lead to recouping lost revenue from tourists if a popular coastline is closed due to pollution.

Article 229 should not be confused with article 232 of UNCLOS, which allows a shipowner to institute proceedings for any economic loss the owner may have suffered due to wrongful enforcement by a State.

E. Monetary Penalties and Observance of Recognised Rights: Article 230

Article 230(1) provides that only *monetary penalties* (fines) may be imposed for violations committed by a foreign ship outside the territorial sea of a (coastal) State. This means that coastal States cannot impose other sanctions, such as imprisonment, for violations that take place in the EEZ, nor can port States sanction (discharge) violations on the high seas with any penalty other than a fine.

There are no exceptions to this, yet it must be emphasised that this provision does not bar flag States from sanctioning such violations with non-monetary penalties, for instance with imprisonment, as article 230 refers to *foreign ships*. Indeed, article 228(3) allows flag States always to supplement a coastal or port State's financial sanction (fine).

As mentioned in chapter 4, some port States use imprisonment as a PSC sanction for what seemingly relates to violations that have occurred in the EEZ or on the high seas. But before one reaches the conclusion that such a sanction is not compliant with article 230(1), two important questions must be asked and answered: What precise violation is being sanctioned? And when is it finalised?

The answers to these questions are important, because if a ship infringes one of the Annexes to the MARPOL Convention in the EEZ or on the high seas, that infringement, by itself, can only be met with a fine. But if the master and/or crew are/is questioned by a Port State Control Officer (PSCO) during a PSC about the violation in the EEZ or the high seas, and they provide false information, that false information could by itself constitute a new violation, which is committed the moment it occurs, which, in some States, can result in imprisonment, as a PSC takes place in port and imprisonment is therefore allowed in accordance with article 230.

The PSC enforcement carried out by the US Coast Guard has used this approach, and it is interpreted to be in accordance with article 230 of UNCLOS, albeit that the US is not party to UNCLOS.[11]

The same conclusion applies if a shipowner, master and/or crew has falsified any of the documentation presented during the initial part of a PSC. This could be perceived as amounting to forgery of documents, which in many States is punishable by imprisonment. Again, the crime is not committed until the moment when the falsified log or document is given to the PSCO while the vessel is in port, thereby allowing imprisonment as a sanction in accordance with article 230.

Article 230(2) stipulates that coastal States may also only impose fines for violations taking place in the *territorial sea*, unless a violation is deliberate and

[11] See CD Michel, 'Vessel Source Pollution and Protection of the Marine Environment' in MH Nordquist, TTB Koh and JN Moore (eds), *Freedom of Seas, Passage Rights and the 1982 Law of the Sea Convention* (Martinus Nijhoff Publishers, 2009) 480. The views expressed in the chapter are US Coast Guard Captain Michel's own, not necessarily representing those of the US Coast Guard.

causes serious pollution. The main rule in article 230(2), relating to the territorial sea, mimics the regulation in article 230(1) regarding the EEZ and the high seas, apart from the fact that article 230(2) allows for an exception if the violation is deliberate (wilful) and serious. This exception makes sense, as violations in the territorial sea are closer to the coast and enclosed areas of the coastal State, so that violations are more likely to cause damage. It furthermore aligns itself with the general principle in article 2 of UNCLOS, which stipulates that the coastal State, at the outset, has the same jurisdiction in the territorial sea as in internal waters.

The reverse conclusion to be drawn from article 230, and article 2(1), is that there are no limitations if a State wishes to sanction violations taking place in *internal waters* with non-monetary penalties, including imprisonment.[12] This also provides the legal basis for the abovementioned PSC sanctioning of untruthful statements or falsified documents during PSCs, as those violations are committed while the ship is at berth in port in internal waters.

Finally, article 230(3) requires that the 'recognized rights of the accused' must be observed if non-monetary penalties are imposed. Such recognised rights must, inter alia, encompass the rights to a fair trial, to be presumed innocent until proved otherwise, to legal representation, to translation and to the protection of minors.[13] It also, in the view of this author, includes ensuring that the accused enjoys the most beneficial safeguarding rights regarding any limitation period. Thus, a national two-year limitation period would trump the three-year limitation found in article 228(2).

F. Notification to the Flag State and Other States Concerned: Article 231

Article 231 of UNCLOS establishes that all States must immediately notify the flag State if they institute any enforcement measure in accordance with section 6, for example when a port State exercises jurisdiction pursuant to article 218 or a coastal State according to article 220. A notification is also made to any State affected by the pollution, such as a coastal State that sustained damage.

Enforcement measures should, recalling that article 231 relates to safeguards, be interpreted in the broadest possible sense, thereby encompassing any stopping and investigation by a coastal State pursuant to article 220, for instance pertaining to ships sailing in the EEZ, covered by article 220(5), and if a port State decides to investigate a high seas violation in accordance with article 218.

As stipulated in the second sentence of article 231, a coastal State is only under obligation to inform the flag State of measures taken concerning a violation that

[12] ibid 479.

[13] See, eg, at https://ec.europa.eu/info/policies/justice-and-fundamental-rights/criminal-justice/rights-suspects-and-accused_en.

took place in the *territorial sea* when those measures form part of legal proceedings. The discussion in and conclusion reached in chapter 9 (section III.C) should be recalled when considering whether coastal States are required to notify a flag State of any enforcement measures relating to a violation committed in their *internal waters*. Adopting a reverse conclusion from article 231, it is suggested that they are not.

The last sentence of article 231 clarifies which flag State authorities should be notified under article 231, by explicitly referring to 'diplomatic agents or consular officers and where possible the maritime authority'. Recognising the technological and communicational advances that have been made since the adoption of UNCLOS in 1982, an article 231 notification sent by email to an embassy or the consulate of the flag State located in a coastal State or port State, and to the official email address of the flag State's maritime authority, would fulfil this requirement.

The article 231 notification can represent the point in time from which the six-month period, referred to in article 228(1) (see chapter 11, section III.A), begins if the notification explicitly informs the flag State of any legal proceedings instituted in the coastal or port State.

It is important to note that the article 231 notification does *not* require any response from the flag State. Enforcement measures that have been instituted, and the six-month deadline, are therefore not affected by a lack of response from the flag State.

It is consequently important that a coastal State or port State can prove that it has fulfilled the requirement under article 231 (*in fine*) to inform the diplomatic or consular officers, and if possible the maritime authority, of the flag State. This could be done by ensuring that an email containing the article 231 notification is logged, recorded and saved in a data system used by the relevant authorities. This is a procedure already commonly applied amongst many coastal and port State authorities.

G. Liability of States Arising from Enforcement Measures: Article 232

As noted in section I.D, in instituting civil proceedings under article 229, article 232 of UNCLOS makes States liable for any loss or damage caused to a shipowner, or to others, by their wrongful application of a section 6 enforcement measure, such as a coastal State's or port State's wrongful enforcement pursuant to article 220 or article 218, or a flag State's wrongful enforcement under article 217. Also, as mentioned in chapter 4, a port State's unlawful PSC detention under article 226 could also lead to such port State liability.[14]

[14] This could be the result of a trial at the ICJ, ITLOS, etc, pursuant to art 292, which focuses on detention.

Article 232 specifies that a State's liability is limited to damage or loss resulting from the unlawful enforcement 'in the light of available information'. If a State has a legitimate reason for believing that a violation has taken place, but further investigation is required, then the State is not liable for any loss following from this investigation. For instance, if a port State detains a ship in accordance with article 226(1)(a), or if a coastal State stops a ship in transit of the EEZ pursuant to article 220(5), due to the State's having *clear grounds* for assuming that a violation had occurred. Examples of what could constitute such clear grounds regarding MARPOL Annex VI enforcement are given in chapters 4 and 9, pertaining to information provided by CEMS, sniffers, authorities in other States, etc.

The words 'in the light of available information' must include also situations where a ship (master and crew) has contributed to the belief that a violation had occurred, for example if a ship, during an initial PSC, cannot present the required valid documents to the port State authorities, or if a ship, during its transit of the EZZ, does not respond to a coastal State's request for further information in accordance with article 220(3) or if it provides incorrect information.

H. Safeguards relating to Straits Used for International Navigation: Article 233

Article 233 is a safeguard that relates exclusively to the right of foreign ships to make a transit passage through a strait used for international navigation in accordance with article 38 and the other provisions of part III of UNCLOS, as described in chapter 3 of this book.

Article 233 emphasises that 'Nothing in sections 5, 6 and 7 affects the legal regime of straits used for international navigation.' This limits the rights for coastal States to prescribe and enforce legislation in their territorial sea, for instance according to articles 211, 212, 220 and 222 of sections 5 and 6.

Yet article 233 also contains an explicit reference to the exception in article 42(1)(a) and (b), which allow a coastal State to adopt regulations relating to navigation and the protection of marine environment, with which foreign ships must comply while transiting the strait.

Article 42(1)(b) limits the scope for protection of the marine environment to a right for coastal States to implement international rules regulating discharges of *oil, oily wastes or other noxious substances* in to the strait, which must be considered a clear and unambiguous reference to MARPOL Annexes I and II. Such implemented international regulations can be enforced by the coastal State, provided it respects the other safeguarding provisions of section 7 of UNCLOS.

Article 233 refers to appropriate enforcement measures being applied *mutatis mutandis*, which, in the view of this author, is a reference to the principles of article 220(2).

As affirmed in chapter 9, because of the clear reference in article 42(1)(b) to international rules on the discharge of oil and noxious substances (ie MARPOL Annexes I and II), the provision does not cover international regulations on air emissions, such as MARPOL Annex VI or future regulations on GHG. Consequently, article 233 does not provide (coastal) States with any wider jurisdiction to act in response to violations of air emissions, including regulation 14 of Annex VI.

II. Responsibility and Liability: Section 9 of Part XII of UNCLOS

Article 235(1) of UNCLOS specifies that all States are responsible for fulfilling their international obligations for protection of the marine environment.[15] This is basically a codification of the international law principle *pacta sunt servanda*, also found in article 26 VCLT.

An example of such an obligation could be the requirement in article 217(3) for a flag State to ensure that all ships under its flag have all required certificates on board, including the International Air Pollution Prevention Certificate (IAPP Certificate) and the International Energy Efficiency Certificate (IEE Certificate) (see regulations 5–9 of MARPOL Annex VI). Another (obvious) example could be the obligation for a flag State to carry out enforcement effectively in accordance with article 217(1), and to inform the IMO and other States thereof pursuant to article 217(7).

If a flag State does not fulfil these requirements it can, besides losing its primary jurisdiction over such violations in accordance with the second exception in article 228(1), become 'liable in accordance with international law' under article 235(1).

Port States and coastal States could also be liable in accordance with article 235, for example if a ship were unlawfully detained contrary to article 226, or if a coastal State were to stop a ship in the EEZ without clear grounds in breach of article 220(5).

It must be underlined that the liability of port, coastal and flag States as regards the shipowner is covered by the previously studied article 232 (see section I.D). Article 235 relates to a State's international responsibility towards other States that are parties to UNCLOS.

The reference to liability in accordance with *international law* could cover violations of general responsibilities codified in the VCLT and application of the International Law Commission's (ILC's) work, the Articles on Responsibility of

[15] Y Tanaka, *The International Law of the Sea*, 2nd edn (Cambridge University Press, 2015) 303.

States for Internationally Wrongful Acts ('the Articles on State Responsibility').[16] Chapter II of the ILC Articles on State Responsibility relates to 'Reparation for injuries' following a State's violation of its international obligations, covering *reparation* (article 34), *restitution* (article 35), *compensation* (article 36), *satisfaction* (article 37) and *interest payable on sums awarded* (article 38).

The concept of *compensation* in particular must be presumed to be an effective legal tool for penalising a State, provided another State has been injured (sustained a loss), for example a coastal State that has been injured by pollution, or a flag State whose vessels have been detained unlawfully by a port State. This assumption is based, inter alia, on case law from the ICJ (and its predecessor the Permanent Court of International Justice (PCIJ)) and ITLOS, as they have used compensation as a measure of adjudication that is described by the ILC in its comments to article 36. The ILC refers[17] to the *Corfu Channel case*,[18] where the United Kingdom (UK) sought compensation from Albania for damage sustained by the UK destroyers *Saumarez* and *Volage*, which the PCIJ awarded.

The ILC also refers[19] to the decision from ITLOS in the second *M/V Saiga case*,[20] where the State of Saint Vincent and the Grenadines sought compensation from Guinea following a wrongful arrest and detention of the vessel *M/V Saiga* flying its flag. ITLOS accepted, in general, this claim for compensation.

If a State therefore violates its obligations pursuant to part XII of UNCLOS, it can result in that State's being liable in accordance with international law, which potentially can result in the State's paying compensation pursuant to article 36 of the ILC Articles on State Responsibility to any State that has sustained damage from the violation.

Article 235(2) of UNCLOS attempts to ensure that the judicial systems of all States will allow a party to bring a claim for compensation or other relief in a fast and uncomplicated way. This could relate to a flag State's being liable for damage resulting from a violation committed by a ship under its flag and which is attributable to the flag State. It could also relate to a port State's being liable for a PSCO's wrongful exercise of power, or a coastal State's being liable for the unlawful stopping of a foreign ship during its innocent or transit passage of the territorial sea.

Article 235(3), inter alia, requires all States to cooperate in the implementation of existing international law and the further development of international law

[16] International Law Commission, *Draft Articles on Responsibility of States for Internationally Wrongful Acts*, November 2001, Supplement No 10 (A/56/10), chp IV.E.1, available at https://www.refworld.org/docid/3ddb8f804.html. It should be noted that references to these Articles sometimes use the full title set out in the ILC Yearbook, ie 'Draft Articles …'. However, given their adoption in 2001, this author will not refer to these Articles as 'drafts'. This is also in line with how the Special Rapporteur behind the work, Cambridge Professor James Crawford, refers to them (see at http://legal.un.org/avl/ha/rsiwa/rsiwa.html).

[17] ibid, 100, section 9.

[18] See *Corfu Channel, Assessment of Amount of Compensation*, Judgment, [1949] ICJ Rep 244, 250.

[19] ILC, n 16, 100, section 10.

[20] See *M/V Saiga (No 2) (Saint Vincent and the Grenadines v.Guinea)*, Judgment, [1999] ITLOS Rep 65, para 170.

pertaining to State responsibility and liability.[21] The reference to 'further development of international law' clearly stresses that international law, including the law of the sea, is dynamic and constantly evolving. This supports the interpretation of article 218(1) that has the term 'discharge' encompassing releases.[22]

III. Settlement of Disputes: Part XV

Part XV (articles 279–299 and Annexes V–VIII) of UNCLOS contains provisions on how disputes regarding the interpretation and application of the Convention should be resolved. Some of these are discussed in this section, as it establishes that there are judicial organs that, for example, can ensure compliance with the protective safeguards described in this chapter.

The interpretations and conclusions set out in this part of the book (Part II), including applying the term 'discharge' under article 218(1) to cover emissions and using article 228(1) as described in chapter 11, can also be brought before one of the judicial bodies under part XV of UNCLOS.

Article 279 of UNCLOS reiterates one of the most basic principles of international law, which stipulates that States shall settle any dispute by peaceful means. This is considered a jus cogens norm,[23] and it is codified in article 2(3) of the UN Charter.[24]

Article 280 allows the participating States to decide which peaceful means (judicial organ) should be used to settle a dispute.

Under article 282, any provisions or procedures for dispute settlement found in another international convention or agreement, to which the States in dispute are also parties, should be applied unless the parties in question agree otherwise.

A State may, in accordance with article 286, submit a request to a court or tribunal described in section 2 of part XV, if it is not possible for the parties

[21] Examples could include the International Convention on the Establishment of an International Fund for Compensation for Oil Pollution Damage (adopted 18 December 1971, entered into force 16 October 1978, as modified by the Protocol of 27 November 1992, entered into force 30 May 1996) 1110 UNTS 57 ('the Fund Convention') and the International Convention on Civil Liability for Oil Pollution Damage (adopted 29 November 1969, entered into force 19 June 1975, as modified by the Protocol of 27 November 1992, entered into force 30 May 1996) 973 UNTS 3 ('the CLC Convention').

[22] That international law is under constant development is also the reason why this author believes that the combination of the continued worsening conditions of the environment due to pollution from GHGs and the fact that ships occupy a unique legal position within international law, means that it is possible to develop principles for future enforcement of GHG regulations relating to ships by application of the jus cogens and erga omnes principles. especially considering ICJ case law and the comments from the ILC pertaining to previous Draft Articles on State Responsibility on this matter. This is analysed in Part IV of this book.

[23] M Dixon, *International Law* (Oxford University Press, 2007) 275.

[24] Charter of the United Nations and Statute of the International Court of Justice (adopted 26 June 1945, entered into force 24 October 1945) 1 UNTS XVI.

to agree on which dispute-settlement mechanism should be used. Section 2 (articles 286–296) contains the *compulsory procedures entailing binding decisions.*

Article 287(1)(a)–(d) list the different judicial organs on which a party to UNCLOS can call to settle a dispute with binding effect.[25] These are;

(a) ITLOS;
(b) the ICJ;
(c) an arbitral tribunal; or
(d) a special arbitral tribunal relating to one or more of the categories specified in Annex VIII to UNCLOS.

Article 287(3) specifies that that arbitration in accordance with article 287(1)(c) shall be deemed to have been accepted if no other measure is decided upon.

Article 288 establishes that all the courts and tribunals listed in article 287(1) have jurisdiction over any dispute pertaining to the interpretation and application of UNCLOS. They can also, where a case so warrants, use experts and invoke provisional measures pursuant to article 289 and article 290.

All States that are parties to UNCLOS have, in accordance with article 291(1), the right to have proceedings brought before these courts and tribunals. However, all *local remedies* (ie in national courts) must have been exhausted beforehand (see article 295).

Article 292(1) relates to the specific situation where a vessel is *detained*, stipulating an expedited procedure. It allows such a dispute to be submitted to a court or tribunal, listed in article 287(1), within 10 days. This could, for example, relate to a port State's detaining a vessel during a PSC in accordance with article 226(1)(c), or a coastal State's detaining a vessel during its direct transit of the EEZ pursuant to article 220(5).[26]

This underlines that detention, as described in chapter 4, is an effective, but also expensive, PSC sanction due to the costs of a commercial vessel's being unable to deliver its cargo on time and meet future cargo deliveries and port slots in a timely manner.

Article 292(2) provides that only the flag State, or a State acting on behalf of the flag State, can submit such a detention dispute as is described in article 292(1). The court or tribunal in question must thereafter, without delay, decide upon the legality of the detention itself (see article 292(3)). This is without prejudice to a subsequent trial and a settlement of the underlying dispute that resulted in the detention.

Article 292(4) affirms that these principles also apply to detention instituted to obtain financial security (a bond) in accordance with article 226(1)(b).

[25] Tanaka, n 15, 426.

[26] Art 297 lists several exceptions to the jurisdiction of the courts or tribunals mentioned in section 2. However, a coastal State's enforcement of international environmental rules for the protection of the marine environment, such as MARPOL Annex VI, is not one of these – see art 297(1)(c).

Article 292(4) requires the detaining (coastal or port) State to release the vessel once the court has reached a decision in this matter and it has been executed.

Article 293(1) specifies that the court or tribunal must apply the provisions of UNCLOS when settling disputes. It can also apply other rules of international law, provided these are not incompatible with the Convention. This must, inter alia, encompass the recognised sources of international law listed in article 38(1)(a)–(d) of the Statute of the ICJ,[27] especially as the ICJ is one of the courts to which a dispute can be referred. One of the recognised sources comprises international conventions, in accordance with article 38(1)(a) of the Statute, which means that the VCLT, the ILC Articles on State Responsibility and the MARPOL Convention, including Annex VI, can be applied by the courts and tribunals. Article 38(1)(b) of the ICJ Statute also includes customary principles of international law, such as the principle of NMFT.

As concluded in chapter 5, the principles of article 18 of the EU Sulphur Directive[28] might also be considered by a court or tribunal when determining whether a flag State's fine has been an effective measure of enforcement in accordance with article 217(1), or whether a port State's claim for financial security is proportionate. An EU regulation, such as article 18 of the Sulphur Directive, could be considered 'subsidiary means for the determination of rules of law', as it was developed and presented by a highly qualified publicist (the EU) within international law (see article 38(1)(d) of the ICJ Statute).

Both article 293(2) of UNCLOS and article 38(2) of the ICJ Statute allow the courts and tribunals to decide a case *ex aequo et bono*, if the parties so agree.

Finally and very importantly, article 296(1) states that any decision rendered by a court or tribunal is *final and binding* on the parties to the dispute. Conversely, article 296(2) proclaims that these decisions have no effect on States other than those parties involved in the dispute.

The wording of article 296(2) should, in the view of the author, be interpreted in a very literal sense, meaning that the legal precedent formed by the case law of the ICJ and ITLOS is still undeniably important, as these rulings provide insight into the interpretation, application and development of international law, including the law of the sea.

A. Cases Relating to Enforcement of MARPOL Annex VI

As mentioned previously, article 282 of UNCLOS stipulates that any dispute settlement provisions or procedures incorporated in another international convention or agreement, to which the States in dispute are also parties, should be applied unless the parties in dispute agree otherwise.

[27] See the UN Charter, n 24.

[28] Directive (EU) 2016/802 of the European Parliament and of the Council of 11 May 2016 relating to a reduction in the sulphur content of certain liquid fuels [2016] OJ L132/58.

When looking into the enforcement of MARPOL Annex VI, it is worth noting that article 10 of the MARPOL Convention, which also applies to the Annexes pursuant to article 1(2), provides that any disputes should be settled by negotiation or, if necessary, in accordance with the dispute-settlement procedures of Protocol II to the MARPOL Convention, which refers to *arbitration*.

It is essential to distinguish what forms the basis of a dispute pertaining to the enforcement of MARPOL Annex VI. If it concerns a dispute directly related to Annex VI, for example whether an IAPP Certificate is issued correctly, or whether an Exhaust Gas Cleaning System (EGCS) is correctly approved by the flag State, then the dispute is subject to arbitration in accordance with article 10 of the MARPOL Convention. But if it concerns the interpretation and application of the provisions of UNCLOS, such as applying article 218(1) to cover emission violations on the high seas, then the dispute should be settled in accordance with part XV of UNCLOS at one of the courts or tribunals listed in article 287(1).

The application of article 228(1) of UNCLOS would also be encompassed by part XV, for example to ensure that the requirements for invoking the main rule were met by a flag State, including determining if evidence of effective enforcement by the bringing of corresponding charges was subsequently furnished to the coastal State or port State. The judicial organ under part XV could also check if there were grounds for invoking any of the exceptions in article 228(1), including the second exception relating to whether a flag State has repeatedly disregarded its obligations of effective enforcement.

Article 292(1) allows the judicial bodies listed in article 287(1) to make an in-depth evaluation of whether financial security, required by a port State as part of a detention for violation of MARPOL Annex VI, is 'reasonable' (ie adequate and proportionate). Otherwise, the court or tribunal can determine what the size of the financial security should be pursuant to article 292(4). Article 292 therefore provides the courts and tribunals with an implicit safeguarding right to determine whether an impending port State fine for violation of the regulation 14 sulphur limit is adequate and reasonable, as the financial security should be proportionate to the possible size of a future fine which the security is intended to cover.

The reference in article 293 to *international law* allows the judicial organs to use the parameters set out in article 18 of the EU Sulphur Directive as guidance on whether a flag State's enforcement has been effective, and thereby meets the requirements of article 217(1), and to determine the proportionate size of any financial security requested by a coastal State or port State for an Annex VI violation.

It is worth noting that ITLOS has the power to convene a *special chamber* when special expertise is needed, for example in environmental matters: see article 15 of the ITLOS Statute, found in Annex VI to UNCLOS. Also, article 2(1) of Annex VIII to UNCLOS ('Special Arbitration') explicitly refers to the 'protection and preservation of the marine environment' as being one of the fields for which a list of experts must be established and maintained.

Article 5(1) of Annex VIII to UNCLOS allows a party to request the special arbitral tribunal to *carry out an inquiry* in order to establish what the facts are in a particular case. The result thereof is binding on the parties to the case pursuant to article 5(2). This could be a valuable tool when establishing whether an infringement of regulation 14 of Annex VI has been sufficiently proved, for instance by a CEMS or a sniffer attached to a drone.

Finally, the ICJ must also be considered extremely capable of handling such disputes, as it often rules in cases with a maritime[29] and environmental[30] interest.

IV. Conclusion

When a coastal State or port State enforces rules and regulations for the protection of the marine environment in accordance with the *lex specialis* provisions of section 6 of UNCLOS, such as articles 220 or 218, it must meet the safeguarding standards and requirements set out in section 7.

Article 223 provides that all States must allow representatives from the IMO, the flag State and any (coastal) State affected by the pollution to attend the legal proceedings instituted. Those representatives are protected under international law.

Article 224–227 cover a (port) State's ability to investigate a foreign ship and the limitations thereon, for instance relating to the PSC enforcement regimes described in chapter 4.

Article 228(2) lays down a three-year limitation period, as well as codifying the principle of *ne bis in idem*. The latter principle only applies if an *effective* penalty (fine) has been imposed. Article 228(3) ensures that flag States are always allowed to supplement a foreign State's sanction with their own sanctions, such as imprisonment.

Article 229 enables a coastal State to institute civil legal proceedings against a shipowner for additional costs and damage not covered by a fine.

Article 230 sets the parameters regarding how non-flag States can sanction foreign ships and shipowners, stipulating that monetary penalties (fines) are the main rule. They are also the only rule (option) for port States when enforcing on the high seas pursuant to article 218(1). Exceptions to the main rule apply for coastal States taking action against damaging and wilful violations within their internal or territorial waters. Port States can also, as part of a PSC, penalise violations with non-monetary penalties if a violation is committed while the vessel is

[29] See, for instance, the *Corfu Channel case* in n 18. See also the judgment of the PCIJ (the ICJ's predecessor) in the *Lotus case* (*SS Lotus (France v Turkey)* PCIJ Rep Ser A No 10), discussed in ch 6.

[30] See the ICJ case law analysed in ch 17 of this book, relating to the *Legality of the Threat or Use of Nuclear Weapons*, *Gabčíkovo-Nagymaros Project* and *The Pulp Mills of the River Uruguay*.

at berth in port, for example by giving a false statement to a PSCO or presenting falsified certificates and official documents.

Article 231 provides that all coastal or port States must promptly inform the flag State authorities of any enforcement instituted pursuant to section 6. These authorities are diplomatic or consular officers and, if possible, the maritime authorities. The article 231 notification should also mark the beginning of the six-month period within which the flag State must react if it wishes to suspend proceedings undertaken by a coastal State or port State (see 'the main rule' in article 228(1), discussed in chapter 11).

Article 232 specifies that a State is liable for any unlawful enforcement measure that results in damage or loss. This liability is limited to what was unlawful in the light of *available information*. This means that a port State is not liable for any delay or loss suffered by a ship(owner) due to a detailed PSC inspection, even though no violation is found, provided the PSC inspection had been instituted on clear grounds.[31]

Article 233 allows a coastal State to proceed against any violation of the rules and regulations adopted in accordance with article 42(1)(a) and (b), which regulate the right of foreign ships to transit an international strait, in matters of navigation and the protection of the marine environment. Article 42(1)(b) limits the scope of the latter to implemented international rules relating to the discharge of oil and other noxious substances, such as MARPOL Annexes I and II, but clearly not Annex VI.

Section 9 (article 235) of part XII of UNCLOS reiterates the basic obligation for all States to comply with any agreements or conventions into which they voluntarily may have entered, in accordance with the principle *pacta sunt servanda*. If not, a State may be held liable by other States that are parties to the same agreement or convention. This can, inter alia, result in the payment of compensation.

Part XV of UNCLOS establishes how disputes between parties to the Convention can be settled. First, article 279 stipulates that any dispute settlement must be peaceful.

Article 286 provides that if agreement cannot be reached on how to settle the dispute, one of the judicial organs listed in article 287(1)(a)–(d) can render a binding decision. These organs are ITLOS, the ICJ, an arbitral tribunal or a special arbitral tribunal.

Article 292 relates specifically to matters of detention, including determining whether financial security should be paid and the amount thereof.

Article 293 creates a tie to article 38(1) of the ICJ Statute, establishing that any decision must be made in accordance with international law, which includes

[31] It is noted that clear grounds for detailed PSC inspections could result from different factors, for instance following a sniffer measurement, faulty documentation, interviews with the crew, the ship's risk profile (previously logged deficiencies and its flag State's white-, grey- or black-list status) or by looking at the overall condition of the ship. Concentrated inspection campaigns can also subject a ship to a detailed inspection.

UNCLOS and other relevant sources of international law. The EU Sulphur Directive can be seen as a secondary source pursuant to article 38(1)(d) of the Statute.

Article 296(1) concludes that any decision by a court or tribunal listed in article 287(1) is final and binding on the parties.

Disputes regarding the interpretation and application of the provisions and regulations of the MARPOL Convention and its Annexes should be settled by negotiation or by arbitration according to article 10 and the procedures found in Protocol II to the MARPOL Convention. However, the judicial bodies listed in part XV of UNCLOS should be applied if a dispute relates to the enforcement of MARPOL Annex VI, through the provisions of part XII of UNCLOS. This means that ITLOS or the ICJ could determine, and approve, the application of article 218(1) to cover emission violations on the high seas, and how the exceptions in article 228(1) of UNCLOS apply.

13

Enforcement of Sulphur Regulations: Conclusion

This chapter seeks to provide a brief summary of some of the aspects discussed in the chapters comprising Part II of this book.

It will be recalled that the chapters in Part I offered a general overview of and insight into the overall problems with enforcing the sulphur regulations in MARPOL Annex VI (chapter 1),[1] the detailed regulations in Annex VI (chapter 2), the basic regulation of UNCLOS (chapter 3),[2] Port State Control (PSC – chapter 4), the EU Sulphur Directive (chapter 5)[3] and the basic jurisdictional principles of international law (chapter 6).

Part II has attempted to clarify relevant parts of the legal basis found in part XII of UNCLOS, particularly the provisions in section 5 on international rules and national legislation to prevent, reduce and control pollution of the marine environment (including articles 211 and 212), section 6 on enforcement (including articles 217, 218 and 220) and section 7 on safeguards (including article 228).[4]

I. Obligations to Implement, Adopt and Enforce

The focus in the first chapter in Part II (chapter 7) was on the provisions relating to joint cooperation between States, for instance within the International Maritime Organization (IMO), for the protection of the marine environment (see articles 197–199).

In this context it was recalled that the term 'protection of the marine environment' encompasses the protection of human health, following the definition in

[1] International Convention for the Prevention of Pollution from Ships (adopted on 11 February 1973, as modified by the Protocol of 17 February 1978, entered into force 2 October 1983) 1340 UNTS 61 (MARPOL), Annex VI, IMO Publication: IMO-520E.

[2] United Nations Convention on the Law of the Sea (adopted 10 December 1982, entered into force 16 November 1994) 1833 UNTS 3.

[3] Directive (EU) 2016/802 of the European Parliament and of the Council of 11 May 2016 relating to a reduction in the sulphur content of certain liquid fuels [2016] OJ L132/58.

[4] Parts I and II have also included brief discussions of, and made reference to, section 8 (ice-covered areas), section 9 (responsibility and liability), section 10 (sovereign immunity) and section 11 (obligations under other conventions on the protection and preservation of the marine environment).

article 1(1)(4) of UNCLOS, covering sulphur pollution. Greenhouse gas (GHG) pollution will also be covered by this description given the direct and indirect adverse impact of it, as described in Part IV of this book.

Chapter 7 then described the provisions conferring obligations on States to implement international rules and regulations for protection of the marine environment, and the possibilities for adopting national laws. The general legal basis for this is found in article 211, with somewhat limited possibilities for coastal States to regulate in the EEZ according to article 211(6), or according to article 234 if the EEZ is ice-covered.

It was noted that article 212 has a fixed scope on adopting national regulations pertaining to air pollution, including establishing national sulphur limits.

Article 222 refers to the enforcement of legislation on air pollution adopted in accordance with article 212. The provision also emphasises the need for all States to implement international regulations for protection of the marine environment, for example MARPOL Annex VI. Article 222 could therefore, in principle, be seen as the primary provision of part XII of UNCLOS for ensuring that coastal and flag States enforce Annex VI.

Nevertheless, article 222 of UNCLOS has a narrow geographical scope that limits a coastal State's right to enforce air pollution regulations in internal and territorial waters. Article 220, on the other hand, supports a coastal State's enforcement of such air pollution regulations, for example pertaining to extraterritorial enforcement of MARPOL Annex VI (or GHG rules) in the EEZ. This is conditional upon the ship's subsequently calling into a port in the coastal State, pursuant to article 220(1). It is also noted that the narrow geographical scope of article 222 has no overlap with the port State's extraterritorial jurisdiction on the high seas pursuant to article 218(1).

Article 222 must further be supported by article 217 pertaining to flag State enforcement of such emissions regulations, for example regulation 14 of Annex VI, as article 222 lacks clarity and detail regarding what flag State enforcement should entail. The generally applicable article 217 on flag State enforcement is therefore the primary provision of part XII for obligating and ensuring a flag State's effective enforcement of such regulations.

II. Flag State Jurisdiction

Chapter 8 sought to analyse article 217(1), which requires flag States to proceed against violations, committed by ships under their flag, of environmental regulations, such as MARPOL Annex VI, irrespective of where these occur, including on the high seas.

This requires flag States to impose sanctions (fines) for violations of regulation 14 of Annex VI, which will confiscate all savings achieved by violating the regulations *and* deter future violations by imposing a discouraging punitive element that will increase in the event of aggravating circumstances, such as repeated violations. This follows the principles set out in article 18 of the Sulphur Directive. Furthermore, the national legislation must not bar such effective enforcement as is specified in article 217(8).

Also, flag States must, in accordance with the first part of article 217(7), immediately report their effective enforcement directly to the IMO and the State that, pursuant to article 217(6), reported the violation. The same obligation is found in article 4(3) of the MARPOL Convention. This information must afterwards, in accordance with the second part of article 217(7), be made available to all States. This could include using the IMO-operated GISIS database, especially when focusing on MARPOL Annex VI enforcement, as GISIS already has a module for such flag State reports given the onus under regulation 11.4 of MARPOL Annex VI, which requires a flag State to report its enforcement of the Annex, including the sulphur limits in regulation 14.

National legislation cannot prevent the flag State providing such article 217(7) information (and regulation 11.4 information), as stated in article 27 of the Vienna Convention on the Law of Treaties (VCLT).[5]

Finally, the flag State must allow representatives from the IMO, and any State affected by the violation, to attend any legal proceedings (see article 223). Any representative attending such proceedings shall, inter alia, enjoy such diplomatic rights of protection as are provided under international law.

III. Coastal State Jurisdiction

Chapter 9 studied the implications of coastal States' proceeding against violations of regulation 14 of MARPOL Annex VI in accordance with article 220 (with article 222). This enforcement relates to violations taking place in internal or territorial waters, or in the State's EEZ. The coastal State has full jurisdiction over all such violations, including in the EEZ, provided the ship subsequently voluntarily calls into a port in the coastal State, pursuant to article 220(1).

Certain exceptions apply to a coastal State's right to enforce in territorial waters, regarding foreign ships making an innocent passage in accordance with articles 17–19, or a transit passage in an international strait in accordance with article 38. Yet environmental rules adopted in accordance with article 21(1)(f) can be enforced over ships making an innocent passage pursuant to article 220(2), and rules adopted in accordance with article 42(1)(a) and (b) can be enforced over ships making a transit passage (see article 233). The adoption and enforcement of regulations relating to transit passage in an international strait are limited to MARPOL Annexes I and II, not covering Annex VI.

[5] Vienna Convention on the Law of Treaties 1969 (adopted 22 May 1969, entered into force 27 January 1980) 1155 UNTS 331 (VCLT).

A coastal State's jurisdiction could, in theory, be extended in accordance with the principle of *hot pursuit* found in article 111. This would allow coastal States to follow a ship, which is not complying with MARPOL Annex VI while sailing in internal or territorial waters, out to the high seas, where it can stop, investigate and prosecute it. This, inter alia, necessitates that the coastal State has clear grounds for believing that a violation has occurred and that the State, at all times, has visual contact with the ship. These requirements could perhaps be met by using a drone with a visual uplink and an attached sniffer.

IV. Port State Jurisdiction

Chapter 10 describes how article 218(1) provides port States with an extraterritorial jurisdictional basis for penalising discharge violations of international regulations committed outside the State's territory, meaning on the high seas or in the waters of another State. This enforcement is conditional upon the foreign ship's afterwards voluntarily going into port in the port State.[6]

The term 'discharge' in article 218(1) should, in accordance with article 31(1) VCLT, be interpreted and applied in the broadest conceivable way, as it refers to 'any' discharge, an approach that was confirmed by the European Court of Justice in the 2018 *Bosphorus Queen case*.[7]

This interpretation clearly shows that 'discharge' can cover emissions, such as sulphur or GHG (CO_2) emissions, as the word *discharge*, according to different English dictionaries and thesauruses, inter alia covers the release of 'fumes', 'particles', 'gas', 'smoke' and, of course, 'emissions'.

This conclusion is reinforced by the definition of 'discharge' in article 2(3)(a) of the MARPOL Convention, which applies to Annex VI,[8] as, inter alia, it refers to *emitting*, thus covering emissions. This also coincides with how the term is already applied in regulation 12.7.3 of MARPOL Annex VI, covering discharge of ozone depleting substances into the atmosphere.

Consequently, port States can, in accordance with article 218(1) of UNCLOS, take action against violations of the 0.5% sulphur limit under MARPOL Annex VI on the high seas, and in the waters of other (coastal) States (see article 218(2)–(4)). This is irrespective of which flag the foreign ship flies due to the principle of 'no more favourable treatment' found in regulation 10.3 of Annex VI, in accordance with article 5(4) of the MARPOL Convention. Such enforcement by the port State

[6] A Kanehara, 'Environmental Protection of Ocean and Flag-State Jurisdiction' Rikkyo University, Faculty of Law, paper presented at the 8th Conference of SCA Qingdao, China, 27–30 May 2008, 14.

[7] Case C-15/17 *Bosphorus Queen Shipping Ltd Corp v Rajavartiolaitos*, ECLI:EU:C:2018:557.

[8] See art 1(2) of the MARPOL Convention.

can only result in a monetary sanction (fine) as it concerns a high seas violation, in accordance with article 230(1) of UNCLOS.

It is noted that the port State must notify the flag State of this enforcement action pursuant to article 231. It must also allow representatives from the IMO and flag State to participate in any legal proceedings (see article 223).

The port State's extraterritorial jurisdiction according to article 218(1) (outside its own waters) leads to this jurisdiction's overlapping with the jurisdiction of a flag State regarding violations on the high seas, and overlapping with the jurisdiction of a coastal State regarding violations taking place within the waters of that State.

Article 218(2)–(4) clarify the latter situation between a coastal State and a port State, stipulating that the coastal State has primary jurisdiction. However, this impliedly allows coastal States to empower/allow port States to take action against such violations on their behalf, as article 218(1) also gives port States jurisdiction in these areas. Such consent can be given on an *ad hoc* or *post hoc* basis, but it could also be given as *ante hoc consent*, for example as part of the coordinated PSC scheme within the different Memorandums of Understanding (MoUs).

This would allow port States to carry out an end-to-end enforcement regarding a foreign ship's entire violation of MARPOL Annex VI, irrespective of where the ship had sailed before coming into port in the port State, including sailing in and out of foreign and own waters and on the high seas.

This extraterritorial jurisdiction for port States to *enforce* would, as described in chapter 6, encompass the right to *adjudicate on* these violations, which includes prescribing fines that remove (confiscate) all savings achieved during the entire violation and imposing a punitive element, which will increase in the event of aggravating circumstances such as repeated violations.

V. Clarifying Overlapping Jurisdictions

Chapter 11 describes the overlapping jurisdiction between flag States (article 217) and port States (article 218(1)) for violations on the high seas, which is resolved in accordance with article 228(1). This provision also deals with a flag State's overlapping jurisdiction with a coastal State (article 220).

That article 228 is the beacon for resolving these matters is accentuated by the direct reference in article 217(4) to the flag State's jurisdiction respecting articles 218, 220 and 228.

The geographical scope of article 228(1) covers violations taking place on the high seas or in a coastal State's EEZ. The reverse conclusion establishes that a coastal State has primary jurisdiction over violations taking place in its internal and territorial waters.

Article 228(1) consists of a main rule and two exceptions for determining which State can assert primary jurisdiction over a violation that takes place in an EEZ or on the high seas.

A. The Main Rule under Article 228(1)

The *main rule* grants the flag State primary jurisdiction over these violations, provided it can meet three preconditions set out in the first and second sentences of article 228(1):

(a) the flag State must within *six months* request the coastal or port State to suspend its proceedings;
(b) the flag State must, at the same time, make assurances as to its intent to bring *corresponding charges.*

The coastal or port State can carry out its own enforcement proceedings if these two preconditions are *not* met by the flag State, but the coastal or port State must suspend its proceedings if these preconditions *are* met. The following condition then stipulates:

(c) the flag State must afterwards present the coastal State with case records, etc, *proving* that it brought corresponding charges that resulted in effective enforcement.

The coastal or port State can, if the flag State does not provide such information or if the information reveals an ineffective or non-corresponding enforcement,[9] *resume* the suspended proceedings and impose a fine, which is covered by claiming any financial security posted.

If a flag State can fulfil all three requirements, it qualifies to invoke the main rule and assert jurisdiction over the violations occurring in the EEZ and on the high seas. The port or coastal State must fully terminate its, up-until-that-point, suspended proceedings and release any security posted.

However, if one of the two exceptions codified in the first sentence of article 228(1) applies, a coastal State or port State may deny a 'main-rule request' from a flag State, regardless of the flag State's observing preconditions for doing so.

B. The First Exception under Article 228(1)

The *first exception* in article 228(1) pertains to a coastal State's suffering *major damage* following a violation in the EEZ.

The coastal State may, in these first-exception cases, where an EEZ violation has caused major damage, instigate proceedings immediately and deny any flag State request for suspension. The coastal State should nonetheless still observe the requirement set out in article 231 and inform the flag State of its enforcement.

It is at the outset for the coastal State to determine if pollution has caused major damage, but the coastal State can be held liable in accordance with article 232

[9] An effective flag State fine for a violation of reg 14 of Annex VI should meet the criteria set out in art 18 of the Sulphur Directive, which also applies when determining if a flag State has brought a 'corresponding charge'.

if its enforcement is unlawful or disproportionate, for example if its assessment of the severity of the pollution is wrong. A flag State can use the judicial mechanisms found in part XV of UNCLOS (such as the International Court of Justice (ICJ) or International Tribunal for the Law of the Sea (ITLOS)) if it wants to question this assessment by the coastal State.

As stated in chapter 11, it cannot be presumed that a violation of regulation 14 of MARPOL Annex VI will result in pollution that can be deemed to have caused *major damage*, as this must refer to tangible pollution.

C. The Second Exception under Article 228(1)

The *second exception* in article 228(1) allows all States, including port States, to deny a flag State's request for suspension pursuant to the main rule, if the flag State in question 'repeatedly has disregarded its obligation to enforce effectively' the international regulations for protection of the marine environment, including MARPOL Annex VI.

This refers to the flag State's obligation to undertake effective enforcement pursuant to article 217, including the requirement (in article 217(1)) to take action against violations irrespective of where these took place, including on the high seas.

An example of a flag State's *disregard* could, relating to enforcement of regulation 14 of Annex VI in accordance with article 217(1), be a flag State's failure to prescribe fines that confiscate all savings achieved by a violation or to impose a dissuasive punitive element. A disregard could also be a flag State's neglecting to inform the IMO and other States of this enforcement in accordance with article 217(7).

The word 'repeatedly' is not defined in UNCLOS, but several possibilities for applying the second exception could be envisioned, for instance by setting a specific number of occasions of disregarded enforcement that, under certain conditions, would constitute acting (or not acting) 'repeatedly'. The term could also be applied on an individual flag State basis, perhaps by following the principles set out for calculating if a flag State should be white-, grey- or black-listed, as used by some MoU systems (see chapter 4).

Thus, all port States may, if a flag State has repeatedly disregarded its obligation to enforce regulation 14 of Annex VI effectively, deny the flag State's request for suspension of any proceedings commenced, and instead conclude these by imposing a dissuasive fine.

The port State must still observe the requirements to: notify the flag State of its enforcement in accordance with article 231; only impose a monetary penalty pursuant to article 230; and allow flag State (and IMO) representatives to attend the legal proceedings (see article 223).

A flag State's loss of primary jurisdiction over certain violations, such as all violations of MARPOL Annex VI committed by ships under its flag, should be limited to a specific period, for example one to two years, followed by a probationary period during which the flag State's enforcement record will be under close scrutiny.

VI. Safeguards

Chapter 12 lists many of the safeguard obligations found in section 7 of part XII of UNCLOS, to which States must adhere when enforcing in accordance with the provisions of section 6. This includes allowing representatives to participate in proceedings (article 223), only imposing fines depending on where the violation took place (article 230) and notifying the flag State of any enforcement (article 231).

Articles 224–227 were also re-examined after a more comprehensive study of them in chapter 4 on PSC enforcement.

Article 228(2) establishes a three-year limitation period and codifies the principle of *ne bis in idem*. Article 228(3) underlines the flag State's prerogative to supplement sanctions imposed by other States with its own penalties, such as imprisonment. The flag State is not subject to the limitations in article 230.

Article 229 enables a coastal State to commence civil legal proceedings for any damage sustained that is not covered by a normal fine.

Article 232 makes a State is liable for any unlawful enforcement if it results in damage or loss for a ship(owner) or others.

Article 233 enables coastal States to enforce violations of rules adopted in accordance with article 42(1)(a) and (b). Article 233 basically provides the jurisdictional basis that allows coastal States to proceed against violations of MARPOL Annexes I and II (but not Annex VI) over foreign ships in transit through an international strait located in the territorial sea.

Article 235, in section 9, stipulates that all States that are party to UNCLOS must meet and fulfil the obligations placed upon them in accordance with the Convention.

Chapter 12 also addresses part XV of UNCLOS, pertaining to how disputes on the interpretation and application of the Convention can be settled. Article 287(1)(a)–(d) refer to ITLOS, the ICJ, an arbitral tribunal or a special arbitral tribunal as being the judicial organs that can settle such disputes with binding effect. Article 292 specifically relates to settling disputes concerning detention, including determining proportionate financial security. A dispute relating to definitions and the application of the provisions of part XII of UNCLOS, such as articles 217, 218, 220, 226 and 228, for instituting proceedings for violations of MARPOL Annex VI, could therefore be brought before these courts and tribunals.

VII. Enforcing MARPOL Annex VI: Conclusion

A port State has, pursuant to article 218(1) of UNCLOS, extraterritorial jurisdiction to bring proceedings for violations of the 0.5% sulphur limit in MARPOL Annex VI, including on the high seas. The flag State also has an obligation to ensure compliance with the rules and standards, in accordance with article 217.

The flag State's high seas jurisdiction over violations supersedes the port State's jurisdiction, provided it meets the conditions of article 228(1) for invoking the main rule.

Even if these obligations are met, a coastal State or a port State can deny the flag State its primary jurisdiction if one of the two exceptions in article 228(1) applies. The first exception relates solely to coastal State proceedings against violations that resulted in major damage, and is not applicable to enforcement of MARPOL Annex VI.

Port States and coastal States can, however, deny a flag State its primary jurisdiction by way of the second exception in article 228(1), which applies if a flag State has repeatedly disregarded its obligation to effectively enforce the rules and regulations applicable to violations of Annex VI in accordance with article 217.

Port States could, according to article 218(2)–(4), enter into agreements with any affected (coastal) States, enabling the port State to carry out end-to-end enforcement against violations of regulation 14 of Annex VI, by imposing a fine that covers the entire violation, irrespective of where the ship sailed, and confiscating all savings achieved, plus adding a dissuasive punitive element. To re-use the example from chapter 1, a port State could thus impose a fine that confiscates the $750,000 saving gained by a ship(owner) from violating regulation 14 of MARPOL Annex VI whilst sailing from Asia to Europe. The fine must also include a dissuasive element, which will increase in the event of aggravating circumstances, such as repeated violations by the same shipowner. This could, for example, amount to one-third of the saved amount ($250,000), so that the port State could punish a first-time violation with a fine of $1 million. Fines of $10 million or above would thus be possible if a shipowner were repeatedly and intentionally to violate these regulations.

Such fines might seem unnaturally high for some port States compared to the normal level of fines imposed during PSCs. Nonetheless, such fines must be imposed to remove the economic incentive to violate the sulphur regulations in Annex VI and deter future violations. This is imperative if the estimated human health benefits are to be achieved from applying the lower 0.5% sulphur limit, including avoiding 137,000 early deaths and preventing 7.6 million children from developing asthma.

Whether these conclusions can be applied to other existing IMO regulations is examined in the next part of the book (Part III), in chapter 14. That Part also contains a chapter (chapter 15) examining how these conclusions in Part II might apply to future regulations on GHG (CO_2), depending on what legislative measures are adopted by the IMO, for example whether it introduces measures regulating zero-carbon or fossil-free fuels, speed reductions, etc. Each of these potential legislative measures presents its own possibilities and challenges for being enforced by non-flag States in accordance with part XII of UNCLOS, including on the high seas.

PART III

Enforcement of Existing IMO Regulations and the IMO GHG Strategy

14

Enforcement of Existing IMO Regulations

This part of the book (Part III) will attempt to clarify to what extent the conclusions in the previous part (Part II) are applicable to other existing environmental International Maritime Organization (IMO) regulations (this chapter) and proposed future regulatory measures on greenhouse gases (GHGs) relating to the IMO's GHG Strategy (chapter 15).

One of the conclusions reached in Part II was that port States, in accordance with article 218(1) of UNCLOS,[1] have extraterritorial jurisdiction over emissions (discharge) violations committed by foreign ships outside the State's own territory. This includes the discharge of sulphur emissions on the high seas, thereby giving port States jurisdiction over infringements of regulation 14.1.3 of MARPOL Annex VI.[2] Port States can also assert jurisdiction over violations of the 0.1% or 0.5 % sulphur limits that occur inside the waters of another (coastal) State pursuant to article 218(2)–(4) of UNCLOS, if so agreed with that State. A coastal States also has jurisdiction over such violations in internal and territorial waters and in the EEZ, provided the foreign ship calls into a port in the coastal State (see article 220(1)). Flag States have, according to article 217 of UNCLOS, jurisdiction over such violations pertaining to ships flying their flags.

The conclusions in Part II also explain that article 228(1), which determines whether a coastal State or port State is able to eliminate a flag State's primary jurisdiction, consists of a main rule and two exceptions. The flag State must fulfil three conditions for invoking the main rule that allows it to suspend any legal proceedings instituted by a coastal State or port State. These conditions are (i) bringing corresponding charges; (ii) within six months of the date on which proceedings were first instituted; and (iii) subsequently informing the coastal or port State of the outcome of these charges. A flag State's request for suspension can be denied, however, even when it meets all three conditions, if one of two exceptions applies.

[1] United Nations Convention on the Law of the Sea (adopted 10 December 1982, entered into force 16 November 1994) 1833 UNTS 3.

[2] International Convention for the Prevention of Pollution from Ships (adopted on 11 February 1973, as modified by the Protocol of 17 February 1978, entered into force 2 October 1983) 1340 UNTS 61 (MARPOL), Annex VI, IMO Publication: IMO-520E.

The first exception enables a coastal State to deny a flag State's request for suspension if the violation resulted in major damage to the coastal State. However, this exception is not deemed applicable to enforcement of MARPOL Annex VI. The second exception is of relevance to port State (and in principle coastal State) enforcement of emission regulations, as it involves a flag State's repeated disregard of its enforcement obligations, for example its article 217 obligations to effectively enforce and inform the IMO and all States thereof.

The following sections examine the applicability of these conclusions and others[3] when enforcing different relevant IMO conventions,[4] including the regulations covering emissions besides sulphur under MARPOL Annex VI, such as ozone-depleting substances (ODSs), nitrogen oxide (NOx), particular matter (PM), volatile organic compounds (VOCs) and carbon dioxide (CO_2), and also MARPOL Annexes I–V, the BWM Convention,[5] the AFS Convention,[6] the London Convention (including certain interfaces with the Nairobi Convention),[7] the Hong Kong Convention[8] and the Intervention Convention.[9] Infringements of these IMO rules might not offer a shipowner the same economic gains as not complying with the sulphur limits in regulation 14 of Annex VI. Nevertheless, the legal basis for enforcing the relevant IMO regulations should still be examined, as the motivation for non-compliance is in part irrelevant when applying the enforcement and safeguarding provisions of sections 6 and 7 of part XII of UNCLOS.

I. MARPOL Annex VI

MARPOL Annex VI regulates, as well as sulphur (SOx) in regulation 14, ODSs in regulation 12, NOx in regulation 13, VOCs in regulation 15, emissions from incinerators in regulation 16 and 'energy efficiency' in regulations 19–23, this last

[3] These other conclusions of Part II include the extended right for coastal States to invoke art 220(5)–(6) relating to the stopping, investigating, detaining and prosecuting of a foreign ship transiting the EEZ. The right for coastal States to exercise such jurisdiction in an international strait in accordance with art 233 and art 42(1)(b) and to invoke the first exception in art 228(1) are also examined.

[4] The term 'relevant IMO conventions' is used in this context to underline that not all of the IMO's environmental legislative measures are discussed in this chapter, but merely those conventions that, in the view of this author, are the most relevant to discuss.

[5] The International Convention for the control and management of ship's ballast water and sediments (adopted 13 February 2004, entered into force 8 September 2017) ('the BWM Convention').

[6] The International Convention on the Control of Harmful Anti-fouling Systems on Ships (adopted 5 October 2001, entered into force 17 September 2008) ('the Anti-fouling Convention' or 'AFS Convention').

[7] Convention for the Prevention of Marine Pollution by Dumping of Wastes and Other Matter (adopted 13 November 1972, entered into force 30 August 1975) 1046 UNTS 120 ('the London Convention'); the Nairobi International Convention on the Removal of Wrecks (adopted 18 May 2007, entered into force 14 April 2015) ('the Nairobi Convention').

[8] The Hong Kong International Convention for the Safe and Environmentally Sound Recycling of Ships (adopted 15 May 2009, not yet entered into force) ('the Hong Kong Convention').

[9] The Convention Relating to Intervention on the High Seas in Cases of Oil Pollution Casualties (adopted 29 November 1969, entered into force 6 May 1975) 970 UNTS 211 ('the Intervention Convention').

pertaining to reducing emissions of CO_2. Each regulation is given a brief introduction in the following subsections, succeeded by consideration of whether the conclusions set out in Part II of this book are applicable when enforcing that regulation.

It may not come as a surprise, however, to learn that these regulations, to a considerable extent, can be directly enforced in accordance with the specific enforcement principles examined in Part II, as these also regulate emissions from ships. It is therefore principally the differences in enforcement that are highlighted.

A. Ozone-depleting Substances: Regulation 12

Regulation 12.2 of MARPOL Annex VI bans the deliberate discharge of ODSs, which, as the name suggests, is a common term used for a category of harmful substances that can destroy the ozone layer. This includes gases such as freon, chlorofluorocarbon (CFC) and hydrochlorofluorocarbons (HCFCs), which can comprise substances such as 'halons', 'methyl bromide', 'carbon tetrachloride', 'hydrobromofluorocarbons' and 'methyl chloroform'.[10] Such substances were often found in different equipment on board, being predominantly used for refrigeration or for extinguishing fires.[11]

International legislation for the protection of the ozone layer has been in place since the 1980s, inter alia in the form of the Vienna Convention for the Protection of the Ozone Layer (1985)[12] and the successive Montreal Protocol (1987).[13] These rules have a particular focus on CFC gases, because these are considered the most harmful as they can remain in the atmosphere for over 50 years (perhaps up to 100 years in the upper stratosphere), after which time they break down and emit their chlorine content, which serves as a catalyst in the decomposition of the ozone layer, leading to increased UV radiation. The global warming potential of CFC gases is also greater than that of HCFC gases.[14] The regulation of ODSs in MARPOL Annex VI therefore distinguishes between HCFCs and *other* ODSs, which includes CFC. Regulation 12.3.1 banned the use of other (non-HCFC) ODSs on ships constructed on or after 19 May 2005, whereas regulation 12.3.2 bans HCFC substances on ships constructed on or after 1 January 2020.[15]

[10] See at https://www.esrl.noaa.gov/gmd/hats/publictn/elkins/cfcs.html.

[11] See at http://apps.sepa.org.uk/spripa/Pages/SubstanceInformation.aspx?pid=119.

[12] The Vienna Convention for the Protection of the Ozone Layer (adopted 22 March 1985, entered into force 22 September 1988) 1513 UNTS 323.

[13] The Montreal Protocol on Substances that Deplete the Ozone Layer (adopted 16 September 1987, entered into force 1 January 1989) 1522 UNTS 3, was adopted as a protocol to the Vienna Convention for the Protection of the Ozone Layer.

[14] World Meteorological Organization, *Scientific Assessments of Ozone Depleting Substances: 2010 – Global Ozone Research and Monitoring Project, Report No 52*, National Oceanic and Atmospheric Administration (NOAA) et al (March 2011), xvi–xviii.

[15] It is noted that reg 12.3.1.2 and reg 12.3.2.2 specify other compliance dates with regard to the contractual delivery dates of the equipment in question.

Regulation 2.16 defines ODSs by referring to their definition in article 1(4) of the Montreal Protocol. Regulation 12.5 stipulates that ships must keep a list of any ODS on board; and if these are part of a system that can be recharged, the ship must keep an 'Ozone-depleting Substance record book', logging all replenishments, repairs or discharges of these systems (see also regulations 12.6 and 12.7). This record book is one of the documents that the ship must be able to present during an initial Port State Control (PSC) inspection (document control), as described in chapter 4.

As noted in chapter 10, regulation 12.7.3 includes an important reference to the 'discharge of ozone-depleting substances to the atmosphere', supporting an inference that the term 'discharge' in article 218(1) of UNCLOS should be interpreted as including emissions. It therefore comes as little or no surprise that the conclusion reached in Part II applies to enforcement of regulation 12 of MARPOL Annex VI. Any release of ODSs in violation of regulation 12 can be enforced by port States outside their territory pursuant to article 218(1) and by coastal States inside their territory according to article 220(1), both conditional upon the foreign ship's subsequently calling into port (or at an offshore terminal) and the flag State's not having primary jurisdiction in accordance with article 228(1).

It could be argued that discussing enforcement of regulation 12 is outdated, as the most harmful ODSs have been banned from use by the Montreal Protocol and MARPOL Annex VI. However, a study in 2018, made by the Canadian Research Institute NOAA, showed a disturbing development. Despite international regulations, a global rise in emissions of the harmful CFC substance Trichlorofluoromethane, also referred to as 'CFC-11', had been registered.[16] This substance, CFC-11, can exist for up to 45 years in the atmosphere after being released, and has the greatest ozone-depleting potential of all ODSs, as well as being one of the ODSs that contribute most to global warming.[17] Ensuring effective enforcement of regulation 12 of MARPOL Annex VI through UNCLOS might therefore still be a relevant focal point in coming years to counter such adverse developments.

B. Nitrogen Oxide: Regulation 13

Nitrogen oxide (NOx) is covered by regulation 13 of MARPOL Annex VI and the accompanying 'NOx Technical Code'. It is a gas that is produced during combustion of fuels and subsequently emitted to the atmosphere. NOx gases can, like sulphur, lead to reduced respiratory functions and may increase the risk of infections in the lungs.[18]

The NOx rules differ from the SOx rules, as regulation 13 only applies to ships built after a certain date with diesel engines having a power output above

[16] Available at https://www.eurekalert.org/pub_releases/2018-05/uoca-nfr051618.php.
[17] See World Meteorological Organization Report, n 14, Q.18, Table Q7-1.
[18] See at https://www.epa.gov/no2-pollution/basic-information-about-no2#Effects.

130 kilowatts. Regulations 13.3–13.5 set different regulatory steps (tiers) for those ships, with regard to which requirements they must meet. The applicable 'Tier' depends on the time of construction of the engine. The most demanding Tier (Tier III), as stipulated in regulation 13.5.3, applies only in Nitrogen Oxide Emission Control Areas (NECAs), described in chapter 2. Areas off the US and Canadian East and West coasts, as well as in the Caribbean and around Hawaii, have since 2016 been designated as NECAs (see regulation 13.6).[19] In addition, the IMO agreed under MEPC 70 in October of 2016 that the Baltic and the North Sea will be designated as NECAs from 1 January 2021.[20]

The ships that are covered by the regulation must, according to regulation 13.8, carry an Engine International Air Pollution Prevention Certificate (EIAPP Certificate) and a so-called 'Technical File', which shows the ship's compliance with the applicable requirements. The EIAPP Certificate must be presented during a PSC, along with the Technical File and other required documents proving compliance with regulation 13.[21]

Whether the enforcement paradigm set out in Part II of this book applies to enforcement of regulation 13 must depend on whether a violation of these rules can result in a 'discharge' of emissions. The provision undoubtedly has a fixed focus on engine performance, size, output and other similar matters, which could indicate that any violation would fall outside the scope of article 218, etc. On the other hand, regulation 13 of Annex VI uses the engine parameters and requirements to ensure that the ship emits (discharges) NOx below a certain calculated limit specified in Tiers I–III, pursuant to regulations 13.3–13.6.

This author would therefore argue that a violation of regulation 13 of MARPOL Annex VI, for example if a ship were to sail in a NECA without meeting the Tier III limits, would constitute a discharge violation that could be proved quite effectively.

If a ship does not operate in accordance with its EIAPP Certificate and Technical File, this will be a strong indication that the engine emits (discharges) more NOx than the certified value. Authorities should be able to prove a ship's non-compliance, for example if it applied a setting outside the range allowed, or used a non-compliant engine component. Further, if an engine has been operating in Tier II inside a NECA – where it should have been operating in

[19] See at http://www.imo.org/en/OurWork/Environment/PollutionPrevention/AirPollution/Pages/Emission-Control-Areas-(ECAs)-designated-under-regulation-13-of-MARPOL-Annex-VI-(NOx-emission-control).aspx.

[20] See at https://worldmaritimenews.com/archives/205936/imo-designates-north-sea-baltic-sea-as-neca/.

[21] See IMO, 'Guidelines for port State control under the revised MARPOL Annex VI', adopted by Resolution MEPC.181(59) of 17 July 2009, point 2.1.1, which states that the an initial PSC (document) inspection should include 'the Engine International Air Pollution Prevention Certificate (EIAPP Certificate) including its Supplement, for each applicable marine diesel engine' (2.1.1.2), 'the Technical File for each applicable diesel engine' (2.1.1.3) and 'the method used for demonstrating NOx compliance' (2.1.1.4), for example 'the record book of diesel engine parameters for each diesel engine' (2.1.1.4.1).

Tier III mode – it would constitute a clear case that the ship has discharged excess amounts of NOx.

Coastal States can thus take measures against such a violation in accordance with article 220(1) of UNCLOS; and port States may act in accordance with article 218(1), provided article 228(1) allows for this.

C. Particulate Matter: Regulation 14

Particulate matter (PM) can be extremely harmful to human health depending on the size of the particles, fine PM particles being the most harmful as they can easily penetrate the human respiratory and capillary systems through the lungs when inhaled.[22] A distinction is therefore often made between 'dust' (size PM10[23]), 'fine particulate matter' (size PM 2.5[24]) and 'ultra-fine particulate matter' (size PM 1[25]).

Particulate matter is, along with SOx, also released during the combustion of fossil fuels.[26]

It is noted that the heading of regulation 14 reads 'Sulphur Oxides (SOx) and particulate matter', despite the fact that the provision contains no further reference to PM.

Although PM is very harmful to human health, regulation 14 includes no limit on PM or other regulation thereof, thus making it impossible to take action against any 'violation' as there are no rules to violate. Consequently, there is no *delictum sui generis* (criminal act) against which enforcement measures can be taken under part XII of UNCLOS.

D. Volatile Organic Compounds: Regulation 15

The release of VOCs is controlled in accordance with regulation 15 of MARPOL Annex VI.

In general, VOCs are substances that have a low boiling point and which therefore easily evaporate. The term 'volatile organic compound' is not defined in Annex VI, but the EU's VOC Directive defines VOCs as being

> any organic compound having at 293,15 K a vapour pressure of 0,01 kPa or more, or having a corresponding volatility under the particular conditions of use. For the purpose of this Directive, the fraction of creosote which exceeds this value of vapour pressure at 293,15 K shall be considered as a VOC …[27]

[22] See at https://www.health.ny.gov/environmental/indoors/air/pmq_a.htm.

[23] 'PM10' refers to PM that has a diameter of 10 micrometres (μm) or less.

[24] 'PM 2.5' refers to PM that has a diameter of 2.5 μm or less.

[25] 'PM 1' refers to PM that has a diameter of 1 μm or less.

[26] http://www.valleyair.org/air_quality_plans/AQ_plans_PM_definition.htm.

[27] Council Directive 1999/13/EC of 11 March 1999 on the limitation of emissions of volatile organic compounds due to the use of organic solvents in certain activities and installations [1999] OJ L085/1, art 2(17).

Many VOC substances can cause respiratory diseases, allergic skin reactions and headaches, but can also cause damage to the liver, kidneys and the central nervous system. Some are also assumed to be potentially fatal to humans and animals, as they are carcinogens, including the substance benzene (C_6H_6).[28]

Regulation 15.1 of Annex VI stipulates that the requirements of the provision only apply to tankers, as it covers those VOCs that can be generated from carrying liquid cargo (oil) in cargo spaces on board, and the chances of these VOCs being released during the loading and unloading of the cargo.[29]

Regulation 15.5 requires tankers to have equipment on board that can collect VOC gases and deliver them to a facility in a port that is officially approved (see regulation 15.3) to receive such VOC emissions.

It is also, pursuant to regulation 15.6, the responsibility of tankers to have a 'VOC management plan' on board. This plan must prove that the ship meets the IMO regulations and guidelines[30] for VOC handling. This includes, according to regulations 15.6.1–15.6.4, having a written procedure to minimise VOC emissions, and having a designated person (crew member) on board who is responsible for ensuring compliance with these requirements. The required information must be provided in a language the master and the responsible crew members understand, and a copy must be kept in English, Spanish or French. This VOC management plan must be ready for presentation during the initial part of a PSC. It must be approved by the authorities of the ship's flag State (see regulation 15.6).

A violation of regulation 15 could, in principle, result in VOCs' being released into the atmosphere on the high seas. However, given the content of the provision, and the practical and technical aspects of VOCs' being potentially released in port during loading or unloading, it is more likely that a violation of regulation 15 would be considered committed while the ship is in port or at an offshore terminal.[31]

A port State would therefore have normal 'PSC jurisdiction', pursuant to article 226 of UNCLOS, over a violation of regulation 15, whether relating to a faulty or missing VOC management plan or to faulty or missing technical equipment that results in the discharge of VOCs into the atmosphere. Any violation may be met with fines and/or detention, as described in chapter 4.

[28] See at https://www.epa.gov/indoor-air-quality-iaq/volatile-organic-compounds-impact-indoor-air-quality.

[29] Many tankers use inert gas to secure cargo tanks. When the inert gas is discharged, it can simultaneously result in the discharge of VOCs.

[30] See, eg, MSC/Circ.585 of 16 April 1992 on 'Standards for vapour emission control systems', available at https://www.transportstyrelsen.se/globalassets/global/sjofart/dokument/imo_dokument/msc/msc_circ_585.pdf.

[31] A possible exception to this could be ship-to-ship (STS) transfers that result in the release of VOCs in an EEZ or on the high seas. In this situation, the application of art 218(1) and art 220(1) of UNCLOS could become relevant, *if* the discharge violation can be proved, which might be difficult. Also, STS transfers are often not performed on the high seas but in designated/approved areas by a coastal State outside shipping lanes and routes, so that it is the use of art 220 that is most likely.

The port State also has, according to article 230 of UNCLOS, an unrestricted right to sanction any violations that result in a measurable release of VOCs while the ship is in port, as the offence is committed in internal waters over which the port State has unrestricted territorial jurisdiction (see article 2(1) of UNCLOS).

E. Ship Incinerators: Regulation 16

Regulation 16 of MARPOL Annex VI contains rules about burning (incinerating) materials, fabrics, etc a on board ships. The scope of the provision can be divided in to two parts, with regulations 16.1–16.4 applying to *all* ships and regulations 16.6–16.9 solely applying to ships built after 1 January 2000 or to ships that had an incinerator installed after that date.[32]

Regulation 16.1 states that all burning of materials, etc on board a ship may only be done in an incinerator designed for this purpose.

Regulation 16.2 lists several substances that may not be incinerated, not even in an approved incinerator. This list includes residues and materials from transporting MARPOL Annexes I–III substances, substances containing polychlorinated biphenyls (PCBs), waste containing heavy metals (provided it is more than just *traces* of heavy metals), refined petroleum products containing *halogens*, and sewage residue and oil residues, in so far as these are not produced on board during the operation of the ship.

Regulation 16.3 also prohibits the incineration of polyvinyl chloride (PVC) materials, unless this is done in an incineration system approved for this purpose by the IMO.[33]

Regulation 16.4 allows for incineration of sewage and oil sludge in the main or auxiliary engines or boilers if this was generated as part of the normal operation of the ship and the incineration occurs outside ports, harbours and estuaries.

Regulation 16.5 clarifies any situations where the rules of this provision and those of the London Convention (and its 1996 Protocol) overlap when disposing of waste. In such cases, the (*lex specialis*) provisions of the London Convention prevail over regulation 16 of Annex VI. Regulation 16.5.2 also underlines that nothing in the regulation must be considered an obstacle to the continued development of technical solutions that ensure the correct disposal (incineration, etc) of waste.

Regulation 16.6.1 specifies that incinerators that are on board ships that are built after 1 January 2000 or which are installed on any ship after that date, must meet the exact specifications set out in Appendix 4 to MARPOL Annex VI.[34]

[32] Reg 16.5 refers to overlaps with the London Convention and future technological developments.

[33] This requires that a 'Type Approval Certificate' be issued in accordance with the guidelines for this. For more details, see Resolutions MEPC.59(33), MEPC.76(40) or MEPC.244(66).

[34] Reg 16.9 also sets specific requirements for how quickly these systems must achieve a minimum temperature of 850°, depending on whether it is a continuous-feed type incinerator or a batch-loaded type incinerator. In the latter case, the incinerator should achieve a temperature of 600° within 5 minutes, after which it must be able to stabilise at 850°.

This must be certified by the flag State. Regulation 16.6.2 contains an exception to this for ships that only undertake domestic voyages in waters under the sovereignty of the flag State.

Ships covered by regulation 16.6 must also, pursuant to regulation 16.7, have an Operating Manual on board, issued by the manufacturer of the incinerator, which details how the incineration system is operated. The crew members who use the system must familiarise themselves with the Manual and undergo training to operate the system (see regulation 16.8).

It must be noted, in terms of applying the conclusions set out in Part II, that most of the requirements of this provision are of a technical nature, relating to the construction, approval and application of such incineration systems. Nonetheless, violations of regulations 16.2 and 16.3 could provide grounds for a port State to assert jurisdiction in accordance with article 218(1) of UNCLOS, and a coastal State in accordance with article 220(1), for instance if a ship incinerates PCB or PVC materials whilst sailing on the high seas or in an EEZ, subsequently going into port in a port State or coastal State. Coastal States would also, pursuant to article 220(1) and article 2(1) of UNCLOS, have unrestricted jurisdiction over violations of regulation 16.4 that occur within ports, harbours and estuaries.

F. EEDI/SEEMP (CO_2): Regulations 19–23

As mentioned in previous chapters, the international community has increasingly focused on air pollution over recent decades. This includes the IMO, as seen with the adoption of MARPOL Annex VI in 1997, which, inter alia, regulates emissions of ODSs, NOx, SOx and VOCs.

This was followed by a strong focus on emissions of GHGs, which include gases such as carbon dioxide (CO_2), methane (CH_4) and nitrous oxide (N_2O).[35] These GHGs contribute to global warming and climate change.

The IMO has initiated three major GHG studies since the start of the new millennium, in 2000, 2009 and 2014, to establish to what extent shipping contributes to the release of GHGs on a global scale. These studies have showed that although shipping is the form of transport that emits the lowest amount of GHGs compared to the quantity of goods transported over long distances, shipping still overall contributes a large quantity of manmade GHGs to the atmosphere, particularly CO_2. The IMO therefore often refers to the reduction of CO_2 emissions as a proxy for reducing the overall contribution of GHGs from shipping.

The IMO has estimated that shipping accounts for 2.2% of the total release of CO_2 emissions on a global scale, where ships of 5,000 GT and above account for 85% of this emissions factor.[36] It was therefore decided at MEPC62 in July of 2011,

[35] The term 'greenhouse gases' can also cover ODS gases such as CFC and HCFC gases covered by reg 12 of Annex VI.

[36] See at http://www.imo.org/en/OurWork/Environment/PollutionPrevention/AirPollution/Pages/Default.aspx.

following the IMO's second GHG study, to amend MARPOL Annex VI by inserting a new chapter in the Annex (chapter 4), containing regulations 19–23, which seek to reduce CO_2 emissions from shipping.[37]

The rules in chapter 4 introduced a mandatory performance-based system that sets requirements for the minimum energy efficiency of new ships of 400 GT or above (see regulation 19.1), unless the flag State postpones the entry into force of chapter 4, which it can do for up to four years, or as regards ships that, within their working life, only sail in waters under the jurisdiction of the flag State (see regulation 19.2.1). Also exempt are ships using non-mechanical propulsion or platforms (see regulation 19.2.2).

A special index, the Energy Efficiency Design Index (EEDI), found in regulations 20 and 21, lays down the minimum energy efficiency levels that new ships must meet, depending on their type, size, etc. The regulations do not specify how these levels are to be achieved, which allows shipowners and shipbuilders to choose the ship designs, engines, materials, equipment, etc they deem appropriate, provided the EEDI requirements are met.

Regulation 20 differentiates between 'new builds' meeting the specific EEDI targets, described as 'attained EEDI',[38] new ships that have undergone a major conversion and existing ships that have undergone a major conversion.

Regulation 21 refers to 'required EEDI' and includes two tables (Table 1 and Table 2) specifying the EEDI requirements in the form of three phases, which aim to ensure the continued reduction of CO_2 from new ships, according to ship type.[39] The first phase entailed a 5–10% reduction from 1 January 2015 to 31 December 2019; phase two a 15-20% reduction from 1 January 2020 to 31 December 2024; and phase three a 30% reduction by January 2025.[40]

Regulation 22.1 requires all ships, including existing ships, to develop, implement and carry on board a ship-specific Ship Energy Efficiency Management Plan (SEEMP) that determines how the ship can be operated in the most energy-efficient way. To that end an Energy Efficiency Operational Indicator (EEOI) has been developed to enable shipowners to monitor the efficiency performance of its fleet over a specific period of time. The EEOI can therefore measure the fuel efficiency of a ship in operation.

The SEEMP must, as stated in regulation 22.3, follow the IMO guidelines for developing a SEEMP, that is, Resolution MEPC.282(70), 'Guidelines for the Development of a Ship Energy Efficiency Management Plan (SEEMP)'. These Guidelines, inter alia, refer to the ship's improving and optimising certain administrative and

[37] See Resolution MEPC.203(62).

[38] See MEPC.245(66), 'Guidelines on the method of calculation of the Attained EEDI for New Ships' (4 April 2014).

[39] There are, in principle, four phases, as there was a 'phase 0' from 1 January 2013 to 31 December 2014, which set the baseline for the calculated reductions.

[40] The IMO's GHG Strategy, described in ch 15, may result in phase 3 being moved forward and a new phase 4 adopted with even higher reduction targets.

operational parameters, such as voyage planning, weather routing, 'Just in Time Arrival' at port, speed, shaft power, ship handling, trim, ballast, hull, propulsion systems, waste heat recovery, fleet management, cargo handling, energy management, fuel type, trade and sailing areas.[41]

Regulation 22.2 stipulates that all ships of 5,000 GT or above must have a SEEMP that also contains a methodology describing how the ship will calculate and collect data on the its energy consumption and subsequent (CO_2) emissions. This is related to MEPC Resolution MEPC.278(70), adopted on 28 October 2016, which amended chapter 4 of MARPOL Annex VI by implementing regulation 22.A.[42] This regulation links to the IMO's 'Data collection system for fuel oil consumption of ships' ('data collection system'), which was briefly mentioned in chapter 4 of this book on PSC.

The data that are collected in accordance with regulation 22.2 must, at the end of each year, be processed and submitted to the flag State within three months of the year's end, in accordance with regulations 22.A.2 and 22.A.3.[43] The ship must keep these data for one year (see regulation 22.A.8).

The flag State must review the submitted data and confirm their validity in accordance with regulation 22.A.7.

Once the flag State has received the data and confirmed their validity under regulation 22.A, it must issue a *Statement of Compliance* related to fuel oil consumption to the ship within five months from the beginning of the year, pursuant to regulation 6.6 – 6.7 of MARPOL Annex VI.

The flag State is required to report these data to the IMO pursuant to regulation 22.A.9. More precisely, the data must be entered into IMO's data collection system. The IMO generates an annual report, based on these reported data, which is forwarded to MEPC and which contains an analysis of the results submitted (see regulations 22.A.10–22.A.12).[44]

Regulation 23 of MARPOL Annex VI requires flag States that are party to Annex VI to cooperate with the IMO, international bodies and other States, including developing States, to create and promote technology relating to improvement of the energy efficiency of ships.

[41] See point 5 on 'Guidance on best practices for fuel-efficient operations of ship' of Resolution MEPC.282(70), 'Guidelines for the development of a Ship Energy Efficiency Management Plan (SEEMP)'.

[42] http://www.imo.org/en/MediaCentre/PressBriefings/Pages/04MARPOLamendments.aspx.

[43] Reg. 22.A.1 stipulates that the data collection obligation described in regulation 22.2 cf. 22.A began 1 January 2019.

[44] It should be noted that the European Union (EU) has adopted a Regulation that, inter alia, demands that all ships of more than 5,000 GT flying the flag of an EU Member State and sailing to, from or between EU ports must undertake separate reporting of such data, including information on departures, arrivals, fuel quantity, distance travelled, time at sea, cargo, etc, pursuant to arts 8–9 of the Regulation. See Regulation (EU) 2015/757 of the European Parliament and of the Council of 29 April 2015 on the monitoring, reporting and verification of carbon dioxide emissions from maritime transport, and amending Directive 2009/16/EC [2015] OJ L123/55 ('the MRV Regulation').

The common denominator of these chapter 4 regulations is thus *energy efficiency*. The EEDI ensures that new ships will be *designed* to meet an energy-efficiency target, and the SEEMP seeks to guarantee that all ships are *operated* in the most energy-efficient way. These optimisations of a ship's energy usage will lead to ships' using less fuel and thereby releasing less GHG, including CO_2. The IMO has calculated that these regulations could reduce CO_2 emissions by up to 75%.[45]

This also reduces the fuel costs of shipowners and should, in principle, result in a win–win scenario for the owners and the environment alike. However, compliance is not a given, as complying with the EEDI and SEEMP requirements can involve significant expenditure for some shipowners. If an owner chooses to employ ship designs, constructions or operations that are cheap but energy-ineffective, this may yield higher short-term profits.

Determining how regulations 19–23 of Annex VI can be enforced in accordance with part XII of UNCLOS is a challenge. Violations of these regulations may result in excessive discharges of CO_2 emissions compared to the emissions that could be achieved if ships were to fulfil the requirements. That being said, the causal link between a violation of the regulations of chapter 4 of Annex VI and establishing that this resulted in a quantifiable discharge of illegal CO_2 emissions, for example whilst sailing on the high seas, is not strong enough.[46] Violations of the regulations in chapter 4 of Annex VI are therefore best found and enforced during PSC, when inspecting documentation and the ship in accordance with regulation 10.3 of MARPOL Annex VI. Pursuant to article 226(1)(a)(ii) of UNCLOS, a port State has jurisdiction over such violations found during a PSC, also recalling that the principle of 'no more favourable treatment' (NMFT) applies in accordance with article 5(4) and article 1(2) of the MARPOL Convention.

It is noted that the *statement of compliance* the flag State issues each year must be considered a document covered by article 217(3) of UNCLOS. Any failure by the flag State to issue such a statement – or, even more relevant, if such a statement is issued on false grounds, for example where no data or data showing non-compliance were submitted and a statement is still issued – is a violation of article 217(3).

It should finally be noted that the IMO has sought to reduce GHG emissions from shipping further by adopting, in 2018, a GHG Strategy that sets certain fixed reduction goals for GHG (CO_2). This GHG Strategy, including its reduction goals and proposed regulatory (candidate) measures for achieving them, is discussed in detail in chapter 15, including in the context of enforcement through part XII of UNCLOS. Part IV of this book examines whether it is conceivable that some of

[45] This is reiterated in para 9 of the Preamble to the MRV Regulation.

[46] It should be noted that the aforementioned EU MRV Regulation (n 44) includes a reference in art 9 to the CO_2 emissions' being calculated for each ship and for each leg of a voyage, in accordance with the methods specified in Annex 1 to the Regulation. It could be argued that if this approach were adopted by the IMO, it would allow a port State or coastal State to prove that a ship had discharged an excess, and therefore illegal, amount of CO_2 emissions whilst sailing on the high seas or in the waters of the State, thereby allowing them to invoke art 218(1) or art 220(1) of UNCLOS respectively.

these proposed (short-, mid- and long-term) candidate measures on GHG could, in future, be perceived as representing norms of a *jus cogens* character, given the extremely adverse effects of GHGs.

II. Annexes I–V of the MARPOL Convention

As noted in chapter 1 of this book, the MARPOL Convention contains six Annexes. Annex VI on air pollution has been analysed at length; this section will therefore focus on enforcement of the other five Annexes. These are:

- Annex I: Regulations for the Prevention of Pollution by Oil
- Annex II: Regulations for the Control of Pollution by Noxious Liquid Substances in Bulk
- Annex III: Regulations for the Prevention of Pollution by Harmful Substances Carried by Sea in Packaged Form
- Annex IV: Regulations for the Prevention of Pollution by Sewage from Ships
- Annex V: Regulations for the Prevention of Pollution by Garbage from Ships.

The Annexes are briefly examined individually, with regard to how violations of them can be subject to enforcement in accordance with the provisions of part XII of UNCLOS.

A. MARPOL Annex I: Regulations for the Prevention of Pollution by Oil

MARPOL Annex I specifies what obligations certain ships, in particular oil tankers, must meet when sailing with oil on board for combustion or as cargo.

All ships are, unless exempt, covered by the provisions of Annex I pursuant to regulation 2.1.[47]

All oil tankers above 150 GT and all other ships above 400 GT must carry an Oil Record Book Part I in accordance with regulation 17.1. All oil tankers above 150 GT must also carry an Oil Record Book Part II, as stated in regulation 36.1.

Oil tankers must also have segregated ballast water tanks and double hulls, in accordance with regulations 18–30 of Annex I. They must also possess an International Oil Pollution Prevention Certificate pursuant to regulations 6–10. All certificates and Oil Record Books must be presented during a PSC (see regulations 17.6 and 36.7).

[47] Not all the technical, construction, administrative and operational requirements found in MARPOL Annex I will be set out in full here.

Regulation 38 of Annex I also requires ports, terminals, yards, etc to have equipment and facilities (*reception facilities*) that can receive oil and oily residues.

As a starting point, regulations 15.1 and 34.1 prohibit oil from being discharged into the sea, except for in situations of force majeure (regulation 4) or in the conditions specified in regulation 15.2 and regulations 34.1.1–34.1.2 and 34.1.4–34.1.6.

As discussed in chapter 3 of this book, MARPOL Annex I also allows for the designation of 'Special Areas' where strict regulations apply, including a prohibition on the discharge of oil in any form (see regulations 15.3 and 34.1.3 and regulation 1.11). These Special Areas include the Mediterranean Sea, the Baltic Sea, the Black Sea, the Red Sea, the Gulf of Aden, Gulf areas,[48] North West European waters, Oman, the Arabian Sea, southern South African waters and Antarctica.[49]

The MEPC adopted Resolution 189(60) in 2010, which implemented a new chapter (chapter 9) in MARPOL Annex I, containing regulation 43. This provision prohibits the use and carriage of oil, both as cargo and fuel, in the area around Antarctica,[50] provided the oil in question meets the characteristics specified in regulation 43:

1. crude oils having a density at 15'C higher than 900 kg 1m3;
2. oils, other than crude oils, having a density al 15'C higher than 900 kg/m3 or a kinematic viscosity at 50'C higher than 180 mm'/s; or
3. bitumen, tar and their emulsions …

These definitions describe oil types that can have a very harmful impact on the fragile marine environment in polar regions.

There have therefore also been discussions in the IMO on whether similar requirements should be established in the Artic.[51] It was decided at the sixth meeting of the Pollution, Prevention and Response (PPR) Subcommittee in February 2019 to forward such a regulation for discussion, acceptance and approval at MEPC level.[52]

Any kind of violation of MARPOL Annex I that leads to a discharge of oil in to the sea is covered by the conclusions in Part II of this book, including allowing a port State to assert extraterritorial jurisdiction over such violations pursuant to article 218(1) of UNCLOS. For example, for violations of regulation 15.1 of Annex I, which directly provides that 'discharge into the sea of oil or oily mixtures from ships shall be prohibited', or for infringements of regulations 15.3 and 34.1.3 pertaining to any discharge within a Special Area.

[48] The term 'Gulf areas' is a reference to the Arabian Gulf, being the sea north-east of the rhumb line between Ras al Hadd (22° 30' N, 059° 48' E) and Ras al Fasteh (25° 04' N, 061° 25' E).

[49] See also MEPC.1/Circ.778/Rev.2 of 6 April 2017 and http://www.imo.org/en/OurWork/Environment/SpecialAreasUnderMARPOL/Pages/Default.aspx.

[50] Emergency situations are exempt, including when participating in a rescue operation. Regulation 43.2 states that ships that have previously used or transported these oils are *not* required to 'flush' their tanks and piping systems before they sail into the waters of Antarctica.

[51] This work was often described within the IMO as the work on 'HFO in the Arctic'.

[52] See at https://www.highnorthnews.com/en/imo-inches-forward-ban-heavy-fuel-oil-arctic.

It will be recalled from chapter 10 that the term 'discharge' in article 218 of UNCLOS includes, besides emissions, the 'usual' applications of the term; applications that fall within the definition of 'discharge' in article 2(3)(a) of the MARPOL Convention, including 'escape', 'disposal', 'leaking', spilling', 'pumping' and 'emptying', thereby also covering the discharge of oil into the sea. Coastal States can therefore also assert jurisdiction over such violations pursuant to article 220(1).

However, several of the conclusions set out in Part II may be different when it comes to taking enforcement measures against (discharge) violations of MARPOL Annex I. Article 220(5)–(6), which refer to discharges that cause (or threaten to cause) significant and major damage to a coastal State, will become relevant when enforcing violations of MARPOL Annex I, as one specific violation of that Annex, unlike a violation of Annex VI, can result in specific, tangible damage to a coastal State. A coastal State could thus, in accordance with article 220(5)–(6), stop, investigate, detain and prosecute a foreign ship transiting the EEZ if the ship has violated MARPOL Annex I. This was also explicitly established in the *Bosporus Queen case*, discussed in chapters 5 and 9.[53]

It will be recalled that the European Court of Justice noted in the *Bosporus Queen* that even though the Baltic Sea is deemed a Special Area in accordance with MARPOL Annex I, a violation of this regulation does not automatically result in major damage pursuant to article 220(6) but it can influence that overall assessment.[54]

If a violation of Annex I does result in major damage, it will also enable a coastal State to invoke the *first exception* in article 228(1) of UNCLOS. The coastal State's article 220 jurisdiction would consequently automatically extinguish the article 217 jurisdiction of the flag State, as described in chapter 11.

Finally, coastal States would also have jurisdiction, according to article 233 and article 42(1)(b) of UNCLOS, to stop, investigate and prosecute foreign ships that infringe MARPOL Annex I while making *transit passage* in an international strait, provided the infringement caused (or threatened to cause) major damage to the marine environment of the strait.

It should also be noted, regarding coastal State enforcement of regulation 43 of MAPROL Annex I (the special requirements for Antarctica), that Antarctica, in international law, is regarded as being *terra nullis*, that is land over which no State can claim sovereignty, and it can only be used for peaceful purposes.[55] The waters off Antarctica should therefore be considered *res communis* (high seas that belong to humanity).[56] It is thus only logical to assume that violations of regulation 43 of Annex 1 should predominantly be enforced by port States

[53] Case C-15/17 *Bosphorus Queen Shipping Ltd Corp v Rajavartiolaitos*, ECLI:EU:C:2018:557.

[54] ibid, paras 107–08. The same conclusion is reached in Advocate General Wahl's Opinion in the case. Advisory Opinion of Advocate General Wahl, delivered on 28 February 2018, paras 106–08.

[55] This principle is repeated in art 1 of the Antarctic Treaty (adopted 1 December 1959, entered into force June 23 1961) 402 UNTS 71.

[56] CC Joyner, *Antarctica and the Law of the Sea* (Martinus Nijhoff Publishers, 1992) ch 3, 'Jurisdiction Offshore Antarctica'.

in accordance with article 218(1) of UNCLOS and, obviously, flag States according to article 217. Any overlapping jurisdiction between these States is settled in accordance with article 228(1), but only through application of the *main rule* and possibly the *second exception*, as the *first exception* relates to coastal State jurisdiction.

The same conclusion might not apply if a similar prohibition is adopted regarding the Arctic, as different States have sovereign rights in that area, thereby also having the right to exercise a coastal State's territorial jurisdiction in Arctic waters pursuant to article 220 of the Convention. These (coastal) States would likely also be able to adopt national environmental legislation in these ice-covered areas according to article 234, as described in chapter 3.

B. MARPOL Annex II: Regulations for the Prevention of Pollution by Noxious Liquid Substances in Bulk

MARPOL Annex II imposes, in much the same way as Annex I does for tankers transporting oil, requirements for tankers transporting *noxious liquid substances in bulk*, such as the transport of harmful chemicals. The rules and regulations in Annex II are therefore only briefly examined here.

Regulation 1.16.1 of MARPOL Annex II defines a 'chemical tanker' as a ship that sails with substances covered by chapter 17 of the International Code for the Construction and Equipment of Ships carrying Dangerous Chemicals in Bulk ('the IBC Code').[57]

Regulation 6 of Annex II refers to harmful substances, chemicals, etc that are classified as category 'X', 'Y' or 'Z' substances, or as 'other substances' according to chapter 18 of the IBC Code.

Regulation 13.1.1 contains the main rule – that all discharges of substances listed as category X, Y and Z substances are prohibited.

In regulation 13.8, MARPOL Annex II designates Special Areas in which strict requirements apply. This includes a complete prohibition on discharging harmful substances (chemicals) into the sea pursuant to regulation 13.8.2. Antarctica is such an Area.[58]

A ship (tanker) must, in accordance with regulations 8–9 of Annex II, be certified to sail with such substances, this being documented by an International Pollution Prevention Certificate for the Carriage of Noxious Liquid Substances in Bulk. This certificate must be presented as part of an initial PSC.

Regulation 18 requires that ports, terminals and shipyards have adequate facilities for receiving such harmful substances.

[57] The International Code for the Construction and Equipment of Ships carrying Dangerous Chemicals in Bulk (adopted 5 December 1985, entered into force 1 July 1986) ('the IBC Code').

[58] See MEPC.1/Circ.778/Rev.2 of 6 April 2017 and http://www.imo.org/en/OurWork/Environment/SpecialAreasUnderMARPOL/Pages/Default.aspx.

The conclusions in Part II of this book apply to enforcement of MARPOL Annex II, with the same expanded jurisdiction as described regarding enforcement of MARPOL Annex I, including allowing coastal States to invoke article 220(5)–(6) and the *first exception* in article 228(1), as a violation of MARPOL Annex II can result in tangible and serious (major) pollution of a coastal State. This also applies with regard to article 233 of UNCLOS, as article 42(1)(b) refers to noxious substances being released into a strait.

Finally, the same principles apply to enforcement of MARPOL Annex II in Antarctica as described in connection with Annex I.

C. MARPOL Annex III: Regulations for the Prevention of Pollution by Harmful Substances Carried by Sea in Packaged Form

In contrast to Annexes I and II, MARPOL Annex III deals with ships transporting harmful substances in packaged form and not in liquid form.

Regulation 1.1 defines such harmful substances as those covered by the International Maritime Dangerous Goods Code ('the IMDG Code').[59] The reference to the harmful substances' being transported in *packaged form* encompasses substances transported in portable tanks, containers, barrels, road or train tank wagons, etc.

Annex III stipulates that such cargo must be packed in a secure manner (regulation 3), have labels affixed specifying the content and hazard it represents (regulation 4) and must be secured on board (regulation 6.)

Regulation 5 requires ships to adhere to the requirements for documentation (list, manifest or stowage plan) set out in the IMDG Code.

Regulation 8.1 provides that packages containing harmful substances must not be *wilfully* thrown overboard, and regulation 8.2 establishes that such packages must not have any leaks, etc that *unintentionally* might pollute the marine environment.

Compliance with these requirements of Annex III can be ensured as part of a PSC (see regulation 9).

Discussing enforcement of MARPOL Annex III can involve walking a tightrope between violations of regulations 8.1 and 8.2, when deciding which provision of part XII of UNCLOS applies. If regulation 8.2 is infringed, for example if a 'package' (barrel, etc) leaks its content into the sea, it must be considered a discharge, as article 2(3)(a) of the MARPOL Convention defines the term 'discharge' to cover, inter alia, 'leaking' and 'spilling'. Violations of regulations 8.2 of MARPOL Annex III

[59] Maritime Solid Bulk Cargoes Code (adopted 4 December 2008, entered into force 1 January 2011) ('the IMSBC Code').

can therefore be enforced in accordance with the enforcement principles discussed in Part II. The conclusion regarding enforcement of MARPOL Annexes I and II, that coastal States can invoke articles 220(5)–(6), 233[60] and the *first exception* in article 228(1), also applies if the leaked substance is so harmful that it potentially can (or did) cause major damage to a costal State.

However, the same does not apply to enforcement of regulation 8.1, as that provision concerns packages that are *deliberately* thrown overboard. This cannot be considered a discharge but is *dumping*, in accordance with the definition thereof in article 1(1)(5)(i) of UNCLOS, which states that 'dumping' includes, inter alia, 'any deliberate disposal of wastes or other matter from vessels, aircraft, platforms or other man-made structures at sea'.

It will also be recalled from chapter 10 that article 2(3)(b) of the MARPOL Convention provides that the term 'discharge' does not include dumping.

Violations of regulation 8.1 of Annex III can therefore not be enforced on the high seas in accordance with article 218 of UNCLOS. Such violations are instead covered by article 216, which was briefly discussed in chapter 9.

Violations of regulation 8.1 of MARPOL Annex III *can* be enforced by coastal States in accordance with article 216(1)(a), by flag States pursuant to article 216(1)(b) and by port States in accordance with article 216(1)(c). Article 216(1)(c) does note, however, unlike article 218, provide a port State with any extraterritorial jurisdiction to enforce outside its own waters (ports).

D. MARPOL Annex IV: Regulations for the Prevention of Pollution by Sewage from Ships

MARPOL Annex IV regulates discharges of waste water (sewage) from ships over 400 GT engaged in international voyages, or from ships that are certified to carry more than 15 passengers (see regulation 2). Regulation 1.3 defines what is covered by the Annex, which includes drainage from toilets, medical facilities, etc.

It should be noted that water that has been used for bathing, washing dishes, washing clothes, etc – often referred to as 'grey water' – is not covered by Annex IV. It has, however, been discussed at IMO level whether such regulation of grey water should be adopted. The following conclusions of this subsection pertaining to enforcement of MARPOL Annex IV would therefore also apply to any future regulation of grey water discharges.[61]

Ships that are covered by Annex IV must, in accordance with regulation 9, have systems and installations on board for treating such sewage, as it is illegal to discharge untreated sewage pursuant to regulation 11.1, with certain exceptions applying that are codified in regulations 11.1.1 and 11.1.2.

[60] Art 42(1)(b) of UNCLOS refers to the discharge of oil and noxious substances into a strait, which encompasses a violation of reg. 8.2 of MARPOL Annex III but not violations of reg. 8.1.

[61] See at https://gcaptain.com/addressing-grey-water-pollution-ships/.

A ship's compliance with the requirements of Annex IV is, according to regulations 4–8, to be certified in the form of an International Sewage Pollution Prevention Certificate, which must be presented during PSC. This certificate is issued by the flag State. Regulation 12 requires ports, etc to have facilities to receive such sewage waste.

Annex IV also allows for the designation of Special Areas, where particularly strict discharge requirements apply to discharges from passenger ships (see regulation 11.3). The Baltic Sea has been designated as such a Special Annex IV Area.[62]

Any deliberate or unintentional 'discharge' of sewage in violation of Annex IV is an infringement that is covered by the conclusions set out in Part II of this book, allowing for articles 218(1) and 220(1) to be used. It must, however, be questionable whether a coastal State can invoke articles 220(5)–(6) and 228(1) (first exception) in response to such violations, as the potential damage caused by the infringement is not likely to be severe enough to cause *major damage* to a coastal State. It is not relevant to discuss the application of article 233, as article 42(1)(b) only refers to the discharge of oil and noxious substances into a strait, which does not encompass MARPOL Annex IV.

E. MARPOL Annex V: Regulations for the Prevention of Pollution by Garbage from Ships

MARPOL Annex V stipulates how ships may, and may not, dispose of waste generated on board.[63]

'Waste' means garbage, residues and leftover products originating from the normal operations of a ship, including kitchen waste, plastic from food or beverage products (plastic bottles, candy wrappers), etc. For example following the definition in regulation 1.9 of Annex V.

According to regulation 2, Annex V applies to all ships, unless otherwise stated.

Regulation 3 prohibits, as a main rule, any disposal of waste from a ship into the sea, but certain exceptions apply in accordance with regulations 4 and 7, depending on the type of waste and where the ship is sailing. For example, food products and leftovers can be ground down, and the result may be disposed of into the sea if the ship is the required minimum distance from any coast/coastal State (see regulation 4.1).

All vessels of 400 GT or more and all ships certified to carry more than 15 passengers must, in accordance with regulation 10 of Annex V, also have a 'Garbage Record Book'. This record book must be presented during PSC in accordance with regulation 9.

[62] See MEPC.1/Circ.778/Rev.2 of 6 April 2017 and http://www.imo.org/en/OurWork/Environment/SpecialAreasUnderMARPOL/Pages/Default.aspx.

[63] See also Resolution MEPC.219(63), 'Guidelines for the implementation of MARPOL Annex V'; and Resolution MEPC.220(63), 'Guidelines for the development of garbage management plans'.

Ports and terminals must, under regulation 8, maintain adequate reception facilities for such waste.

Special Areas can also be designated in accordance with MARPOL Annex V pursuant to regulation 6, in which strict requirements for the disposal of waste apply.[64] The Mediterranean Sea, the Baltic Sea, the North Sea, the Black Sea, the Red Sea, the outer Caribbean area, Antarctica and the 'Gulf areas' are all Special Areas.[65]

The need for effective enforcement of MARPOL Annex V has become very apparent within recent years due to repeated reports of enormous and ever-growing floating islands of plastic in the world's oceans,[66] for instance in parts of the North Pacific[67] and in the Caribbean.[68] This pollution can by itself lead to fish or mammals consuming it or getting caught in it. This waste can also dissolve into very small particles, which may be consumed and absorbed by even smaller fish, which can have toxic consequences for the marine environment, including the health of humans and animals if those contaminated fish enter the food chain.

The disposal of waste, including plastic in oceans, is also a subject that has repeatedly been addressed at the IMO under the heading 'marine litter'. The MEPC adopted an 'IMO Action Plan' in 2018, to address and combat marine litter in the oceans.[69] The Action Plan is due to be completed by 2025.

When looking in to enforcement of Annex V, a distinction must be made depending on whether the term '*disposal* of waste' can be characterised as involving *discharge* or *dumping*. Two arguments indicate that disposal should be viewed as being a discharge within the realm of part XII of UNCLOS. First, article 2(3)(a) of the MARPOL Convention defines 'discharge' as also being a form of 'disposal', which also applies to the Annexes to the Convention pursuant to article 1(2). Second, although article 1(1)(5)(a)(i) of UNCLOS defines 'dumping' as 'any deliberate disposal of wastes or other matter from vessels, aircraft, platforms or other man-made structures at sea' – which could indicate that disposal of waste amounts to dumping – article 1(5)(b)(i)–(ii) go on to state that dumping does *not* include (i) 'the disposal of wastes or other matter incidental to, or derived from the normal operations of vessels' and (ii) 'placement of matter for a purpose other than the mere disposal thereof'. Disposal of waste that is generated on board as part of the *normal operations* of a ship, which includes food waste, empty plastic bags, plastic bottles, etc, is therefore not considered dumping, also because this waste is not placed onboard with the sole purpose of being disposed/dumped.

[64] See MEPC.1/Circ.778/Rev.2 of 6 April 2017 and http://www.imo.org/en/OurWork/Environment/SpecialAreasUnderMARPOL/Pages/Default.aspx.

[65] For the meaning of 'the Gulf areas', see n 49 above.

[66] http://www.imo.org/en/MediaCentre/HotTopics/marinelitter/Pages/default.aspx.

[67] https://www.nationalgeographic.org/encyclopedia/great-pacific-garbage-patch/.

[68] http://www.independent.co.uk/environment/un-ocean-plastic-waste-resolution-us-china-india-reject-pollution-sea-united-nations-environment-a8095541.html.

[69] http://www.imo.org/en/MediaCentre/HotTopics/marinelitter/Pages/default.aspx.

Therefore, given the positive demarcation of the term 'discharge' in article 2(3)(a) of the MARPOL Convention and the negative demarcation of the term 'dumping' in article 1(1)(5)(a)(i)–(ii) of UNCLOS, it can be concluded that 'disposal of waste' covered by Annex V can be regarded as a 'discharge'.

The conclusions set out in Part II of this book therefore apply to enforcement of Annex V, giving port States extraterritorial jurisdiction over MARPOL Annex V violations outside their own territory, for example on the high seas, pursuant to article 218(1) of UNCLOS. Coastal States have jurisdiction over all infringements of Annex V within their territory, including in the EEZ, pursuant to article 220(1) of UNCLOS. Both articles are conditional on the violating ship's subsequently calling voluntarily into port in the port State or coastal State.

Section II.D stated that it was considered questionable whether a violation of MARPOL Annex IV (sewage) would allow a coastal State to assert the wider jurisdictional basis found in article 220(5)–(6) and invoke the first exception in article 228(1).[70] Pertaining to enforcement of MARPOL Annex V, there can be little or no doubt that *one* violation of Annex V, as a main rule, cannot result in potential significant or major damage to the coastal State. It is, as was concluded in Part II (chapter 9) regarding enforcement of the sulphur regulations of Annex VI, the accumulated amount of plastic waste, etc in the oceans that inflicts the required damaging effect.

However, a possible exception to this main rule could perhaps be envisaged if *one* ship were to dispose of such substantial amounts of waste at *one* time that it might result in *major damage*, allowing the coastal State, in accordance with articles 220(5)–(6) and 228(1), to stop, investigate and detain the ship in the EEZ and deny the flag State its request for suspension of legal proceedings instituted by the coastal State.[71] It must to this end be recalled, however, that the disposed of waste must originate in normal operations on board, meaning that if waste is taken on board *as* waste, with the purpose of its being dumped during the voyage, then that is *dumping*, covered by the London Convention and article 216 of UNCLOS.

III. The Ballast Water Management Convention

The IMO's Convention for the Control and Management of Ships' Ballast Water and Sediments, is often referred to as the 'Ballast Water Management Convention' or simply the 'BWM Convention'.[72] It seeks to regulate the discharge of ballast water from ships.

[70] It is not relevant to discuss the application of art 233, as art 42(1)(b) only refers to the discharge of oil and noxious substances in to a strait, which does not encompass MARPOL Annex IV.

[71] Art 233 is not relevant to this discussion, as art 42(1)(b) only refers to the discharge of oil and noxious substances.

[72] See n 5 above.

Some ships have ballast tanks that they fill with seawater to stabilise the vessel during a voyage. The amount of ballast water needed is assessed depending on the ship's cargo (weight), voyage plan, weather conditions, etc.

This seawater in the ballast water tanks is logically referred to as *ballast water*. The ship can discharge the ballast water back into the sea when needed, but this discharge may occur in regions far away from the sea where the ballast water was originally taken in.

The biodiversity, ecology and marine biology of different sea areas may not be the same, different sediments (containing biological organisms) and species existing in each of them. Some of these organisms and species (including their eggs) can be very invasive and harmful when introduced into a new marine environment. An example of such an invasive species is the Chinese wool-handed crab, which, as the name suggests, originates in the South China Sea, but today is found in a number of European waters, including in Denmark.[73]

The BWM Convention seeks, as laid down in article 2(1), to 'prevent, minimize and ultimately eliminate the transfer of Harmful Aquatic Organisms and Pathogens through the control and management of ships' Ballast Water and Sediments'. The BWM Convention regulates, in short, a ship's discharge of ballast water through two standards: D-1 and D-2. Those standards are incorporated into the Annex to the Convention, which 'forms an integral part of this Convention' pursuant to article 2(2). This provision also states that, 'Unless expressly provided otherwise, a reference to this Convention constitutes at the same time a reference to the Annex.'

The first standard, the *D-1 Standard* (regulation D-1) provides, pursuant to regulation B-4.1, that all ships covered by the regulation may only discharge ballast water on the high seas or in areas over 50 nautical miles (nm) from any coast, in waters with a depth over 200 metres or, pursuant to regulation B-4.2, in specified areas designated for the discharge of ballast water.

Ships covered by the second standard, the *D-2 Standard* (regulation D-2), must ensure, pursuant to regulation B-5 and regulation B-3.3, that the ship 'treats' the ballast water before discharging it, which means that the ballast water is cleansed of any sediments, etc.

There are several different technologies that can be applied to meet the D-2 Standard, including systems that expose the ballast water to UV-lighting, which kills off organisms living in the water. All ships covered by the regulation must meet the D-2 Standard from 8 September 2024, pursuant to regulation B-3.8.[74]

[73] The Chinese wool-handed crab has, for example, been observed in streams near the city of Aunslev and in Nyborg Fjord, see https://www.fyens.dk/indland/Frygtet-kinesisk-krabbe-fundet-i-oestfynsk-aa/artikel/2158594 and https://www.tv2fyn.dk/artikel/uoensket-uldhaandskrabbe-har-indtaget-nyborg-fjord.

[74] Reg B-3, including reg B-3.8, and the entry into force in 2024 were amendments to the Annex of the BWM Convention made by Resolution MEPC.297(72) on Amendments to the International Convention for the Control and Management of Ships' Ballast Water and Sediments, 2004, adopted on 13 April 2018.

Those ships must also have a 'Ballast water management plan' (regulation B-1) and a 'Ballast water record book' (regulation B-2), plus an *International Ballast Water Management Certificate* (regulation E-2–E-5), which must be presented during a PSC, as stated in article 9(1).

As the BWM Convention relates to the *discharge* of ballast water, the different principles and conclusions discussed in Part II of this book apply to the enforcement of this Convention, including on the high seas, as the D-2 Standard also will prohibit discharges of untreated ballast water in these areas.

Coastal States can, in principle, invoke article 220(5)–(6)and the first exception in article 228(1), as sediments potentially can cause major damage to the ecosystems of a State.[75] But it must nonetheless be considered unlikely that a coastal State will be able to prove that a specific violation of the BWM Convention resulted in specific damage, that is, that a specific invasive species was introduced to the marine environment after a specific ship discharged its ballast water. It may be recalled that the general enforcement principles in articles 218(1) and 220(1) have no such requirement. The mere infringement of the BWM Convention is sufficient to invoke those jurisdictions for port States and coastal States.

It should be noted in this context that article 16 of the Convention explicitly refers to the BWM Convention's respecting the principles of UNCLOS and international law in general.

IV. The Anti-Fouling Systems Convention

'Fouling', in this context, refers to unwanted marine organisms, sediment, animals (algae and molluscs), etc that become attached to a ship's keel below the waterline. This fouling can reduce the aerodynamic function of the hull, thereby increasing its water resistance, thus requiring more engine power to achieve a certain speed, which increases the ship's fuel consumption.

'Anti-fouling system' (AFS) is a term used to describe material, often a special coat of paint or chemicals, that is applied to a ship's keel, which prevents fouling from occurring, or at least makes it more difficult. Article 2(2) of the Convention on the Control of Harmful Anti-fouling Systems on Ships ('the AFS Convention')[76] defines an 'anti-fouling system' as being 'a coating, paint, surface treatment, surface, or device that is used on a ship to control or prevent attachment of unwanted organisms'.

The AFS Convention regulates which anti-fouling systems can and cannot be used, as some systems (paints, etc) contain substances that are harmful to the marine environment. Such systems can slowly release those substances into the sea while a ship is sailing, thereby poisoning and polluting the marine biodiversity.

[75] It is not relevant to discuss the application of art 233, as art 42(1)(b) only refers to the discharge of oil and noxious substances into a strait.

[76] See n 6 above.

Enforcement of the AFS Convention is therefore covered by the regulatory scope of part XII of UNCLOS.[77] It is worth noting, with regard to this, that article 3(1)(a) of the AFS Convention requires flag States, which are party to the Convention, to ensure that ships under their flags comply with its rules and regulations; while article 3(1)(b) requires participating coastal States to enforce the Convention within their waters, as it refers to ships operating 'under the authority of a party', for example in the territorial sea; likewise for port States within their ports in accordance with article 3(1)(c).

Drawing a parallel with other IMO conventions and article 236 of UNCLOS, the AFS Convention does not apply to State-owned ships (see article 3(2)).

Under article 4(1), flag and coastal States have a duty, in accordance with article 3(1)(a) or (b), to ensure that no ship installs, applies or uses an AFS that does not meet the requirements for such systems set out in Annex 1 to the AFS Convention. Ensuring that ships do not install, apply or use an unlawful AFS during their stay in port pursuant to article 3(1)(c) is explicitly referred to in article 4(1)(b) as the responsibility of port States. The jurisdiction of port States is linked to PSC and the NMFT principle codified in article 3(3).

Article 10 provides that ships must have an International Anti-fouling System Certificate, proving that the ship complies with the AFS Convention. Regulation 1.1. of Annex 4 to the AFS Convention states that this Certificate is required for vessels of 400 GT and above. The Certificate must be presented as part of a PSC (see article 11(1)(a)).

Article 11(1)(b) specifies that is possible for Port State Control Officers (PSCOs), during the initial part of a PSC, to take a sample of the AFS from the ship to ensure that it complies with the Convention. This means that it is not necessary for a PSCO to have *clear grounds* before taking a sample, as the sample is taken as part of the initial PSC inspection and not as part of a detailed PSC. However, article 11(2) refers to more thorough inspections of the AFS where there are clear grounds, that is, as part of a detailed PSC inspection. This can, in accordance with article 11(3), lead to detainment of the ship until such surveys are completed, and possibly until the test results from any survey and/or sampling are returned.

Articles 6–9 of the AFS Convention contain provisions relating to the need for continued scientific research to determine if an AFS is harmful, and whether the use thereof should be banned or restricted. For example, it was discussed at PPR6 in 2019 whether the substance *cybutryne* should be banned.[78]

When determining the applicability of the Part II conclusions to the enforcement of the AFS Convention, it must first be determined what a violation comprises, that is, whether it is a discharge, dumping or something else. The IMO refers to toxic AFSs slowly 'leaching' into the sea from the hull of ships.[79] It must therefore initially be concluded that the term 'dumping', and thus article 216 of

[77] IMO, 'Implications of the United Nations Convention on the Law of the Sea for the International Maritime Organization' (LEG/MISC. 8, 30 January 2014) 57.

[78] See at http://www.imo.org/en/MediaCentre/MeetingSummaries/PPR/Pages/PPR-6th-Session.aspx.

[79] See at http://www.imo.org/en/OurWork/Environment/Anti-foulingSystems/Pages/Default.aspx.

UNCLOS, has no relevance pertaining to enforcement of the AFS Convention, as this (the term 'leaching') falls outside the definition in article 1(1)(5)(a)(i)–(ii) of UNCLOS.

Synonyms of 'leach' include, according to the Oxford Electronic Thesaurus, 'discharge' and 'leak'.[80] It can therefore be assumed that a violation of the AFS Convention may, in principle, be enforced in accordance with the different conclusions set out in Part II, including using articles 218(1) and 220(1). However, a coastal State would probably not be able to invoke article 220(5)–(6) and the first exception in article 228(1) of UNCLOS, as the continued, slow release (leaching) of an AFS from a specific (violating) ship's hull into the sea cannot potentially cause specific significant or major damage to the coastal State.[81]

An interesting perspective on anti-fouling has been discussed within the IMO at MEPC level regarding the possible mandatory use of AFSs to avoid 'biofouling'.[82] As mentioned initially, fouling impairs the aerodynamic function of the hull, which increases water resistance, resulting in the need for more engine power and greater fuel consumption. Fouling that is released from the hull can contain invasive species, which, like ballast water, may contaminate a new sea area. It has therefore been discussed whether ships should be required to use AFSs, provided these comply with the AFS Convention. An IMO Guideline was adopted as early as 2011 addressing this issue – MEPC Resolution MEPC.207(62), Guidelines for the control and management of ships' biofouling to minimise the transfer of invasive aquatic species.

If such regulations on biofouling were to be adopted, they would, depending on their final wording, probably be enforceable in accordance with the conclusions in Part II of this book, as the result of not complying would be the discharge of more harmful substances into the air (due to increased fuel consumption) and into the sea (due to the continued release of sediment from the hull).

V. The London Convention

The Convention for the Prevention of Marine Pollution by Dumping of Wastes and Other Matter[83] is often referred to as 'the London Convention', as it was adopted in London in November 1972.

Article I of the London Convention states that its purpose is to

> prevent the pollution of the sea by the dumping of waste and other matter that is liable to create hazards to human health, to harm living resources and marine life, to damage amenities or to interfere with other legitimate uses of the sea.

[80] See at https://en.oxforddictionaries.com/thesaurus/leach.

[81] It is not relevant to discuss art 233 and art 42(1)(b), as the AFS Convention does not pertain to discharge of oil or noxious substances.

[82] See at http://www.imo.org/en/OurWork/Environment/Biofouling/Pages/default.aspx.

[83] See n 7 above.

The participating parties to the Convention must, in accordance with article II, take steps to prevent illegal dumping.

Article III(1)(a)(i)–(ii) of the London Convention, as previously described, contain a positive definition of what the term 'dumping' includes – a definition almost identical to the definition of 'dumping' in article 1(1)(5)(a)(i)–(ii) of UNCLOS.

Article IV(1)(a) provides that it is the responsibility of all States party to the Convention to prohibit the dumping of any waste and other substances in any form or mode, if these substances or waste are covered by Annex 1 to the London Convention.

The dumping of types of waste or substances listed in Annex 2 to the Convention requires a *prior special permit* from coastal States (see article IV(1)(b)). All other forms of wastes and substances, not covered by Annex 1 or Annex 2, merely require general permission from the coastal State for dumping these, in accordance with article IV(1)(c).

The factors that (coastal) States must consider before granting permission in accordance with article IV(1)(b) or (c) are listed Annex 3 to the Convention. It is a requirement under article VI that a (coastal) State must designate a special authority for issuing such permits.

Article VII requires the participating parties take the necessary steps to ensure compliance with the London Convention. Article VII(1)(a) stipulates that flag States must ensure that the requirements of the Convention are met by ships under their flags. Article VII(1)(b) obliges States to ensure that the requirements of Convention are fulfilled by foreign ships dumping waste and hazardous substances in ports or at offshore terminals. All States must also, in accordance with article VII(1)(c), enforce the provisions of the Convention within their territory.

Situations of *force majeure* may exempt ships from complying with the London Convention (see article V).

Needless to say, violations of the London Convention cannot be considered discharge violations; they are 'dumping violations', and the two terms do not overlap (see, for example, the positive and negative demarcations of the term 'discharge' found article 2(3)(a)–(b) of the MARPOL Convention). Violations of the London Convention are therefore not covered by the different enforcement principles discussed in Part II of this book. Violations are instead covered by article 216 of UNCLOS, which was briefly described in chapter 9 and in section II.C of this chapter regarding enforcement of MARPOL Annex III.

Dumping violations can be made the subject of enforcement measures by coastal States pursuant to article 216(1)(a), by flag States in accordance with to article 216(1)(b) and by port States in accordance with article 216(1)(c).

Article 216(1)(a) of UNLCOS refers to the coastal State's enforcement of international rules on dumping (ie the London Convention) within its territorial sea, EEZ or on the *continental shelf*. It will be recalled from chapter 9 that the clear reference to the *continental shelf*, without any specified restrictions, must allow coastal States that have extended their continental shelf to 350 nm from the baseline, in accordance with article 76(4)–(8) of UNCLOS,[84] to exercise jurisdiction under article 216(1)(a) over that entire 350 nm area.

A. The London Convention's Interface and Overlap with Other IMO Regulations

It should be noted that the regulations in the London Convention sometimes overlap with other IMO regulations, such as regulation 8.1 of MARPOL Annex III pertaining to the wilful dumping of hazardous substances in packed form into the sea. The London Convention must supersede these other regulations from a *lex specialis* standpoint, which also is provided for in regulation 16.5 of MARPOL Annex VI regarding the use of ship incinerators. This provision specifies that in the event of any overlap between regulation 16 and the London Convention, the latter prevails.

However, sometimes regulations might not directly overlap but merely supplement one another. For example, regarding the applicability of the London Convention and the Nairobi Convention.[85] There can be several interfaces at which these two conventions might overlap, for example where containers are lost at sea, which is a subject that also has been discussed at IMO level.[86] If a container is deliberately dumped at sea then it is covered by the London Convention. If, on the other hand, a container is unintentionally lost at sea, for instance during a storm, then this matter is covered by the Nairobi Convention, as it would be encompassed by the definition of a 'wreck' in article 1(4)(a)–(d) of that Convention, that is:

(a) a sunken or stranded ship; or
(b) any part of a sunken or stranded ship, including any object that is or has been on board such a ship; or
(c) any object that is lost at sea from a ship and that is stranded, sunken or adrift at sea; or
(d) a ship that is about, or may reasonably be expected, to sink or to strand, where effective measures to assist the ship or any property in danger are not already being taken.

A container, or other equipment or cargo on board, would be considered an 'object that has been on board' ((b)), as well as an 'object that is lost at sea from a ship' ((c)).

A coastal State that is party to the Nairobi Convention may, if a wreck (eg a lost container) constitutes a hazard or danger to marine traffic or the environment, require that the wreck be located, pursuant to article 7 of the Convention. It may also require that the wreck be *marked*, in accordance with article 8, if this

[84] See ch 3 for more on art 76 and the delimitation of the continental shelf.
[85] See n 7 above.
[86] See http://www.imo.org/en/MediaCentre/HotTopics/container/Pages/default.aspx.

is necessary. Finally, the wreck can also be required to be *removed* under article 9, if this is crucial for safety and/or for environmental reasons. A shipowner can, in accordance with article 10(1) of the Nairobi Convention, be held liable for the costs that follow the application of the measures listed in articles 7–9.[87]

VI. The Hong Kong Convention on Ship Recycling

The Convention for the Safe and Environmentally Sound Recycling of Ships[88] is often referred to as 'the Hong Kong Convention', as it was adopted at a conference held in Hong Kong from 11–15 May 2009. The Hong Kong Convention has, at this point in time, not yet entered into force, article 17 providing that it will do so two years after the three cumulative conditions in article 17(1)(1)–(3) have been fulfilled.[89]

It is the purpose of the Convention to ensure that ships that are decommissioned are scrapped in a safe, sustainable and environmentally-sound manner. Some ships, particularly larger merchant ships, are often scrapped at a yard or near a beach. The concept of *scrapping* covers ships' being dismantled piece by piece, which can be a dangerous process for the people who do the work as well as for the environment, as the ship could contain chemicals, oils and other residues.

The Hong Kong Convention does not apply to government ships or ships under 500 GT (see article 3(2)–(3)).

Article 4(1) of the Hong Kong Convention requires flag States to ensure that ships flying their flags adhere to its provisions, which, pursuant to article 5, must be demonstrated by the surveying and certification of all ships covered by the regulations. A flag State must maintain a list covering all certified ships and submit this list to the IMO annually in accordance with article 12(4). Chapter 2 of the Annex to the Convention (containing the 'Regulations for safe and environmentally sound recycling of ships') sets out detailed requirements for ships and shipowners.

[87] An interesting overlap could also occur between MARPOL Annex III and the Nairobi Convention if a container containing noxious substances were unintentionally lost at sea, resulting in the contents' leaking/escaping from their packaging. The two regulatory instruments will in this case supplement each other, as this could be considered a violation of reg 8.2 of MARPOL Annex III, incurring legal responsibility in accordance therewith, but also constitute a violation of the Nairobi Convention, which would allow the relevant articles (arts 7–10) to be invoked as regards locating, marking and potentially removing the 'wreck' (container) at the expense of the shipowner.

[88] See n 8 above.

[89] These three conditions require that: (i) 'not less than 15 States have either signed it without reservation as to ratification, acceptance or approval, or have deposited the requisite instrument of ratification, acceptance, approval or accession in accordance with Article 16'; (ii) 'the combined merchant fleets of the States mentioned in paragraph 1.1 constitute not less than 40 per cent of the gross tonnage of the world's merchant shipping'; and (iii) 'the combined maximum annual ship recycling volume of the States mentioned in paragraph 1.1 during the preceding 10 years constitutes not less than 3 per cent of the gross tonnage of the combined merchant shipping of the same States'.

Article 4(2) requires all participating States to ensure that any ship recycling yards within their territory, at a minimum, meet the conditions laid down in the Convention, including the specific requirements found in chapter 3 of the Annex.

Article 12(4) and regulation 11.11 of the Annex provide that a ship that has been surveyed and certified by the flag State (or by a classification society) and found to be in compliance must be issued with an International Ready for Recycling Certificate, which the flag State reports to the IMO. This Certificate must be presented during PSCs (see article 8).

It is explicitly stated in article 9(3) that port States may detain or refuse access to ships that do not fulfil the conditions of the Convention, including being able to present an International Ready for Recycling Certificate. A detention may not be longer than necessary, however, following article 11. This is in alignment with the principles of (PSC) detention found in article 219 and article 226(1)(b) and (c) of UNCLOS, as described in chapter 4 of this book. A detention that violates these detention principles in UNCLOS and the Hong Kong Convention can be subject to the expedited judicial procedure of article 292 of part XV of UNCLOS, as described in chapter 12.

Determining *enforcement* of the Hong Kong Convention in accordance with part XII of UNCLOS can be somewhat complicated. First, it must be stated that the conclusion in Part II of this book, relating to enforcement in response to emission violations by port States and coastal States, is not relevant in this context, as ships do not discharge (or dump) any polluting substances if they violate the Hong Kong Convention whilst still in commission (sailing).

When the ships cease to operate and are being dismantled at a yard or on a beach, they likewise cease to be a vessel covered by the enforcement principles of part XII of UNCLOS. Ships, in service, that do not meet the conditions of the Hong Kong Convention are subject to the previously described principles of PSC pursuant to articles 219 and 226 of UNCLOS. Also, the obligation for flag States to survey and certify ships under their flag must be an obligation covered by article 217(2)–(3). If States that are party to the Convention fail to meet the requirements contained in the Convention, for example as a flag State to survey, certify, register and inform the IMO or as a State with a shipwrecking facility within its territory, that State can be held liable according to article 235 and the principles of State responsibility described in chapter 12. However, certain interfaces with the London Convention and the Nairobi Convention should briefly be underlined.

If a vessel is deliberately 'dumped' (sunk) at sea to avoid complying with the Hong Kong Convention, this is not covered by this Convention but is instead a violation of the London Convention, given the definitions of 'dumping' in article 1(1)(5)(i)–(ii) of UNCLOS and article III(1)(a)(i)–(ii) of the London Convention, which (at (ii)) refers to 'any deliberate disposal of vessels, aircraft, platforms or other man-made structures at sea'. This can subsequently be enforced in accordance with article 216 of UNCLOS.

A ship that accidentally sinks at sea is covered by the Nairobi Convention, relating to whether the wreck is to be found, marked and removed. Again, this situation is not covered by the Hong Kong Convention.

VII. The Intervention Convention

Going from studying enforcement of the Hong Kong Convention to examining enforcement of the Convention Relating to Intervention on the High Seas in Cases of Oil Pollution Casualties ('the Intervention Convention'[90]) is like time-travelling through the history of the IMO's environmental legislation. From a convention not yet in force to a convention that was adopted in 1969[91] and which entered into force in 1975.

Embedded in the Convention is also an amending protocol that was adopted as early as 1973, pertaining to a coastal State's right to intervene on the high seas in the event of accidents involving dangerous substances and materials. The regulatory scope of the Intervention Convention thereby overlaps with MARPOL Annexes I–III regarding harmful substances (oil and chemicals).

The Intervention Convention and the Protocol give coastal States special extra-territorial jurisdiction to take necessary measures on the high seas against foreign ships, to prevent or reduce a substantial risk of pollution of the Coastal State with oil or hazardous substances. See, for example, article I(1) of the intervention Convention.

The scope and jurisdiction of the Intervention Convention is thus limited to situations where a *grave and imminent danger to the coastline* is present and it is a *result of an accident at sea*. The coastal State must also contact the ship's flag State and the shipowner, and consult with experts nominated by the IMO *before* making an intervention, pursuant to article III.

The Convention does not apply to foreign State ships or warships, in accordance with article I(2).

Article V of the Intervention Convention requires that any precautionary measure taken by a coastal State must be proportionate, which includes being necessary and not going beyond the purpose of the intervention. A coastal State can become liable in accordance with article VI if it does not adhere to the requirement for proportionality set out in article V. A coastal State that takes such precautionary measures is also under obligation to inform all relevant parties thereof, including the ship's flag State and the IMO (see article III).

The conclusions set out in Part II do not apply to enforcement of the Intervention Convention as it relates to pollution, or preventing pollution, which stems from an accident (collision, stranding etc) at sea. However, enforcement of the

[90] See n 9 above.

[91] The loss of the tanker *Torrey Canyon* in 1965 prompted the development of the Intervention Convention.

Intervention Convention may be achieved by invoking article 221 of UNCLOS, which, as described in chapter 9, provides coastal States with an extraterritorial jurisdictional basis for enforcing laws and regulations for the protection of the coastal State's coastlines and interests beyond the territorial sea, including on the high seas.

The wide-ranging geographical scope of the article is, as also stated in chapter 9, curtailed by the reference to its covering pollution that is a consequence of a maritime accident, which in article 221(2) of UNCLOS is defined as a 'collision of vessels, stranding or other incident of navigation, or other occurrence on board a vessel or external to it resulting in material damage or imminent threat of material damage to a vessel or cargo'.

Article 221 does not therefore entail any wider jurisdiction for coastal States to enforce MARPOL Annex VI, or IMO regulations in general, on the high seas. But article 221 does provide a specific extraterritorial basis for coastal States to enforce the Intervention Convention outside their territory, including on the high seas.

VIII. Conclusion

This chapter has sought to apply the enforcement principles that were discussed in Part II of the book to enforcement of other relevant environmental IMO regulations, that is, the other regulations of MARPOL Annex VI, MARPOL Annexes I–V, the BWM Convention, the AFS Convention, the London Convention (including its ties to the Nairobi Convention), the Hong Kong Convention and the Intervention Convention.

It must initially be noted that many (or all) of these IMO instruments contain requirements to keep special certificates, etc on board that document compliance with the different regulations. These are certificates that may be presented during the initial part of a PSC. If these mandatory certificates are not presented or if they are incorrect, this can lead to detailed inspections, detention and fines, as described in chapter 4. All these Conventions also codify the principle of NMFT.

These formal requirements are not the focal point, however, when examining whether the conclusions reached in Part II are applicable to violations of the material requirements of these Conventions, including determining whether those violations can be described as *discharge* violations, or *dumping* or something else.

It was recalled that the enforcement principles set out in Part II related to the special port and coastal State jurisdictions pursuant to articles 218(1) and 220(1), and how those jurisdictions are resolved in accordance with article 228(1) when they overlap with a flag State's jurisdiction under article 217. It was then established that MARPOL Annex VI, regulation 12 (ODSs), regulation 13 (NOx), and regulations 16.2 and 16.3 (incineration of PCB or PVC materials) can be enforced in accordance with the conclusions in Part II.

Violations of MARPOL Annexes I–II and regulation 8.2 of Annex III can also be acted upon in accordance with the conclusions reached in Part II, and coastal

States may invoke article 220(5)–(6), article 233 pursuant to article 42(1)(b) and the *first exception* in article 228(1), as a violation of MARPOL Annexes I–III can result in major damage to a coastal State.

Violations of regulations 8.1 of Annex III are dumping violations that are enforced in accordance with article 216 of UNCLOS.

Infringements of MARPOL Annex IV and Annex V can also be the subject of enforcement measures in accordance with the conclusions in Part II, but they do not allow coastal States to invoke article 220(5)–(6) of UNCLOS, nor the first exception in article 228(1), as it is unlikely that the potential damage from one specific violation will result in major damage to a coastal State. Article 233 does not apply, as article 42(1)(b) only refers to the discharge of oil and noxious substances.

Violations of the D-2 Standard of the BWM Convention can be enforced in accordance with the conclusions reached in Part II of this book, but it is doubtful whether a coastal State could prove that major damage had occurred as a result of a specific violation from one specific ship, rendering the application of articles 220(5)–(6)and 228(1) (first exception) very unlikely. The same conclusion applies to enforcement of the AFS Convention.

Violations of the London Convention are considered dumping violations, and therefore enforcement is in accordance with article 216 of UNCLOS and not articles 218 or 220. Where a ship, container, etc is accidentally lost at sea is, however, that is not considered to involve *dumping* and is therefore covered by the Nairobi Convention.

A ship's violation of the Hong Kong Convention is not covered by the Part II enforcement principles, but if a State fails to ensure that ships under its flag or ship recycling facilities on its territory comply with the Convention, it can result in that State's becoming liable in accordance with article 235 of UNCLOS. Ships that are intentionally sunk at sea to avoid compliance with the Hong Kong Convention are covered by the London Convention.

The enforcement principles covered in Part II do not apply to enforcement of the Intervention Convention, but it can be enforced by coastal States in accordance with article 221 of UNCLOS.

This chapter has also sought to highlight some of the issues that are under discussion at IMO (MEPC/PPR) level, and which might result in future regulations on, for example, a ban on HFO in the Arctic, grey water, marine litter (plastics in oceans), biofouling and containers lost at sea.

The IMO is also, as described in section I.F on the enforcement of regulations 19–23 of MARPOL Annex VI, working on creating and adopting new regulations to reduce GHG (CO_2) emissions from ships. This work is being undertaken to achieve the overall GHG reduction goals set out in the IMO's GHG Strategy from 2018.

Several legal candidate measures to meet these reduction goals have been discussed, for example in relation to adopting rules on *speed optimisation*, use of *on-shore power*, use of *alternative fuels*, etc. Each legal measure may present its own possibilities and challenges for enforcement in accordance with the provisions of part XII of UNCLOS, and in accordance with the conclusions set out in Part II of this book.

The GHG Strategy and the different legal measures, including the potential enforcement of these through UNCLOS, are analysed in further detail in chapter 15.[92]

[92] Part IV (chapters 16–18) of this book examines whether future IMO rules on GHG could be considered legislation that represents international norms of a *jus cogens* character.

15

Enforcement of Candidate Measures Relating to the IMO GHG Strategy

On 13 April 2018 the International Maritime Organization (IMO) adopted an initial strategy for reducing greenhouse gas (GHG) emissions from ships on a global scale, particular CO_2.[1] The strategy was adopted by Resolution MEPC 304(72) on Initial IMO Strategy on Reduction of GHG Emissions from Ships.[2]

The GHG Strategy is, as the name of the Resolution suggests, an *initial* strategy, which is set to be revised at MEPC 80 in 2023,[3] but this chapter will refer to it as the 'IMO GHG Strategy', the 'GHG Strategy' or simply 'the Strategy'. Any references to any new IMO GHG Strategy adopted at MEPC 80 are to the 'revised GHG Strategy'.

The IMO GHG Strategy sets out several reduction goals that the shipping industry must meet in the coming decades. How these goals are to be achieved is not finally decided upon in the Strategy, but it presents several different 'candidate measures', which have been singled out by the IMO, Member States and organisations, etc as being feasible regulatory solutions for reducing GHG emissions from ships.

These IMO candidate measures are categorised in the GHG Strategy as being short-, mid- or long-term candidate measures, depending how fast they could be adopted, as some measures, for example alternative fuels (zero-carbon fuels or fossil-free fuels, etc) are not sufficiently developed at this point in time to form the basis of regulations for their obligatory use.[4] The GHG Strategy defines possible 'short-term candidate measures' as measures that could be 'finalized and agreed by the Committee between 2018 and 2023'; 'mid-term candidate measures' are measures that could be 'finalized and agreed by the Committee between 2023 and 2030'; and 'long-term candidate measures' are measures that could be

[1] This work on developing and adopting the GHG Strategy was done in the IMO's Marine Environmental Protection Committee (MEPC).

[2] IMO GHG Strategy, in Resolution MEPC 304(72) on Initial IMO Strategy on reduction of GHG emissions from ships (adopted 13 April 2018), available at http://www.imo.org/en/OurWork/Documents/Resolution%20MEPC.304%2872%29%20on%20Initial%20IMO%20Strategy%20on%20reduction%20of%20GHG%20emissions%20from%20ships.pdf.

[3] ibid point 6.

[4] Such fuel would also have to be available in sufficient amounts before its use could be made mandatory.

'finalized and agreed by the Committee beyond 2030'. Once such regulatory measures enter into force they would have to be enforced in accordance with international law, including UNCLOS[5] where applicable.

The IMO GHG Strategy refers directly in point 1.5.1 to the Strategy and measures' falling within a broader legal context, which includes 'other existing instruments related to the law of the sea, including UNCLOS'. It also refers in point 3.2.1.1 to application of the principle of 'no more favourable treatment' (NMFT) as it is 'enshrined in MARPOL[6] and other IMO conventions'. It is therefore essential to examine how some of these proposed short-, mid- and long-term measures potentially could be enforced in accordance with the provisions of part XII of UNCLOS and through Port State Control (PSC), including determining if some of the different enforcement principles discussed in Part II of this book apply. This chapter will address this after setting out an overall introduction to the IMO GHG Strategy in section I. It will also be discussed, in section III, whether the IMO's adoption of the regulatory GHG measures, including the long-term measures on the use of alternative fuel, eventually could lead to a complete ban (prohibition) on the use of fossil fuels.

I. The IMO GHG Strategy

The final content of the GHG Strategy had been discussed and negotiated into place at previous MEPC meetings leading up to its adoption at MEPC 72, including at three Intersessional Working Group (ISWG) meetings. The report from the third ISWG, held in the week before MEPC 72, formed, to a considerable extent, the basis for the text and principles adopted in the final Strategy, including as regards the proposed short-, mid- and long-term candidate measures.[7]

The GHG Strategy was developed since the Paris Climate Agreement[8] did not include shipping. This led to work on an IMO GHG Strategy that would seek to achieve reduction levels consistent with the goals of the Paris Agreement,[9] pursuing the global temperature goal of limiting/stabilising the increase in global temperatures (global warming) to well below 2.0°C, and aiming to further limit

[5] United Nations Convention on the Law of the Sea (adopted 10 December 1982, entered into force 16 November 1994) 1833 UNTS 3.

[6] International Convention for the Prevention of Pollution from Ships (adopted on 11 February 1973, as modified by the Protocol of 17 February 1978, entered into force 2 October 1983) 1340 UNTS 61 (MARPOL), IMO Publication: IMO-520E.

[7] See *Report of the Third Meeting of the Intersessional Working Group on Reduction of GHG Emissions from Ships* (ISWG-GHG 3).

[8] The Paris Climate Agreement was adopted at the 21st Conference of Parties (COP21) on 12 December 2015 by 196 States as an agreement within the United Nations Framework Convention on Climate Change (adopted 9 May 1992 entered into force 21 March 1994) 1771 UNTS 107 (UNFCCC), available at https://unfccc.int/process-and-meetings/the-paris-agreement/what-is-the-paris-agreement.

[9] See point 1.7 of the IMO GHG Strategy.

the increase to 1.5°C compared to pre-industrial levels.[10] Any greater increase in global temperatures could have severe adverse effects on the environment and, if global warming is not stopped, for the possibility of the environment's sustaining human life.[11]

The IMO GHG Strategy in point 3.1 sets out specific goals for reducing the global release of GHG emissions from ships, to help achieve this global limitation/stabilisation of global warming:

- The cornerstone of the Strategy is the overall goal to *reduce GHG emissions by a minimum 50% by 2050*, compared to the GHG emissions level in 2008.
- The Strategy also singles out the reduction of CO_2 emissions by setting a *40% reduction goal per transport work, as an average across international shipping, by 2030*,[12] pursuing efforts to achieve a 70% reduction by 2050. These goals are also measured against the 2008 level.
- Finally, the GHG Strategy also explicitly refers to the *implementation of further Energy Efficiency Design Index (EEDI) phases for new ships* as a goal to reduce 'carbon intensity' (ie reduce CO_2 emissions).

The GHG Strategy is also, in part, a consequence of the results of three major GHG studies that the IMO initiated, and which were presented in 2000, 2009 and 2014. These GHG studies, inter alia, revealed that even though shipping is the form of transportation that emits the lowest levels of GHG compared to the quantities of goods that are transported over vast distances, the shipping industry nonetheless accounts for a substantial part of all GHG, particular CO_2, emitted on a global scale. The latest study estimated that commercial shipping accounts for over 2.2% of the total global release of CO_2, and that it is ships of 5,000 GT or more that are responsible for 85% of this (2.2%) release.[13]

It should be noted that point 6 of the GHG Strategy refers to a fourth GHG study's being undertaken in 2019 (using data collected from 2012–18), with publication set for the latter part of 2020 at MEPC 76. This is stated as being part of the follow-up actions to be taken with a view to developing a *revised GHG Strategy*,

[10] The term 'pre-industrial levels' is not clearly defined in the Paris Agreement. For more on this, see https://theconversation.com/what-is-a-pre-industrial-climate-and-why-does-it-matter-78601.

[11] This is also the premise behind Part IV of this book, exploring if the principles of jus cogens and erga omnes would apply if it were to be established that the stabilising of the increase in global temperatures to 1.5°C had failed, in part due to lack of enforcement of internationally adopted GHG regulations, ie the regulatory measures adopted by the IMO in accordance with its GHG Strategy.

[12] 'Transport work' is a term/unit used to establish how much CO_2 was released when a ship transported a certain amount of cargo over a certain distance. This is achieved as a calculated result by multiplying the distance travelled by the amount of cargo carried. This shows how effective, or ineffective, a transport was regarding the amount of cargo carried compared with how much CO_2 was released. It is this efficiency that the international shipping industry must, on average, improve by 40% by 2030, pursuing efforts towards 70% reduction by 2050.

[13] See points 1.3.3 and 1.6 of the IMO GHG Strategy. See also, for further information, http://www.imo.org/en/OurWork/Environment/PollutionPrevention/AirPollution/Pages/GHG-Emissions.aspx and http://www.imo.org/en/OurWork/Environment/PollutionPrevention/AirPollution/Pages/Default.aspx.

which is set to be adopted at MEPC 80 in 2023. This is specified in point 6.2 of the Strategy, which contains a *road map* for adopting a revised strategy.

It is also stated in point 6 of the GHG Strategy that the revised GHG Strategy should include the adoption of short-, mid- and long-term regulatory measures in accordance with the previous implementation schedule, meaning that the short-term candidate measures should, at the latest, be adopted and enter into force as part of a revised IMO strategy in 2023.

The mid- and long-term candidate measures should also be addressed in a revised strategy in a similarly revised implementation plan setting out the different regulatory measures and how these might be implemented, and when the regulations (in the form of new conventions or amendments to existing legislation) implementing these might be adopted and enter into force.

II. Enforcement of the Proposed Short-, Mid- and Long-term Measures

The short-, mid- and long-term candidate measures that potentially may form the basis of new IMO regulations are listed in point 4 of the GHG Strategy.

It is crucial, when developing such regulatory measures, to ensure that the regulations can be enforced in accordance with UNCLOS, as explicitly referred to in point 1.5.1 of the GHG Strategy, in order to achieve the environmental benefits from the adopted regulations. Or rather, to avoid the adverse environmental impacts that would follow if global warming were to continue to rise.

This can be compared to enforcement of the sulphur regulations in MARPOL Annex VI as described in chapter 1, and it would seem logical that everyone, including shipowners, would seek to comply with such regulations, as the environmental advantages, be it reduced amounts of sulphur oxide (SOx) or a reduction in GHGs, are beneficial to all. However, as is the case with the sulphur regulations, shipowners will probably have a huge economic incentive not to comply with some of these proposed short-, mid- and long-term regulatory measures, as they would come at a prohibitive cost, in the form, for example, of higher fuel costs for alternative (zero-carbon) fuels, or excessive costs in building new, compliant ships or having existing ships undergo major conversions to make them compliant. Increased costs for the installation of new equipment, the training of crew and more administrative burdens pertaining to new demands for documentation and certificates are all also relevant cost factors.

All these additional costs can influence a shipowner's decision to comply with such new regulations or not, especially as it must be expected that *open registry flag States*[14] will continue to offer protection under their flags to non-compliant

[14] As stated in ch 1, section VII.B, 'open registry flag States' are sometimes also referred to as offering 'flags of convenience'.

shipowners. It is therefore important to establish whether the short, mid- and long-term candidate measures proposed in the GHG Strategy can be enforced by non-flag States, such as coastal States and port States, in accordance with part XII of UNCLOS.

In this context it is important to establish, as the discussions in Part II and chapter 14 showed, what obligation a regulatory measure exactly imposes, for example whether a violation thereof can be considered a *discharge violation*. Each potential short-, mid- and long-term regulatory measure of the GHG Strategy must therefore be evaluated separately.

It should be noted that not all short-, mid- and long-term regulatory measures proposed in the GHG Strategy will be studied in this section, as some of these measures are unlikely to result in regulations to which shipowners must adhere and for breach of which they will be penalised. Some of the proposed measures rather encourage cooperation and incentivise further developments on a voluntary basis, for instance through cooperation in capacity building or through the Integrated Technical Cooperation Program (ITCP) to assist developing countries.

Market-based measures to incentivise voluntary reduction of GHG emissions, and instituting research and development (R&D) activities and funds, are also measures that do not necessarily place explicit regulatory obligations on ships and shipowners. Agreements and voluntary guidelines implementing such non-regulatory measures will likely not be enforced in accordance with the provisions and principles of part XII of UNCLOS, so they do *not* form part of the following analysis. Only the short-, mid- and long-term regulatory measures proposed under point 4 of the GHG Strategy, which potentially may result in actual IMO rules and regulations for breach of which shipowners can be sanctioned, are examined in the following subsections.

But before examining these measures individually, it must generally be underlined that such provisions on GHG will be covered by part XII of UNCLOS as they regulate harmful, manmade substances that are 'likely to result in such deleterious effects as harm to living resources and marine life, hazards to human health', pursuant to the definition of 'pollution of the marine environment' found in article 1(1)(4) of the Convention.[15] This is based on GHGs causing global warming, resulting in 'deleterious effects' to the environment and directly and indirectly to human health, as described in chapter 17 in Part IV of this book.

A. Short-term Candidate Measures

Point 4.1.1 of the GHG Strategy stipulates, as previously noted, that the short-term candidate measures should be finalised by MEPC between 2018 and 2023.

[15] Y Tanaka, 'Regulation of Greenhouse Gas Emissions from International Shipping and Jurisdiction of States' (2016) 25 *Review of European, Comparative and International Environmental Law* 337.

Point 6 also refers to such measures' being adopted at MEPC 80 in 2023, which, given the wording of point 4.1.1, must be an absolute deadline for adoption of short-term measures. It could more accurately be said that these short-term measures should, at the latest, be adopted at MEPC 77 in 2021, to account for the 16-month implementation period following final IMO (MEPC) adoption before entry into force. This primarily relates to the implementation of further EEDI phases for new ships, which also addresses the third goal of the GHG Strategy.

Other short-term candidate measures listed in point 4.7 of the GHG Strategy are also studied in the following subsections, including: speed optimisation and speed reduction; further measures to address emissions of volatile organic compounds (VOCs); provisions pertaining to shore-side/on-shore power (cold ironing) and port responsibilities (optimisation of the logistic chain and planning, including 'Just in Time Arrival').

i. Implementation of Further EEDI Phases for New Ships

As discussed in chapter 14, the EEDI specifies the energy efficiency target a new ship is expected to meet, based on the ship's specifications.

There are three phases (phases 1–3) setting EEDI requirements, which continually increase in strength pursuant to regulation 21 of MARPOL Annex VI. 'Phase 0' sets the baseline from which these continual improvements are measured.

Phase 3 is set to enter into force on 1 January 2025, but the demand for adoption of a new, stronger EEDI phase (phase 4) before 2023 would lead to the need for phase 3 to enter into force at an earlier time (2022[16]). This would allow the results of improved energy efficiency, meaning lower emissions of CO_2, to be attained at an earlier stage. This early entry into force of phase 3 will not influence the enforcement possibilities (or lack thereof) described in chapter 14. In fact, these findings and conclusions can also be applied to a new EEDI phase 4 (and further phases for that matter), as the content of these regulations – improvement of the energy efficiency of ships specified in percentages compared to the baseline – remains unchanged.

Such new EEDI requirements implemented in a phase 4 would be enforced in accordance with the conclusion in the previous chapter on enforcement of regulations 19–23 of Annex VI, including that the *causal link* between a violation of EEDI requirements and the excess amount of CO_2 released as a result is too weak to establish that a *discharge violation* has occurred in accordance with article 218(1) of UNCLOS. See the analysis of this in chapter 14, section I.F.

However, port States will always be able to enforce the EEDI requirements on a formal basis through PSC by examining whether the ship has the mandatory documentation on board, including a Ship Energy Efficiency Management

[16] An MEPC-instituted correspondence group on 'EEDI review beyond phase 2' was tasked with developing a proposal to move phase 3 forward to 2022.

Plan (SEEMP) and a statement of compliance in accordance with regulation 10.3 of MARPOL Annex VI and article 226 of UNCLOS. Flag States are, in accordance with article 217(3) of UNCLOS, under obligation to ensure that such documentation is in order.

So, even though it is one of the most important and 'simple'[17] short-term candidate measures pertaining to reducing CO_2 emissions from shipping, this measure still presents some difficulties when looking at enforcement through UNCLOS, since some 'entrepreneurial' shipowners and open registry flag States in conjunction will undoubtedly concoct ways to circumvent the regulation and/or falsify the data and documentation needed to show compliance with chapter 4 of Annex VI. Effective document control during PSCs will be the best solution for non-flag States to counter such attempts.

ii. Speed Optimisation and Speed Reduction

'Speed optimisation', often also referred to as 'speed reduction', is a short-term candidate measure that relates to IMO rules setting a regulatory maximum for a ship's speed.

Several proposals have been discussed pertaining to how such a short-term measure might be transposed into legislation. The most favoured solutions have proposed either an 'absolute maximum speed' or an 'average maximum speed'.[18] The absolute speed maximum sets a speed limit (as seen on roads for cars) by which ships are required to abide. The average maximum speed, as the name suggests, sets an average speed limit to which ships must adhere, either *per voyage* or *per year*.

The reasoning behind such requirements is that there is a direct connection between a ship's speed and its release of CO_2. The higher the speed, the higher the level of release of CO_2. For instance, studies have shown that reducing a ship's speed by 10% could result in a corresponding 27% reduction of the ship's demand for power,[19] thus also lowering the ship's fuel consumption.[20]

Setting an average limit might accommodate the needs of ships and shipowners more than an absolute limit, as it gives ships more flexibility to operate and adjust their speed when desirable in a given situation. For example, if adverse weather

[17] The word 'simple' is used in a legal context to indicate that the adoption of this measure would merely require a change of existing IMO regulations (MARPOL Annex VI). It does not therefore include the political, economic and other interests that might complicate or prolong such adoption of a phase 4 EEDI.

[18] See at https://www.transportenvironment.org/sites/te/files/publications/Joint%20industry_NGO%20speed%20letter%20to%20the%20IMO.pdf.

[19] J Faber, T Huigen and D Nelissen, *Regulating speed: a short-term measure to reduce maritime GHG emissions*, Report prepared for CE Delft, Publication code: 17.7L90.155 (2018) 5, available at https://www.cedelft.eu/en/publications/download/2399.

[20] It is noted that the IMO guidelines for the SEEMP, describe in ch 14, already refer to 'speed optimization' as a method to improve the energy efficiency of a ship.

conditions require this, or if the ship needs to make a scheduled call into port. Yet from an enforcement point of view, the *absolute maximum speed* might be easier for non-flag States to enforce, as it could be argued that there is a direct correlation between a ship's violating an absolute maximum speed and the excess release of CO_2 that results from this infringement. If this excess release of CO_2 could be calculated, it would establish the necessary *causal link* between the violation and the excess *discharge* of CO_2, which would allow the enforcement paradigm described in Part II to apply, including assertion of jurisdiction on the high seas by port States in accordance with article 218(1) of UNCLOS.[21]

This conclusion would also apply to an average maximum speed *per voyage*, as port States and coastal States would, relatively easily, be able to calculate and prove (beyond any reasonable doubt) a violation of such an obligation by using Automatic Identification System (AIS) data, which continuously show a ship's route and speed during the entire voyage (ie each leg).

However, if an average maximum speed *per year* were to be adopted, meaning that over the course of a year a ship must stick to an average limit (quota), it might lead to enforcement difficulties for non-flag States. It would be logical to model such legislation on the provisions found in regulations 22 and 22.A of MARPOL Annex VI, on how ships (shipowners) must report data showing their compliance on an annual basis to the flag State, which in return must verify the data and afterwards issue a certificate or document of compliance. This would, in a perfect world, not present a problem, but if some (open registry) flag States are *unable* or *unwilling* to verify such data effectively (but still issue the certificate), it would be challenging for non-flag States to assert jurisdiction over such infringements. The initial part of a PSC would likely, once again, prove to be the best method for enforcing such regulation when inspecting all relevant documentation pursuant to article 226(1)(a) (ii) of UNCLOS.

The exception might be if the IMO, or the different MOUs, through the use of AIS data, were able to track and calculate the speed of all ships covered by the regulation continuously, as this would enable port States to take action against ships that violate the permissible yearly average but whose infringements are not met with dissuasive sanctions from their flag States.

Finally, it should be noted that a third regulatory measure has been discussed pertaining to speed optimisation as a short-term measure. *Power limitation* is a concept the regulatory focus of which is on how much 'power' a ship needs to create, and therefore how much fuel it needs to combust, to attain a certain speed. This approach would, like the principles behind the EEDI regulation, accommodate

[21] The flag State would still have primary jurisdiction over such violations in accordance with art 217 and art 228(1) of UNCLOS, unless it fails to meet the requirements of the latter article for invoking the main rule or the second exception in the provision applies (discussed in ch 11). Such a violation would not, however, allow coastal States to invoke art 220(5)–(6) and the first exception of art 228(1), as a single violation would not result in tangible, measurable major damage to the coastal State.

newer and more energy-efficient ships, as they have effective engines that need less power, and therefore use less fuel, to reach and maintain a certain speed.[22] This reduces the amount of CO_2 released in comparison to other, less effective ships sailing at the same speed. Such regulation could serve as an incentive for shipowners to ensure that their fleets are well maintained or replaced with more effective, and thereby more environmentally friendly, ships.

Enforcement of such regulation by non-flag States would be feasible, as it would be relatively uncomplicated[23] to examine the engine during PSCs to establish whether it corresponds with any formal engine documentation, and to use this information in conjunction with AIS data from the ship's previous voyage regarding its speed.

A violation of rules on power limitation could therefore be regarded as a discharge (emission) violation, as a sufficient causal link between the violation and the excess release of CO_2 could be established. The principles set out in Part II of this book would apply, such as port States' exercising extraterritorial jurisdiction over such a violation pursuant to article 218 of UNCLOS.

iii. Further Improvements to the Existing Regulation of Methane and VOCs

The IMO GHG Strategy refers in point 4.7.5 to a possible short-term candidate measure, being to 'consider and analyse measures to address emissions of methane and further enhance measures to address emissions of Volatile Organic Compounds'.

If these considerations result in the adoption of further, enhanced measures to reduce VOC emissions, including methane, it would most likely be done through a strengthening of the existing provisions in regulation 15 of MARPOL Annex VI. It is therefore also presumed that the strengthened regulation would be modelled on the same requirements that, as described in the previous chapter, relate to VOCs that can be generated from carrying liquid cargo on board a ship, and the chance of these VOCs being released during the loading and unloading of that cargo.[24] The regulation would thus focus on ensuring that the relevant ships,

[22] Such power limitation regulations would probably, like the EEDI requirements, differentiate between types of ship.

[23] The words 'relatively uncomplicated' are used in this context to underline that Port State Control Officers (PSCOs), as described in ch 4, are people with a professional maritime background who also receive continual training. This training could therefore have a heightened focus on examining engines to a degree that would allow PSCOs to detain a vessel for further, detailed (engine) inspections if they believe that there are *clear grounds*, pursuant to art 226(1)(a)(i) of UNCLOS, indicating a mismatch between the formal description of an engine's parameters in the ship's documentation and the actual state of the engine.

[24] Many tankers use inert gas to secure cargo tanks. When the inert gas is discharged, it can simultaneously result in the discharge of VOCs.

predominantly tankers, have equipment onboard that can collect VOC gases and deliver them to an approved reception facility in port. Such ships must also carry a flag State-approved *VOC management plan* and a *written procedure* pertaining to minimising VOC emissions on board. These obligations could be strengthened, relating both to the requirements ships must meet and to the requirements ports must fulfil as regards having adequate reception facilities, matters that also fall to the (port) State to ensure and approve.

The enforcement of such a short-term candidate measures, which change (strengthen) existing IMO regulations on VOC, must therefore predominately rest on the same enforcement principles as are set out in the previous chapter regarding regulation 15 of MARPOL Annex VI. It should be recalled that although VOCs could be released in to the atmosphere at any time, including on the high seas, the content of regulation 15 focuses on the practical and technical aspects of VOCs' being released in port during the loading or unloading of cargo, so that it is considered more likely that a violation of this provision would be judged to have occurred while the ship is in port or at an offshore terminal. A port State would therefore, pursuant to article 2(1) and article 230 of UNCLOS, have unrestricted territorial jurisdiction to sanction any violations that result in a measurable release of VOCs while the ship is in port, as this occurs (the infringement is committed) within internal waters. The port State will also, as part of a PSC and in accordance with article 226, be able to ensure compliance with the regulation.

Methane (CH4) is also characterised as a VOC. However, it is also the main component of Liquefied Natural Gas (LNG), which is a natural gas that can be used by ships for propulsion. It can also be cooled down to a liquid state for storage or transport.

Methane is singled out in point 4.7.5 of the GHG Strategy alongside VOCs in general, as it has severe adverse effects on the environment.[25] If this is to be understood as an intention to regulate methane separately in the same way as other substances in MARPOL Annex VI, for example SOx and nitrogen oxide (NOx,) it could change the applicable enforcement paradigm. If special regulations are adopted on the release of methane while sailing, and violations thereof could be proved through the use of technical appliances, etc, it could lead to the conclusions reached in Part II of this book becoming applicable. However, if such methane regulation were to mimic the VOC rules of regulation 15 in Annex VI, focusing on any release that occurs during loading or unloading (either as cargo or fuel), then the above-mentioned PSC enforcement principles for VOC (also discussed in chapter 14) would need to be utilised.[26]

[25] Methane has a very high potential for global warming, approximately 21 times higher than CO_2, which makes it one of the worst GHGs. See at http://apps.sepa.org.uk/SPRIPA/Pages/SubstanceInformation.aspx?pid=65.

[26] See at https://www.lngworldshipping.com/news/view,lng-fuel-and-the-ship-emissions-debate_53722.htm.

iv. Shore-side Power and Port Responsibilities

When a ship is at berth in port, it often continues to run its engine, perhaps an auxiliary engine, to generate power for operating equipment, appliances, etc on board. This could include operating communications, safety and security equipment, machines for the cooling of cargo, and machines for the loading and unloading of cargo. Power is also essential for the crew and passengers living on board, for instance for illuminating and heating or cooling rooms, and in general to operate and charge electronic devices, such as televisions and computers. Passenger ships in particular can require huge amounts of power to service numerous facilities, rooms, etc on board while in port.[27]

While the ship is running its (auxiliary) engines in port, it burns fuel, leading to the discharge of emissions, including sulphur and GHG (CO_2) emissions. The IMO GHG Strategy thus refers, in point 4.7.8, to the need for development of provisions on the use of *shore-side/on-shore power*. Utilising this form of power enables ships to operate their equipment, appliances, living spaces, etc on board with electricity supplied from shore, allowing them to shut down their engines and not emit any GHGs or sulphur. This is also the reason behind the requirement in the EU Sulphur Directive, as described in chapter 5, that ships at berth for over two hours in EU ports use 0.1% fuel or 'shore-side electricity'.[28] Such an operation, where a ship at berth in port receives electrical power from shore through large cables, is often referred to as 'on-shore power supply' (OPS), or 'shore-to-ship power' (SSP) or 'alternative maritime power' (AMP). However, the popular term for the use of on-shore electricity is 'cold ironing'.[29]

The use of on-shore power is often more expensive for ships (shipowners) than combusting fuel.[30] It can also entail practical difficulties, including the risk of delaying the ship, as attaching and removing the large electrical cables can be very time-consuming.[31] Thus, there may be several reasons why shipowners are reluctant to switch to on-shore power.

[27] Some (port) States have already implemented national rules on so-called 'cold ironing', which is possible provided the requirements set out in art 211(3) of UNCLOS, and described in ch 7 of this book, are followed. The term 'cold ironing' has its origins in the time where many ships used coal as a fuel. When such a ship was in port, its iron engines would cool down, 'cold ironing' describing how these cold iron engines needed to be warmed up again before the ship departed from port. California's Air Resources Board has adopted requirements for the mandatory use of shore power (cold ironing) for ocean-going vessels at berth in port, available at https://ww3.arb.ca.gov/ports/shorepower/shorepower.htm. See, eg, the port at Long Beach, at http://www.polb.com/environment/air/shorepower.asp.

[28] Directive (EU) 2016/802 of the European Parliament and of the Council of 11 May 2016 relating to a reduction in the sulphur content of certain liquid fuels [2016] OJ L132/58, art 7(1)–(2).

[29] See n 27.

[30] See at https://www.porttechnology.org/technical_papers/the_economics_of_cold_ironing.

[31] International standards have been developed by international standardisation organisations (ISO, etc) to ensure the unified, effective and safe connection of shore power; shore grids standards have been developed, eg standard IEC/IEEE DIS 80005-1 for high voltage systems and standard IEC/PAS 80005-3 for low voltage systems.

Although the IMO has not yet regulated cold ironing in terms of requiring the use of on-shore power, it (MEPC) has adopted a Resolution on OPS,[32] in which the attention of all States is drawn to the environmental problems associated with ships running their engines while at berth. The Resolution also contains a global list of ports that offer land-based power to larger ships.

Adopting IMO regulations on cold ironing, in accordance with the GHG Strategy, will not be easy, as many practical obstacles must be overcome beforehand, including ensuring compatibility between different ship types and different power units in different countries. Also, it will be necessary to ensure the swift attachment and removal of power cables without relaxing safety, and to meet the requirement of the Strategy for the on-shore power (electricity) to come from *renewable sources*.

If, or when, such regulations are adopted, this may also present certain enforcement challenges as to whether it is the ship or the port that fails to meet the regulatory cold-ironing obligations. First, *ships* that do not comply with a requirement to use on-shore power while in port may be sanctioned by the port State as the violation is committed in port within internal waters, where a (port) State has unlimited territorial jurisdiction pursuant to article 2(1) of UNCLOS. The port State can also, in accordance with article 230, impose non-monetary penalties, which, in principle, could include imprisonment, as described in chapter 12.

Second, *ports* that are located within the jurisdiction of States that become parties to these IMO regulations, would also have to comply with any requirements for them to supply ships at berth with on-shore power. Failure to do so could be an infringement by the ports. Irrespective of their being independent legal entities (private companies) or government-owned, ports are solely subject to the jurisdiction of the State in which they are located.

Conversely, this means that other States cannot sanction a port's violation of its obligation to make available the required on-shore power options.[33] Only the (port) State can penalise a port for not complying with IMO regulations, provided the State has implemented the IMO regulations into national law, thereby obliging its ports to comply with them.

A State that voluntarily has become a party to such IMO (GHG) regulations must implement these in its national legislation pursuant to article 211(2) of UNCLOS, and it must also, in accordance with the requirement in article 26 of the Vienna Convention on the Law of Treaties (VCLT),[34] ensure that these rules are complied with, sanctioning any violations.

If a State, which is party to the IMO regulation on cold ironing, fails to implement the IMO regulation or to enforce this in ports located within its territory

[32] See MEPC 1/Circ 794 of 9 October 2012.

[33] The same would apply if a port (State) that is party to MARPOL Annex VI failed to meet the obligations in the Annex for having the required adequate reception facilities for receiving VOCs (reg 15), or ozone-depleting substances (ODSs) or discharge water from closed loop scrubber systems (reg 17).

[34] Vienna Convention on the Law of Treaties 1969 (adopted 22 May 1969, entered into force 27 January 1980) 1155 UNTS 331 (VCLT).

(jurisdiction), this could be a violation of the State's obligations to implement and enforce pursuant to UNCLOS and/or the VCLT. This might lead to other participating States' invoking principles of State responsibility in accordance with international law, including article 235 of UNCLOS and the ILC Articles on State Responsibility[35] as described in chapter 12, in response to the (port) State's failure to abide by its commitment.

It should be noted that point 4.7.8, besides referring to the use of shore-side power (cold ironing), also specifies that ports should develop measures to 'further optimize the logistic chain and its planning'. This could be a link to the concept of 'Just In Time Arrival', ' often abbreviated to 'JIT Arrival'.

JIT Arrival addresses situations where merchant ships are forced to anchor and wait outside a port until the port has an open slot and is ready to receive the ship, where it can unload and load cargo, bunker fuel, etc as planned. Ships can wait outside ports for hours, days or perhaps even weeks, which causes long delays and severe economic losses for shipowners. Such delays are also damaging to the environment, as ships must continue to use their engines, perhaps auxiliary engines, while at anchor to support the necessary operations and functions on board the ship, some of which were noted previously. However, unlike ships at berth in port, the use of on-shore power (cold ironing) is not a possibility for ships at anchor outside ports. Those ships must use their engines to generate power, which creates harmful emissions of different polluting substances, such as GHGs and sulphur.

The principles of JIT Arrival dictate that ports must seek to ensure that the port's infrastructure and logistics, and the planning of ship arrivals and departures, etc, are organised in the most effective and realistic way to minimise any delays. If the IMO were to adopt regulations on JIT Arrival, requiring ports to ensure that certain regulatory measures were followed to reduce the harmful GHG emissions from ships anchored outside ports, the same enforcement principles just stated to apply to cold ironing would also apply here. This means that only the (port) State in which the port is located could sanction any non-compliance with JIT Arrival regulations, but other States that were party to the same regulations could hold that (port) State accountable for not enforcing.

B. Mid-term Candidate Measures

Point 4.1.2 of the GHG Strategy emphasises that the mid-term candidate measures should be finalised by the MEPC between 2023 and 2030. Those mid-term measures are set out in point 4.8 of the Strategy. Some of them focus on developing market-based measures to incentivise GHG emissions reduction, promote

[35] International Law Commission, *Draft Articles on Responsibility of States for Internationally Wrongful Acts*, November 2001, Supplement No 10 (A/56/10), chp IV.E.1, available at https://www.refworld.org/docid/3ddb8f804.html.

and enhance technical cooperation, and develop feedback mechanisms on lessons learned from the implementation of measures. These are all candidate measure that are unlikely to entail any rules for breach of which shipowners can be penalised.

i. Implementation of Programme for the Effective Uptake of Alternative Low-carbon and Zero-carbon Fuels

Point 4.8 also lists an 'implementation programme for the effective uptake of alternative low-carbon and zero-carbon fuels' as a mid-term measure. The use of alternative (zero-carbon) fuel, etc is, in principle, a very important regulatory measure, as it could potentially represent one of the most effective solutions for the shipping industry to achieve the greatest GHG/CO_2 reduction and thereby meet the GHG reduction goals of the Strategy. Nonetheless, the mid-term measure described in point 4.8.1 refers to the development and implementation of programmes for such alternative fuel types, which by itself does not include requirements for ships to use such fuels. Adopting provisions on alternative (zero-carbon) fuel types, etc is nevertheless described as a long-term measure in point 4.9.1 of the GHG Strategy, which is therefore examined in section II.C.

ii. Strengthening Operational Energy Efficiency Measures for New and Existing Ships

One mid-term candidate listed in point 4.8.2 refers to creating 'operational energy efficiency measures for both new and existing ships'.

From an UNCLOS enforcement point of view, this measure introduces no new regulatory concepts compared to the principles already described in chapter 14 regarding enforcement of the regulations in chapter 4 of MARPOL Annex VI and those described in this chapter pertaining to the adoption of new EEDI phases as a short-term candidate measure. There is therefore no need to examine the enforcement of this proposed mid-term regulatory measure, as it relates to issues already clarified.

C. Long-term Candidate Measures

Point 4.1.3 of the GHG Strategy underlines that the long-term candidate measures 'could be measures finalized [by the MEPC] beyond 2030'.

Point 4.9 proposes two long-term candidate measures. The first long-term measure (4.9.1) pertains to developing provisions on 'zero-carbon' or 'fossil-free fuels'. The second measure (4.9.2) encourages the general adoption of 'other possible new/innovative emission reduction mechanism(s)'.

Although the latter long-term candidate measure potentially could result in the adoption of new regulations, the reference to 'possible new/innovative emission reduction mechanism(s)' is too vague to establish exactly what this measure regulates. It is likely that the revised GHG Strategy in 2023 will elaborate on this measure, but the current wording of point 4.2.9 is not precise enough to discuss any enforcement of it in accordance with UNCLOS. This section will therefore focus on the enforcement of future regulatory measures adopted in accordance with the long-term candidate measure set out in point 4.9.1 of the Strategy, that is, enforcement of future IMO regulations on the mandatory use of *alternative fuels* in the form of *zero-carbon* fuels or *fossil-free* fuels.

i. Regulations on Using Alternative Fuel Types: Zero-carbon or Fossil-free Fuels

One of the key measures for achieving the 50% GHG reduction goal by 2050, as set out in the IMO GHG Strategy, is to make a transition within the shipping industry from using fossil fuel to using fossil-free fuel, ideally zero-carbon fuels (either decarbonised fuel, or fuel types that contain no carbon), especially if further GHG reductions are needed to keep global warming to well below 2.0°C and to further pursue a maximum increase of 1.5°C in alignment with the goals of the Paris Climate Agreement.

Several general aspects will have to be addressed before such provisions on the use of alternative (zero-carbon) fuels can come into force, including ensuring their safety, availability, reliability, etc. This could, it is hoped, be tackled as part of the mid-term candidate measure pertaining to the implementation of a programme for the effective uptake of alternative low-carbon and zero-carbon fuels (section II.B.i).

Other issues relating to these alternative fuel types will also have to be dealt with individually, for example which *generation* of biofuel (a fossil-free fuel) should be used, as some forms of biofuel are extracted from plants or crops that could have been used for food, or which were grown in areas that could have been used for farming edible crops. The production of some biofuels may also leave a heavy *CO_2 footprint* depending on how the biofuel material is grown, refined and transported, and from the fact that the material (plant mass, etc) will not contribute to absorbing CO_2 from the atmosphere once harvested.[36]

It is also a problem that biofuels, compared to other alternative (zero-carbon) fuels, actually have a CO_2 output when combusted, although much smaller than that resulting from combusting fossil fuels, such as oil, irrespective of whether this fossil-based fuel oil is low sulphur fuel or not.

This section will therefore refer to *zero-carbon fuels* as the most viable long-term candidate measure, provided they are manufactured and transported in ways that do not deplete the benefits gained from using them.

[36] See further at https://www.nationalgeographic.com/environment/global-warming/biofuel/.

IMO regulations on the mandatory use of such zero-carbon fuels could potentially lead to some of the same problems that are seen, and described in chapter 1, regarding enforcement of the sulphur limits in MARPOL Annex VI, as such alternative (zero-carbon) fuels could be vastly more expensive than fossil fuels, giving shipowners an incentive not to comply. Enforcement of this GHG regulation through UNCLOS could therefore become vital to achieving the reduction goals of the Strategy.

It must initially be assumed that such IMO regulations on zero-carbon fuel would include formal requirements for ships to have specific documentation proving their compliance. For example, certificates issued by the flag State (or a classification society on the flag State's behalf) could show that the particular ship (its engines, etc) is able to use zero-carbon fuels and thereby comply with the rules, which would be flag State-issued certificates covered by article 217(3) of UNCLOS, as described in chapter 8. Another document proving compliance would be a new form of Bunker Delivery Note, showing that the fuel purchased and bunkered was zero-carbon compliant. Relevant certificates would evidently be part of the documentation that should be presented during the initial part of PSCs and enforced by port States in accordance with article 226 of UNCLOS.

Ensuring that sufficient amounts of zero-carbon fuel would be available in a State or region could also be a requirement imposed upon the participating States, following the principle in regulation 18.1 of MARPOL Annex VI.

Lack of compliance with such IMO GHG regulations stipulating the use of alternative (zero-carbon) fuels would be deemed a *discharge violation* in accordance with the conclusions reached in Part II of this book, including the application of article 218(1), following the arguments and interpretation set out in chapter 10. So as a result, port States would be allowed to assert extraterritorial jurisdiction over such violations that occur outside its territory, including on the high seas, pursuant to article 218(1) of UNCLOS.[37]

Knowing this, a port State could consequently exercise full jurisdiction over the infringement (end-to-end enforcement) as described in Part II, which allows the port State to sanction the entire violation. Every flag State would still have primary jurisdiction in accordance with article 217, unless it failed to correctly invoke the main rule of article 228(1), or if the port State could invoke the *second exception* in the provision, as discussed in chapter 11. Repeated disregards by a flag State to ensure effective compliance with such regulations would lead to the second exception applying. Deeming whether a flag State has enforced effectively could presumably also rest on some of the same principles previously discussed, including ensuring that violations are penalised so that all economic gains are removed, and a dissuasive punitive element is imposed.

[37] Proof of non-compliance with such regulations, including on the high seas, could be established by using many of the different means of detection described in ch 1 regarding enforcement of the sulphur regulations, including the use of fuel testing, Continuous Emission Monitoring Systems (CEMSs), sniffers attached to drones, etc and satellite data. See section III.A of this chapter for more on this.

It should be recalled, in this context, that coastal States would likewise have full jurisdiction over violations of such IMO regulations on alternative (zero-carbon) fuels taking place within their territory pursuant to article 220(1) of UNCLOS, including in the EEZ, as described in chapter 9. Given that the release of GHGs can be considered air pollution, this would also allow violations in internal or territorial waters to be covered by article 222, but as discussed in chapter 7, article 220(1) is the primary provision for enforcing such regulations on air pollution. Using article 220(1) requires the ship to have voluntarily called into a port or at an offshore terminal in the State after the violation. A coastal State will have few or no grounds for stopping, detaining and prosecuting a ship in accordance with article 220(5)–(6) if the ship does not go into port, as a ship's excess release of GHG/CO_2, much like emissions of SOx, ODSs, etc, does not result in tangible or measurable major damage to the coastal State. This confirms that a coastal State is barred from invoking the *first exception* in article 228(1) in such situations.[38]

Violations of future IMO regulations demanding the use of zero-carbon fuels, or other alternative fuels, could thus be enforced in accordance with the enforcement paradigm set out in Part II of this book, allowing non-flag States (port States and coastal States) to ensure compliance by foreign ships when those vessels call into a port or at an offshore terminal in the States, irrespective of where the violation occurred.

III. IMO Prohibition on the Use of Fossil Fuels

This section will focus on the IMO's adoption of a specific GHG regulation that is not imbedded in the GHG Strategy. That is, a global prohibition (ban) on the use of fossil-based marine fuels.

This, although not directly mentioned in the IMO Strategy, would be the natural outcome of well-implemented and well-enforced regulatory measures of the Strategy, as these would initially limit the release of GHGs from shipping, until the use of fossil fuels is finally halted by using alternative (zero-carbon or fossil free) fuels. A ban on fossil fuels in shipping would be the endgame of the IMO GHG Strategy.[39]

Discussing adoption of IMO regulations on alternative (zero-carbon or fossil-free) fuels and discussing the adoption of an IMO regulation prohibiting the use of fossil fuels may seem like two sides of same coin. However, a prohibition on the

[38] Art 233 and art 42(1)(b) are not relevant to the discussion in this context, as the latter provision pertains to the discharge of oil and noxious substances in an international strait.

[39] The IMO's adopting such a prohibition on the release of a particular substance would not be unheard of, recalling from ch 14 that, for example, reg 12.3.1 of MARPOL Annex VI completely bans the release of non-hydrochlorofluorocarbon gases, including freon and chlorofluorocarbons (CFCs), on all ships constructed on or after 19 May 2005.

use of fossil fuels leaves the door open for shipowners to decide how they wish to fulfil this requirement. This could obviously be by using alternative fuels, but it could in theory also be achieved by applying other solutions not described in the GHG Strategy. Such solutions could include the use of solar power,[40] wind,[41] electricity (batteries),[42] etc.[43]

This makes a prohibition on the use of fossil fuels a viable alternative to (or to be used in conjunction with) the regulatory measures of the IMO GHG Strategy, since shipowners, as mentioned, would have a discretionary right to decide how to comply with the regulations.

However, despite this right for shipowners to choose their own means of compliance, it could still prove to be very expensive for them to adhere to a ban on fossil fuels, thus giving shipowners a clear economic incentive not to comply, just as with the sulphur regulation, particularly if open registry flag States provide relaxed conditions in terms of enforcement of the rules.[44]

Two points must thus be given special attention in this section, pertaining to the detection and enforcement of such a prohibition.

A. Detecting Violations of a Prohibition on Fossil Fuels

Many of the detection measures described in chapter 1, section VIII regarding detection of SOx could be adapted to detect violations of a ban on fossil fuels.

First, PSCOs could check documentation on the fuel used on board. The PSCOs could also take fuel samples and test the fuel to establish whether it is non-compliant (ie fossil-based).

Second, the ship's GHG (CO_2) emissions could be monitored by a CEMS,[45] satellites,[46] or sniffers[47] attached to drones, bridges, etc.

Third, focusing on certification of alternative fuels and creating licensing schemes for fuel suppliers could also help strengthen enforcement of a fossil fuel ban.

[40] See at http://www.climateaction.org/news/major-shipping-company-tests-solar-for-cargo-ships.

[41] See at https://www.theguardian.com/sustainable-business/2016/aug/16/shipping-emissions-low-carbon-wind-power-climate-change.

[42] See at https://safety4sea.com/are-electric-vessels-the-future-of-shipping/.

[43] This could inadvertently also limit (or completely remove) other pollution sources described in this book, such as SOx, NOx, particulate matter, depending on the solution applied.

[44] The term 'sulphur regulation' refers to the 0.5% sulphur limit that comes into force on 1 January 2020 in accordance with reg 14.1.3 of MARPOL Annex VI. See ch 1 for an analysis of this regulation and discussion of its economic aspects, as well as the role of open registry flag States (ie flags of convenience).

[45] See at https://www.atmos-meas-tech.net/4/2809/2011/.

[46] See at https://www.nature.com/articles/d41586-018-06963-4.

[47] See at https://www.atmos-meas-tech-discuss.net/amt-2019-29/amt-2019-29.pdf.

B. Enforcing a Prohibition on Fossil Fuels

The conclusions set out in Part II of this book, pertaining to enforcement of the sulphur regulation in MARPOL Annex VI through UNCLOS, would also be applicable to enforcing an IMO prohibition on fossil fuels, as the prohibition relates to the same delict (type of crime) – the *use of non-compliant fuel*. This is irrespective of the crime's being the use of fuel with a high sulphur content (over 0.5%), or of the fact that the fuel is fossil-based.

The application of the conclusions in Part II mean, in short, that:

- Flag States will be obligated to take action against (and penalise) violations of such a ban committed by ships under their flags, in accordance with articles 217 and 222 of UNCLOS.[48]
- Coastal States can take measures against violations in all parts of their national waters (including in the EEZ) if the violating ship afterwards voluntarily goes into port (see article 220(1) of UNCLOS). Coastal States will not be able to stop ships transiting the EEZ or an international strait, however, as article 220(5)–(6) and article 233 (pursuant to article 42(1)(b)) are not applicable.[49]
- Port States can assert jurisdiction over violations of the fossil fuel ban when they occur outside their territory, for example on the high seas, in accordance with article 218(1) of UNCLOS, due to an interpretation of the term 'discharge' in accordance with article 31 VCLT. Port States could also enter into agreements with other (coastal) States that allowed them to sanction any violations of the fossil fuel ban occurring in the waters of these States, as specified in article 218(2)–(4). This would also allow port States to exercise full jurisdiction over the entire violation (ie end-to-end enforcement), including sanctioning by the imposition of fines that remove all savings and impose a dissuasive penalty.[50]
- The overlapping jurisdiction between a flag State and a coastal State or a port State would be clarified in accordance with article 228(1) of UNCLOS. The flag State can thus assert primary jurisdiction over the violation, provided it meets all the criteria for invoking the main rule and provided the second exception in the provision does not apply.[51]

IV. Conclusion

This chapter has sought to establish whether some of the relevant regulatory GHG measures that are likely to be adopted at IMO level in the coming years – many of

[48] As described in chs 7 and 8.
[49] As described in ch 9.
[50] As described in ch 10.
[51] As described in ch 11.

these based on the reduction goals and proposed candidate measures set out in the IMO's (initial) GHG Strategy – could be enforced in accordance with UNCLOS, particularly since the short-term candidate measures should be finalised before a revised GHG Strategy is adopted at MEPC 80 in 2023.

It was concluded that pertaining to enforcement of new and/or improved EEDI and SEEMP regulations, these could be enforced primarily through PSC and the measures described in chapter 14 relating to enforcement of the EEDI requirements in regulations 19–23 of MARPOL Annex VI.

Other short-term candidate measures examined included 'speed optimisation and speed reduction', where the possible enforcement relies on whether such regulations adopt an absolute maximum speed (or average maximum speed *per voyage*) or an average maximum speed *per year*. The latter leaves few possibilities for non-flag States to enforce, whereas regulations covering an absolute maximum speed and an average maximum speed *per voyage* (and potential regulations on 'power limitation') allow port States and coastal States to enforce violations to a greater degree, in accordance with articles 218 and 220 of UNCLOS.

Further improvements to the existing regulations covering methane and VOCs will presumably be enforced in accordance with how the current regulation of VOCs (regulation 15 of MARPOL Annex VI) is enforced, described in chapter 14.

Foreign *ships* that violate IMO regulations on shore-side power (cold ironing) may be sanctioned by the port State where the infringement occurs. Violations by *ports* of these regulations – and possible regulations on 'port responsibilities' (including JIT Arrival) – cannot be acted upon by States other than the (port) State in which the port is located. That (port) State can, however, be subject to international liability, for example in accordance with article 235 of UNCLOS, the VCLT and the ILC Articles on State Responsibility, if it fails to implement and enforce any such GHG regulation to which it is party.

The proposed mid-term candidate measure on 'strengthening operational energy efficiency for both new and existing ships' was also briefly described, and it was concluded that the same enforcement principles applicable to new EEDI phases as a short-term measure would also apply to this.

One of the most promising regulatory measures for achieving the 2050 GHG reduction goal of at least 50% (and thereby also contributing to the much-needed reduction in global warming to well below 2.0°C, and aiming to further limit the increase to 1.5°C) is the total transition in the shipping industry away from fossil fuels to fuel types that are fossil-free, preferably zero-carbon fuels. Enforcement of the long-term candidate measure on the use of such alternative fuels was therefore considered, by this author, the most important to be analysed.

Finally, section III entailed a discussion on whether the IMO's GHG regulations could eventually lead to a complete prohibition on the use of fossil fuels.

It was concluded that both regulations on alternative fuels and a regulation implementing a prohibition on the use of fossil fuels could be enforced in accordance with the principles set out in Part II, including allowing coastal States and port States to sanction violations pursuant to articles 220(1) and 218(1) of UNCLOS. This is conditional upon the ship's subsequently calling into a port and the flag State's not being able to assert primary jurisdiction in accordance with the main rule of article 228(1), either due to the flag State's not fulfilling the necessary requirements or because the second exception in the provision applies.

A. Alternative Enforcement of GHG Rules through International Law

It is imperative that the reduction goals of the Paris Climate Agreement and the IMO GHG Strategy are reached, as the consequences of increased global warming are becoming abundantly clear, both through new scientific research and from continued news of the adverse effects of global warming, including extreme weather phenomena (droughts, floods, storms, etc), the melting of the polar ice caps, and the destruction of ecosystems and living conditions for certain species. All, basically, endangering human living conditions.[52]

The focus of Part IV of this book is on the legal possibilities within international law that, in the view of this author, may become relevant and applicable if continued scientific research suggests that the climate goals of the Paris Climate Agreement and of the GHG Strategy are *not* being achieved, or if they prove to be insufficient.[53] Chapters 16–18 therefore examine whether a future IMO regulation prohibiting the use of fossil fuels, as discussed in section III of this chapter, could be considered a *peremptory norm* of a *jus cogens* character and enforceable through the principle of *erga omnes*.

[52] Examples of such scientific research explaining current climate developments and how they affect the conditions for sustaining human life and habitation are reiterated in ch 17.

[53] For example, through a new IMO-instituted GHG study.

PART IV

Can IMO's Future GHG Regulation be Protected by Jus Cogens?

16

Jus Cogens and Erga Omnes: General Description

If the International Maritime Organization (IMO) adopts a greenhouse gas (GHG) regulation prohibiting the use of fossil fuels, as discussed in chapter 15, section III, can it be regarded as an international regulation that represents a peremptory norm of a jus cogens character? And if so, what are the consequences thereof and in light of the erga omnes principle? These, in short, are the questions that this part of the book (Part IV) will try to answer.

Applying the jus cogens principle to help protect the environment is not necessarily a new concept.[1] But these discussions have often not included any practical analysis of what sort of pollution might be deemed 'harmful enough' to be covered by the principle, nor what the legal implications of it could be, especially for non-State actors.

The intention behind this part of the book is to establish not only whether the jus cogens and erga omnes principles are applicable to an IMO regulation that protects the environment from GHG pollution by banning the use of fossil fuel, but also what the consequences would be for States and non-State actors, such as ships (shipowners), ports, etc, including examining how this might influence enforcement of this IMO GHG regulation through UNCLOS.[2]

Part IV of this book consists of chapters 16–18. This chapter provides an introduction to the legal concepts of jus cogens and erga omnes. The discussion of the erga omnes principle will also include references to the possible application of universal jurisdiction for such enforcement. The distinct legal role of ships within international law (ie the flag State principle) will also be addressed, as, in the view of this author, ships represent the perfect legal entities to link the theoretical discussions on environmental jus cogens principles with how they are applied in practice.

Chapter 17 will seek to establish that case law from the International Court of Justice (ICJ) and official conclusions from the UN's International Law Commission (ILC) clearly indicate that the jus cogens principle can encompass environmental

[1] See, eg, Special Rapporteur Dire Tladi, *Fourth Report on Peremptory Norms of General International Law (Jus Cogens)*, ILC, 71st session, A/CN.4/727 (April 2019) 61–62, para 136.

[2] United Nations Convention on the Law of the Sea (adopted 10 December 1982, entered into force 16 November 1994) 1833 UNTS 3.

protection if it relates to pollution that can threaten human existence. The ICJ and ILC are two of the most important sources of international law regarding its development, codification and application, and the conclusions and decisions of these legal bodies will have a binding effect on the international community as a whole.

Chapter 17 will also include a section referring to different scientific studies that show the deleterious effects of global warming on the environment – due to manmade GHG pollution – and how this has direct adverse effects on human health and on the surrounding conditions needed for sustaining human life. This review of the studies is necessary, as application of the jus cogens principle (and erga omnes) is reserved for protecting the most basic of human rights and for enforcing crimes of the most serious nature. A link must therefore be established between the negative scientific predictions on global warming (including the potentially adverse effects on human life) and the seriousness of violating future IMO regulations on GHG.

Finally, chapter 18 will attempt to tie together the findings of the two previous chapters, to establish whether a future IMO prohibition on the use of fossil fuel could be considered a peremptory norm of a jus cogens character and what the legal consequences of that would be, also by virtue of the regulation's being an erga omnes norm.

I. The Jus Cogens Principle

It was briefly stated in chapter 6 that the jus cogens principle represents recognised international customary law. 'Jus cogens' is a Latin term that describes 'compelling law' that cannot be set aside or waived by anyone, including by States, individuals, organisations, companies, etc. This principle can be traced back to Roman law in the form of *jus strictum* ('strict law'), which must always be adhered to.[3]

Some basic (human) rights within international law are deemed to be *peremptory norms*, which the jus cogens principle safeguards. No derogation is allowed from peremptory (jus cogens) norms.

The jus cogens principle is recognised, inter alia, in case law of the ICJ, and in the commentaries, reports, considerations, etc of the ILC, as well as in international law literature. But, most importantly, it is also explicitly codified in article 53 and article 64 of the Vienna Convention on the Law of Treaties (VCLT),[4] which links violations of the principle to practical consequences pertaining to the validity of treaties, agreements and conventions. These articles argue that no international treaty, etc may contradict jus cogens norms, relating both to existing norms (article 53)

[3] See at https://www.merriam-webster.com/dictionary/strictum%20jus.

[4] Vienna Convention on the Law of Treaties 1969 (adopted 22 May 1969, entered into force 27 January 1980) 1155 UNTS 331 (VCLT).

and to norms that have been developed and acknowledged after such treaties have entered into force (article 64). Any treaty, or part thereof, that violates such norms will immediately become void in accordance with article 71 VCLT.

This clearly shows the 'sovereignty' of the jus cogens principle, as it supersedes normally accepted principles within international law, including the principle of *pacta sunt servanda* found in article 27 VCLT, which states that the conditions of a treaty, which are accepted in a legally binding way, must be fulfilled by participating parties.

The jus cogens principle extends beyond the law of treaties and the VCLT,[5] as it also prohibits physical actions, orders, plans, etc that can result in violations of jus cogens norms.

If a jus cogens norm is violated by a State, it can result in that State's incurring 'State responsibility', as laid down in the ILC's Articles on the Responsibility of States for Internationally Wrongful Acts,[6] which were adopted in 2001. In the following paragraphs these are referred to as the 'Articles on State Responsibility' or simply as 'the Articles', depending on the context.

Articles 40 and 41 of the Articles on State Responsibility refer directly to violations of peremptory norms and how these may be considered serious breaches of international law that can result in different legal consequences. It should be noted that the ILC's commentaries on these two articles will be closely examined in the next chapter, pertaining to whether they, and thereby peremptory norms in general, could include environmental norms.

Individuals, organisations and companies can also be held liable for violations of jus cogens norms, for example individuals such as military leaders who are involved in crimes of genocide, torture, etc. The Rome Statute, which is the Statute of the International Criminal Court (ICC),[7] refers in its articles 6–8 to such violations of the prohibition on genocide (article 6), crimes against humanity (article 7), war crimes (article 8) and crimes of aggression (article 8 *bis*).

It should be noted that the Rome Statute does not explicitly refer to these being peremptory (jus cogens) norms, but the clear reference to and definition of these violations leaves no room for doubt as to whether these are covered by the jus cogens principle, as they represent quintessential peremptory norms.

[5] See, eg, this statement made by the ILC: 'It was also observed that jus cogens norms extended, beyond treaty law, to nonconventional instruments and other fields of law, such as the law on responsibility of States for internationally wrongful acts.' *Report of the International Law Commission* ILC, 69th session, A/72/10 (1 May–2 June and 3 July–4 August 2017) 196, para 163.

[6] International Law Commission, *Draft Articles on Responsibility of States for Internationally Wrongful Acts*, November 2001, Supplement No 10 (A/56/10), chp IV.E.1, available at https://www.refworld.org/docid/3ddb8f804.html.

[7] The Rome Statute of the International Criminal Court (adopted 17 July 1998, entered into force 1 July 2002) 2187 UNTS 3.

A. Norms that are Recognised as Being Jus Cogens Norms

Which international norms are of a jus cogens character is a non-static concept, as they are subject to constant evolution and the development of the international community, exemplified by article 64 VCLT, as existing treaties become void because of newly developed norms. In the eighteenth century *piracy* was deemed a violation of a peremptory norm so that all States (also by virtue of the erga omnes principle) could prosecute and punish pirates, irrespective of where the offence took place and which ships they had plundered.[8] Today, piracy should not be perceived as an action violating jus cogens norms, as the term has evolved to encompass the most basic of human rights, including the right to life.[9]

These protected rights are therefore often expressed as prohibitions on actions that violate these rights and which are therefore deemed the most severe and damaging to human life – thus being so completely unacceptable to the international community that no entity, be it a State, a company or an individual, is allowed to violate these norms. Hence the discussion in chapter 18 on whether an IMO *prohibition* on the use of fossil fuels could be considered a jus cogens norm.

Exactly which peremptory norms are covered (protected) by the jus cogens principle is often debated, as no official, exhaustive list of such norms has been codified.[10] It has been recognised by the ILC, however, in a 2019 report on the subject,[11] that, inter alia, the norms unequivocally cover prohibitions on 'crimes against humanity',[12] genocide, torture, slavery, denying the right to claim self-determination,[13] apartheid and a State's illegal use of force (ie the prohibition on aggression).[14]

The ILC has previously elaborated on this in its *Report on Draft Articles on Crimes Against Humanity* (2017)[15] – also by referencing the ICJ's application of the principle – where it (the ILC) stated that such acts are deemed 'crimes against humanity' when they are perpetrated on a *widespread or systematic basis*:

> The International Court of Justice has indicated that the prohibition on certain acts, such as torture, has the character of jus cogens, which a fortiori suggests that a prohibition

[8] E de Vattel, *Le Droit des Gens* (1758).

[9] Although piracy is no longer considered a violation of jus cogens norms, it could still be viewed as such a serious crime that States are allowed to exercise *universal jurisdiction* over it, given the provisions in arts 100–105 of UNCLOS, which allow all States to exercise extraterritorial jurisdiction on the high seas over such violations.

[10] It is noted that ILC decided, at its 67th session in 2015, to include the topic of 'jus cogens' in its work programme, and appointed Mr Dire D Tladi as Special Rapporteur on this topic. See at http://legal.un.org/ilc/reports/2017/, 192, para 142.

[11] ILC, *Fourth Report on Peremptory Norms*, n 1, pt IV, section B, 24–54 and 63, para 137.

[12] *Report of the International Law Commission Sixty-ninth session*, n 5, 22–23.

[13] The right to self-determination is referenced directly in art 1(2) of the UN Charter (*Charter of the United Nations*, (adopted 26 June 1945, entered into force 24 October 1945), 1 UNTS XVI).

[14] The prohibition on the illegal use of force by a State is directly codified in art 2(4) of the UN Charter.

[15] *Report of the International Law Commission Sixty-ninth session*, n 5, 23–127.

of the perpetration of that act on a widespread or systematic basis amounting to crimes against humanity would also have the character of jus cogens.[16]

The ILC has also indicated that the term 'crimes against humanity' could cover murder, the deportation of populations, rape, forced pregnancy and prostitution, and persecution of an identifiable group of people as a result of their political, ethnic, religious, cultural, sexual, etc affiliation.[17]

The aforementioned reference by the ILC to such crimes' being perpetrated on a *widespread or systematic basis*, for example by a government, is also important, as single cases of murder, persecution, etc, although being despicable and a violation of an individual's basic human rights, cannot be considered violations of a jus cogens (or erga omnes) norm. It is the extensive (widespread) and methodical (systematic) nature of the violation, in combination with the adverse effects caused by the violation to many people, that leads to its acknowledgement as a peremptory (jus cogens) norm.

Nonetheless, these considerations also clearly show that the jus cogens principle is constantly developing, regarding how society views certain crimes and how these are deemed more appalling by modern society than they were by previous generations.

B. Case Law of the ICJ Accepting the Jus Cogens Principle

The ICJ has accepted and referred to the existence of the jus cogens principle in its case law, including in the *Nicaragua case*,[18] where the Court stated that article 2(4) of the UN Charter, prohibiting the illegal use of force (prohibition on aggression), probably had achieved the status of being a jus cogens norm.[19]

The ICJ also referenced the principle in the *Gabčíkovo-Nagymaros project case* (1997)[20] (which is discussed in greater detail in chapter 17) and in the *Arrest Warrant case* (2002).[21]

II. The Erga Omnes Principle

'Erga omnes' is a Latin term that means 'towards all'. The erga omnes principle dictates that certain basic rights (norms) should never be violated, and all States

[16] *Report of the International Law Commission Sixty-ninth session*, n 5, 23.
[17] ibid 28, points (a)–(k).
[18] *Military and Paramilitary Activities in and against Nicaragua (Nicaragua v United States)* [1986] ICJ Rep 14.
[19] ibid para 190.
[20] *Gabčíkovo-Nagymaros project (Hungary v Slovakia)* [1997] ICJ Rep 7.
[21] *The Arrest Warrant case (Democratic Republic of Congo v Belgium)* [2002] ICJ Rep 3, 23, para 56.

are required to prevent violation of such rights. This is often referred to as an *obligation erga omnes* or *obligation owed towards the international community as a whole.*[22]

The violation of an erga omnes obligation, as described in chapter 6, allows all States to intervene to try to stop the violation. It is not conditional on the violation's having any effect on, or connection to, the State that seeks to stop the violation; it refers to stopping specific actions or omissions that are deemed damaging to the international community as a whole.

The erga omnes principle was accepted by the ICJ in the *Barcelona Traction case* (1970).[23] The Court established that the principle exists and that it protects certain fundamental rights that all States, and the international community as a whole, have a legal interest in protecting. Violations of such fundamental erga omnes obligations give all States the right to seek judicial assistance to ensure that any infringement is stopped and subsequently sanctioned. This means that all States may bring violations of erga omnes norms before an international court, such as the ICJ or the ICC.

In the *East Timor case* (1993),[24] the ICJ referred to the right of colonised people to *self-determination* as having an erga omnes character.[25]

The ICJ repeated its view on the erga omnes principles in 2004 in its Advisory Opinion in *Legal Consequences of the Construction of a Wall in the Occupied Palestinian Territory.*[26]

A. The Link between Erga Omnes and Jus Cogens Norms

The right for States to act on violations of erga omnes norms (obligations) is – to some extent – linked to violations of jus cogens norms, as the latter also relate to violations of the most basic, universal human rights. This has also been noted by other authors.[27]

The *East Timor case* stated that right to self-determination has the character of an erga omnes norm. It was previously stipulated that this right is also an accepted jus cogens norm,[28] which underlines the clear link between jus cogens and erga omnes norms, which often overlap. For example, erga omnes norms are, like

[22] UN, *Materials on the Responsibility of States for Internationally Wrongful Acts*, ST/LEG/SER B/25 (2012) 260.

[23] *Barcelona Traction case (Belgium v Spain)* [1970] ICJ Rep 3, para 33.

[24] *East Timor (Australia v Portugal)* [1995] ICJ Rep 90.

[25] ibid para 29.

[26] *Legal Consequences of the Construction of a Wall in the Occupied Palestinian Territory*, Advisory Opinion [2004] ICJ Rep 136, para 155.

[27] The reference in ch 6 is to Anders Henriksen's stating that 'In practice, erga omnes obligations will include the norms of a peremptory/jus cogens character.' See Anders Henriksen, *International Law* (Oxford University Press, 2017) 36.

[28] *Materials on the Responsibility of States for Internationally Wrongful Acts*, n 22, 260.

jus cogens norms, believed to include the prohibitions on genocide, torture, a State's illegal use of force, slavery, etc.[29]

This overlap has also been noted by the ILC, which, in the following extract, refers to jus cogens norms as *peremptory norms* and to erga omnes norms as *obligations to the international community as a whole*. The ILC states:

> Whether or not peremptory norms of general international law and obligations to the international community as a whole are aspects of a single basic idea, there is at the very least substantial overlap between them. The examples which ICJ has given of obligations towards the international community as a whole all concern obligations which, it is generally accepted, arise under peremptory norms of general international law. Likewise the examples of peremptory norms given by the Commission in its commentary to what became article 53 of the 1969 Vienna Convention involve obligations to the international community as a whole.[30]

It cannot with absolute certainty be claimed that all erga omnes norms are similarly to be considered jus cogens norms, as each norm must be individually acknowledged by the international community, for example by the ICJ or ILC. Although, it must be presumed that if a norm is acknowledged by the international community as a jus cogens norm, unless the opposite is clearly stated or indicated, that jus cogens norm will overlap with erga omnes norms. Recalling from chapter 6 that these concepts constantly change and evolve in accordance with the development and progress within the international community. Evidently it is thus logical to assume that society deems one form of violation as being so atrocious that the prohibition hereof is deemed a peremptory norm, that society will also deem the same prohibition as an obligation owed to the international community as a whole.

Nevertheless, despite the clear overlap, the two principles confer different rights and obligations on the States who intend to invoke them. Deeming a norm as a jus cogens norm entails that these norms are not to be violated and that they cannot be waived. Several explicit legal consequences may follow if a jus cogens norm is violated. Exemplified by termination of a treaty in accordance with the VCLT, a State's incurring State responsibility in accordance with the ILC's Articles on this, or an individual's being held criminally responsible in accordance with the Rome Statute.

The erga omnes principle dictates that a violation of any of these norms (obligations) gives all States the right to act and bring the matter before a court to ensure that the infringement ends and that it is sanctioned, for example through the abovementioned legal measures. This right can be invoked by all States, despite their not having any connection to, or suffering any impact from, the violation.

Thus, a *symbiotic relationship* seems to exist between the two principles, as a violation of a jus cogens norm results in the violating party's becoming legally

[29] ibid 260–61 and 262, fn 640 [884].
[30] ibid 262.

liable in accordance with international law, while the violation of the erga omnes norm (the same norm) enables all States to bring the violation before a court to have the violating party sanctioned as it has violated principles owed to the international community as a whole.

B. The Link between Erga Omnes Rights and Universal Jurisdiction

A State's right to seek to have violations of erga omnes norms (obligations) stopped, has, by its very nature, a close connection to the principle of universal jurisdiction (see chapter 6), as the common denominator of both legal principles is that the prosecuting State can react to, and assert extraterritorial jurisdiction over, certain crimes that may not relate to, or have no influence on, the State.

The erga omnes principle allows States, perhaps on a more theoretical level, to take certain legal measures when establishing that certain heinous and/or very damaging crimes have been committed. Universal jurisdiction provides the actual, practical jurisdictional basis that States can use to *prescribe* (enact) legislation prohibiting such violations of basic norms (human rights), and to *enforce* and *adjudicate upon* this legislation.[31]

If an environmental norm is considered an erga omnes norm, due to its being an obligation owed to the international community as a whole, it could indicate that any violation would also allow a State to invoke the principle of universal jurisdiction, as this jurisdiction pertains to the most heinous international crimes.[32] This premise would be in alignment with the United Nations Environment Programme (UNEP), which explicitly states that 'universal jurisdiction can play a significant role in bridging the gaps in the enforcement of international environmental law', which was supported by different States.[33] This is also supported by Matthias Reuß and Jakob Pichon, who state that the enforcement of certain international environmental rules can be carried out on the basis of the principle of universal jurisdiction They assert, inter alia:

> In general, the universality principle allows States to exercise jurisdiction for the defence of legal values that the international community considers to be particularly worthy of protection. Such values can result from the identification of international community interests, which are public goods so fundamental in nature that they are of immediate concern to all States …

[31] See, eg, *Arrest Warrant case*, n 21, paras 41–42. For further discussion of a State's right to prescribe, enforce and adjudicate, see ch 6.

[32] *Report of the International Law Commission Seventieth session*, (30 April–1 June and 2 July–10 August 2018) A/73/10 (2018) Annex, 307, para 3.

[33] Report of the Secretary-General, *The scope and application of the principle of universal jurisdiction*, A/72/112, 72nd session, item 86 (2017) 10, para 43.

> The protection of the natural environment has been widely acknowledged as an international community interest. In fact, international environmental law is one of the areas where the international community has first asserted the legal significance of international community interests. At present, numerous international treaties refer to natural resources as global 'common goods' or 'common concerns of mankind'. Such references corroborate the exceptional interest that international law attributes to the protection of the natural environment. There already exist precedents in international case law that illustrate the broad consensus regarding the qualification of environmental protection as an international community interest.[34]

The practical implications of States being able to claim universal jurisdiction over violations of environmental peremptory norms (ie the IMO's future regulation on GHGs) are described in chapter 18.

However, the special role of ships within international law is worth reiterating in this chapter, as it ties into application of the abovementioned principles of jus cogens and erga omnes, including enforcement through universal jurisdiction.

C. The Special Legal Status of Ships in the Context of Jus Cogens, Erga Omnes and Universal Jurisdiction

Enforcement, including penalisation, of violations of jus cogens norms may in general entail practical challenges, as the State that is responsible for the violation, by either its actions or its omissions, enjoys sovereignty over its territory.

Any ICJ judgment, etc that aims to stop a State's illegal activities within its own territory, for instance if the State violates the prohibition on genocide or not allowing an indigenous group of people the right to self-determination, may be difficult for the outside international community to enforce. The use of force through the UN Charter may often prove to be the last and only way to ensure this, especially if non-military actions, such as economic and other sanctions, have been ineffective.[35]

Further, if an individual is to be convicted and sanctioned (imprisoned) by the ICC for violating a jus cogens norm, that person must be apprehended, which can be a challenge if he or she is residing in a State that refuses to extradite him or her.

The same problems would, in principle, arise relating to stopping and punishing polluters, if were deemed that a regulation stopping some types of serious pollution, such as a prohibition limiting the release of GHGs, could be considered

[34] M Reuß and J Pichon, 'The European Union Exercise of Jurisdiction Over Classification Societies – An International Law Perspective on the Amendment of the EC Directive on Common Rules and Standards for Ship Inspection and Survey Organisation and for the Relevant Activities of Maritime Administrations' (2007) 67 *Zeitschrift für ausländisches öffentliches Recht und Völkerrecht* 119, 135, available at https://www.zaoerv.de/67_2007/67_2007_1_a_119_144.pdf.

[35] Art 42 of the UN Charter allows for the use of military force, if sanctioned by the UN Security Council, if non-military sanctions (pursuant to art 41) are ineffective or cannot be applied.

a peremptory norm. Polluters that violate such a regulation are often private companies, and the sources of the pollution are often land-based and stationary, such as factories, plants, etc located within the territory of a State. If that State is unable or unwilling to stop the pollution, it would again be difficult for the international community to enforce any verdict over this polluter or the State, including stopping the polluting activity, without any assistance from the State in which the source of pollution is situated. Any UN-authorised use of force can hardly be imagined as a response to such violations.[36]

However, if it were to be deemed that an IMO rule prohibiting the use of fossil-based marine fuel had the character of a jus cogens norm, it would present the international community with different legal tools to ensure that any violation by a ship (shipowner) was properly sanctioned and the pollution stopped. First, it is the very nature of merchant ships that they sail between ports located in different States. Second, the international coordinated Port State Control (PSC) regime (through the different Memorandums of Understanding (MoUs)), described in chapter 4, creates a tightly woven web through which it is difficult for ships engaged in international trade to slip.

To this end, it is also worth recalling the special principle of *no more favourable treatment* (NMFT), codified in IMO conventions, which allows all (port) States to ensure compliance by foreign ships with IMO regulations (presumably also on GHGs) when they call into a port in the State, irrespective of the ship's flying the flag of a (flag) State that is not party to the regulation.[37] It must therefore be assumed that a ship's violation of an environmental jus cogens norm would more easily be enforced compared to violations of such norms committed by other (land-based) entities.

It was concluded in Part II of this book that the widened jurisdiction of port States over foreign ships in port (article 226 of UNCLOS), and on the high seas pertaining to discharge violations (article 218 of UNCLOS),[38] provides a strong counterpart to the flag State's jurisdiction under the flag State principle codified, inter alia, in article 217 of UNCLOS. This also applies to enforcement of some of the proposed GHG candidate measures and of a ban on fossil fuels described in chapter 15. This could raise the logical question: if a future IMO regulation prohibiting use of fossil fuel can be enforced through the *normal* enforcement paradigm of UNCLOS as described in Part III of this book, why seek to establish

[36] This does not mean, however, that establishing that environmental norms might be considered jus cogens is without effect when looking to enforcement of pollution from land-based pollution sources. It would, inter alia, allow the international community to place stronger political pressure on such countries, eg imposing economic sanctions and embargoes. This would therefore also have an influence and impact on enforcement of international agreements on reducing air pollution, such as the Paris Climate Agreement, which was briefly noted in ch 15.

[37] See also the discussion in ch 4 on whether the NMFT principle can be considered a principle of customary law within the law of the sea.

[38] See the analysis in ch 10 of how the term 'discharge' in art 218(1) of UNCLOS can encompass 'emissions', including sulphur and GHG emissions.

that this IMO regulation also can be deemed to be a (peremptory) norm that could be enforced through universal jurisdiction? The answer is, in short, that this would give the international community, including port States, several unique legal possibilities (rights) and obligations to ensure compliance by all ships. This is examined more closely in chapter 18.

III. The ILC's Principles for Establishing a New Jus Cogens Norm

The previous subsections of this chapter stated that a violation of a *jus cogens norm* is presumably also to be considered a violation of an *erga omnes norm*, which often will also allow (and obligate) States to seek to stop the infringement and sanction it by virtue of *universal jurisdiction*. This leads back to the starting point of this chapter, as it underlines the importance of establishing whether certain environmental norms, for example prohibiting or limiting the release of GHGs from ships, could be considered peremptory (jus cogens) norms.

This is examined further in chapter 17, but first it must be determined which criteria in general need to be met in order for an international norm to be deemed to have obtained the status of a peremptory norm.

The ILC attempted to clarify this point in its report from 2017, in which the Commission made explicit reference to the text of a set of draft conclusions of the Special Rapporteur on this matter of peremptory norms.[39] Those draft conclusions set out, in conclusions 4 to 9, guiding principles for determining when an international norm is considered a jus cogens norm.

Draft conclusion 4 sets out two cumulative criteria for when an international norm can be considered a jus cogens norm:[40]

> To identify a norm as one of jus cogens, it is necessary to show that the norm in question meets two criteria:
>
> (a) It must be a norm of general international law; and
> (b) It must be accepted and recognized by the international community of States as a whole as a norm from which no derogation is permitted.

The first condition ((a)) may prove to be the least complicated to fulfil. A 'norm of general international law' could, in line with draft conclusion 5(4), be a provision or regulation (prohibition) codified in an international agreement, for example in an IMO convention.[41]

[39] Report of the International Law Commission Sixty-ninth session, n 5, ch VIII, 192–202.

[40] ibid 192, fn 808.

[41] ibid 193, fn 808 (continued). It is noted that draft conclusion 5(4) states that a rule found in a treaty (convention) may reflect a norm of general international law – a norm that is capable of rising to the level of a jus cogens norm.

Fulfilling the second condition ((b)), establishing that an international norm has been recognised by the international community as a jus cogens norm, is probably the more difficult condition to meet, as it leads to the questions: What is the international community? And when has it accepted a norm as jus cogens? Some indications and conditions with reference to this are listed in draft conclusions 6–9.[42] Regarding the first question ('What is the international community?'), draft conclusion 7(3) states that 'Acceptance and recognition by a large majority of States is sufficient for the identification of a norm as a norm of jus cogens. Acceptance and recognition by all States is not required.'

The writings and conclusions of the ILC and findings of the ICJ must, in the view of this author and as stated in draft conclusions 9(3) and (4), represent the best sources of international law to establish what the majority international community has and has not accepted as jus cogens.[43] This is also underlined by article 66(a) VCLT, which provides that questions of whether a norm is of a jus cogens character encompassed by article 53 or article 64 of the Convention, may be brought before the ICJ.

Also, as may be recalled from previous chapters, article 38(1) of the ICJ Statute[44] lists four recognised sources ((a)–(d)) that the Court can apply in its judgments. These are generally perceived as representing the primary sources of international law. Article 38(1)(c) refers to 'the general principles of law recognized by civilized nations'. This means that the ICJ can deem whether a principle (norm) has been recognised by 'civilized nations' (ie the international community). This includes determining if an international norm has become recognised by the international community as jus cogens norm from which there can be no derogation, which also corresponds with draft conclusion 8(2).

Given the widespread international support of the ICJ through the acceptance of the Court's Statute by 193 nations,[45] and the widespread participation by States in the work of the ILC,[46] it may be concluded that the judgments of the ICJ and the writings of the ILC represent the views of the international community. It will therefore fulfil the requirement in draft conclusion 7(3) if the ICJ and ILC both recognise an international norm as being a peremptory norm.

The next chapter will therefore, by analysing the relevant case law of the ICJ and studying the findings and conclusions of the ILC (eg in the ILC's commentaries on previous and final drafts of the Articles on State Responsibility), examine

[42] ibid.

[43] Draft conclusion 9(3) refers to 'judgments and decisions of international courts and tribunals as evidence of acceptance and recognition for the purposes of identifying a norm as a jus cogens norm', which undoubtedly includes the ICJ. Draft conclusion 9(4) refers explicitly to the work of the ILC as also providing a basis for identifying the acceptance and recognition of a jus cogens norm.

[44] The ICJ Statute is available at https://www.icj-cij.org/en/statute.

[45] See at https://www.icj-cij.org/en/states-entitled-to-appear.

[46] The reports and findings of the ILC are often presented to the General Assembly of the UN, which also speaks in favour of viewing such ILC conclusions as representing *recognition by a large majority of States*.

whether the *international community*, that is the ICJ and ILC, has accepted the possibility of environmental norms as jus cogens norms. And, if so, how these environmental peremptory norms are to be defined, referring to the requirements that must be met in order to determine what type of pollution can be covered by the norm. This will be supplemented by studying a report from the ICC prosecutor relating to the prioritisation of cases that are brought before the ICC.

IV. Conclusion

In this chapter it has been established that the jus cogens principle in general is recognised by the ICJ and the ILC.

The peremptory norms covered by the jus cogens principle include, but are not limited to, the most basic human rights, such as prohibitions on genocide, slavery, torture, the refusal of self-determination, and likely also widespread and systematic discrimination and prosecution due to political, sexual, religious, etc orientation.

If such a principle is violated, it can result in different sanctions, including the termination of treaties, State responsibility, criminal liability and others.

It is the right of every State, in accordance with the erga omnes principle, to ensure that such violations are brought before the relevant court, and prosecuted and sanctioned in accordance with the abovementioned responsibilities and liabilities. This is because these are infringements of obligations that are owed to the international community as a whole.

Breaches of international norms that have a jus cogens character and which are considered erga omnes obligations can probably also be enforced by invoking universal jurisdiction over such violations. Such universal jurisdiction is presumably more easily enforced over such legal entities as (merchant) ships, given that they constantly sail into the territory and ports of different States, and by virtue of the NMFT principle.

Which international norms are deemed jus cogens norms is a constantly fluctuating concept, given the continual evolution and development of international society, exemplified by article 64 VCLT, which allows existing treaties to become void following the acceptance of new jus cogens norms. For example, the prohibition of piracy was previously a jus cogens norm, but is presumably no longer deemed to be so. Conversely, there are presently deliberations on whether the term can encompass other norms, such as prohibitions on forced pregnancy, murder, deportation of populations, rape and forced prostitution, and persecution of an identifiable group of people, if these violations occur on a widespread and systematic basis. The common denominator of some or all of these prohibitions (recognised as jus cogens or not) is that they aim to protect people's lives and health from hazards and threats that they have not sought out themselves but to which they are inadvertently exposed.

The ILC has set out two conditions that must be met to establish whether an international norm has achieved status of jus cogens:

(a) it must be a norm of general international law; and
(b) it must be accepted and recognised by the international community as a norm from which no derogation is permitted.

Case law from the ICJ, and reports, commentaries, etc from the ILC, will, as main sources of international law from recognised bodies of the international community, serve as sources for determining whether an international norm has been accepted as a jus cogens norm. These are supplemented by a report made by the ICC prosecutor.

The next chapter (chapter 17) will attempt to clarify whether it is possible for environmental norms (regulations) prohibiting very damaging pollution (GHGs) to be considered peremptory norms accepted by the international community, that is by the ICJ and ILC.

Finally, the adverse effects of GHG pollution and the continued negative impact from increased global warming are also described in the next chapter, as they directly relate to the conditions set by the ILC and the ICJ for recognising environmental jus cogens norms.

17

Whether Some International Environmental Norms Can be Considered Jus Cogens

Chapter 16 established that the International Law Commission (ILC), through the report of its Special Rapporteur on this matter, has drafted a set of conclusions codifying the requirements for determining whether an international norm has obtained the status of jus cogens.[1]

Draft conclusion 4 sets out two conditions that an international norm must meet to have the character of jus cogens:

(a) it must be a 'norm of general international law'; and
(b) this norm 'must be accepted and recognized by the international community of States as a whole as a norm from which no derogation is permitted'.

The second condition in particular will be analysed in this chapter. Case law from the International Court of Justice (ICJ) and ILC reports and their commentaries, specially on previous drafts and the final draft of the Articles on State Responsibility for Internationally Wrongful Acts ('the Articles on State Responsibility' or 'the Articles'),[2] will, alongside a report from the International Criminal Court (ICC) prosecutor, be examined in sections II–IV to establish if the international community, via the ICJ, ILC, ICC, etc, has recognised that environmental norms may be accepted as jus cogens norms and, if so, under what conditions.

Before the second condition is examined in further detail, the first condition is briefly studied in section I.

I. A Norm of General International Law

The first condition in ILC draft conclusion 4(a) dictates that a jus cogens norm must be a 'norm of general international law'. It must be presumed that this

[1] UN, *Report of the International Law Commission Sixty-ninth session* (1 May–2 June and 3 July–4 August), A/72/10 (2017) 192–93, fn 808.

[2] International Law Commission, *Draft Articles on Responsibility of States for Internationally Wrongful Acts*, November 2001, Supplement No 10 (A/56/10), chp IV.E.1, available at https://www.refworld.org/docid/3ddb8f804.html.

requirement can be considered fulfilled if it is a norm (regulation/prohibition) codified in an international agreement, such as a treaty or convention.

Draft conclusion 5 lists different examples of norms that can be seen as representing general international law, including norms codified in a treaty or convention. Conclusion 5(4) provides that 'A treaty rule may reflect a norm of general international law capable of rising to the level of a jus cogens norm of general international law.'[3] Thus, if an environmental norm is codified in a convention, for example in an IMO convention regulating GHGs by prohibiting the use of fossil fuels,[4] it should satisfy the first requirement in conclusion 4(a) for deeming that norm a peremptory norm from which there can be no derogation.[5]

II. The ICJ's Recognition of Environmental Norms as Jus Cogens Norms

As stated in chapter 16, ILC draft conclusion 9(3) referred to case law of the ICJ as an important source of international law, and also for establishing whether the international community has accepted a new jus cogens norm, as the ICJ, with its 193 participating States, represents 'recognition by a large majority of States' (see draft conclusion 7(3)).

It is also worth noting that article 66(1)(a) VCLT establishes that any party to a dispute concerning the application or the interpretation of article 53 or article 64 (whether a treaty becomes void due to existing or new jus cogens norms) may bring the matter before the ICJ. This further underlines that the ICJ is the foremost 'source' of international law to determine which peremptory norms exist at a given time.

The ICJ has previously dealt with questions and cases pertaining to protection of the environment, also referring to the jus cogens principle.[6] This link between

[3] Draft conclusion 5(2)–(3) also refer to principles of customary international law and general principles of law pursuant to art 38(1)(c) of the ICJ Statute as norms that should be viewed as being norms of general international law.

[4] It is deemed likely by this author that the IMO GHG regulation, at some point, will impose a complete prohibition on the use of fossil fuels but otherwise allow shipowners to choose which form of compliance best suits them, eg alternative fuels, solar power, wind power, electricity (batteries) or something else. See ch 15, section III for more on this.

[5] It should be recalled that art 53 of the Vienna Convention on the Law of Treaties 1969 (adopted 22 May 1969, entered into force 27 January 1980) 1155 UNTS 331 (VCLT) defines a peremptory (jus cogens) norm as follows: 'a peremptory norm of general international law is a norm accepted and recognized by the international community of States as a whole as a norm from which no derogation is permitted and which can be modified only by a subsequent norm of general international law having the same character'.

[6] The ICJ website states that 'it should be noted that in 1993 the Court created a Chamber for Environmental Matters, which was periodically reconstituted until 2006. However, in the Chamber's 13 years of existence no State ever requested that a case be dealt with by it. The Court consequently decided in 2006 not to hold elections for a Bench for the said Chamber.' See at https://www.icj-cij.org/en/chambers-and-committees.

the violation of environmental regulations (prohibitions) and invocation of the jus cogens principle is of particular relevance to the examination of this subject in this chapter. If it can be established that the ICJ, in principle, has accepted that jus cogens norms may encompass environmental issues, this would be a giant step towards fulfilling the second condition in draft conclusion 4(b), regarding the international community's acceptance of this.[7]

The ICJ case law that will be examined in the following subsections comprises the decisions on the *Legality of the Threat or Use of Nuclear Weapons*, the *Gabčíkovo-Nagymaros project case* and *Pulp Mills on the River Uruguay*.

A. *Legality of the Threat or Use of Nuclear Weapons* (1996)

In 1996 the ICJ gave an Advisory Opinion in the case *Legality of the Threat or Use of Nuclear Weapons*,[8] covering whether the use, or threat of use, of nuclear weapons was incompatible with international law. The 28 States that had brought the case before the ICJ emphasised that the use of nuclear weapons contradicted several fundamental principles of international law, for example concerning the 'use of force' and 'the protection of civilians, children and non-combatants during military operations', etc. The parties also referred to the use of nuclear weapons as having disastrous effects on the environment (air, water, wildlife, etc), which, by itself, would have an indirect negative impact on human health and living conditions. The parties therefore, inter alia, argued that the use of such weapons would conflict with international rules for the protection of the environment, including the principles of the Stockholm Declaration of 1972[9] and the Rio Declaration of 1992.[10] The parties directly referred to compliance with these principles being incumbent upon all States, including 'to ensure that activities within their jurisdiction or control do not cause damage to the environment of other States or of areas beyond the limits of national jurisdiction'.[11]

The ICJ did not find that there was any relevant international law that prohibited the use, or threat, of nuclear weapons.[12] The Court did recognise, however,

[7] This is also important in context of the next chapter – where it is questioned whether the IMO's future regulations on greenhouse gases (GHGs) could be considered prohibitions covered by the jus cogens principle – as it should be recalled from ch 12 of this book that the ICJ is one of the judicial organs that can settle disputes on the interpretation and application of UNCLOS and the law of the sea in general: see art 287(1) of UNCLOS. So although the cases referenced in the following subsections all concern pollution from land-based sources, it must be assumed that the ICJ will adopt the same principles to determine a case brought before it in accordance with art 287(1)(b), pertaining to pollution from ships.

[8] *Legality of the Threat or Use of Nuclear Weapons*, Advisory Opinion [1996] ICJ Rep 226.

[9] See, eg, principle 21 of the Stockholm Declaration, A/CONF.48/14/Rev.1, available at https://undocs.org/en/A/CONF.48/14/Rev.1.

[10] See, eg, principle 2 of the Rio Declaration, UN Doc A/CONF.151/26 (vol I); (1992) 31 ILM 874.

[11] *Legality of the Threat or Use of Nuclear Weapons*, n 8, para 27.

[12] ibid para 20.

a fundamental obligation for all States to ensure that the environment is protected, including in areas outside of a State's jurisdiction, stating that the environment is not an abstract entity but forms the conditions and basis for human habitation and livelihood (health, survival), for future generations too:

> The environment is not an abstraction but represents the living space, the quality of life and the very health of human beings, including generations unborn. The existence of the general obligation of States to ensure that activities within their jurisdiction and control respect the environment of other States or of areas beyond national control is now part of the corpus of international law relating to the environment.[13]

This means that the ICJ established that there is a fundamental principle (obligation) within international law for all States to reduce or prohibit transboundary pollution. This could arguably cover air pollution, especially transboundary air pollution that can exist in the atmosphere for extended periods, such as CO_2.[14]

It should also be noted that the ICJ explained this obligation by reference to the direct and indirect adverse consequences such pollution can have on human existence (health and living conditions), including on future generations. This description would also apply to the consequences resulting from manmade GHG (CO_2) pollution, as the global warming that follows could lead to the destruction of the environment, which has a direct influence on human survival, for example through rising sea levels, drought, floods and storms, which, inter alia, can destroy food (crops, animals), water and habitation needed for human survival, especially for future generations. These impacts are described in greater detail in section VI of this chapter.

So, although the ICJ did not explicitly refer to jus cogens norms in this respect,[15] it did unequivocally recognise the existence of a norm within international law that prohibits all States from releasing transboundary pollution that has such adverse effects on the environment that it also has a negative impact on the conditions needed for human survival, including for future generations. The ICJ clearly stated that this principle 'is now part of the corpus of international law relating to the environment', which allows us to start to assert that the ICJ can accept that the jus cogens principle can cover environmental norms.

B. *The Gabčíkovo-Nagymaros Project* (1997)

The ICJ passed judgment in 1997 in the case of the *Gabčíkovo-Nagymaros project*, also known as the *Danube Dam* case, between Hungary and Slovakia.[16]

[13] ibid para 29.

[14] CO_2 can stay in the atmosphere for up to 200 years; see at https://archive.ipcc.ch/ipccreports/tar/wg1/016.htm.

[15] *Legality of the Threat or Use of Nuclear Weapons*, n 8, para 83. The ICJ did, in that case, mention the jus cogens principle in connection with the application of art 53 VCLT in the context of humanitarian law.

[16] *Gabčíkovo-Nagymaros project (Hungary v Slovakia)* [1997] ICJ Rep 7.

The proceedings concerned a bilateral agreement from 1977 between Hungary and Slovakia, the latter being Czechoslovakia at that time. The agreement related to the establishment of canal, dam and border-crossing projects for the exploitation of the River Danube, especially between the two cities on opposite banks of the river: Gabčíkovo in Slovakia and Nagymaros in Hungary.

Complications arose following economic challenges and changes within the different States after the fall of the Soviet Union. Hungary tried to terminate the agreement in 1992, which led the parties in dispute to bring the case before the ICJ.

The case is of relevance to this part of the book as Hungary argued that *subsequently imposed requirements of international law in relation to the protection of the environment precluded performance of the Treaty*, supporting this by referring to the ILC's commentaries on the (at that point) not yet adopted (draft) Articles on State Responsibility, in particular article 33 pertaining to circumstances precluding wrongfulness due to a 'state of necessity'.[17] Hungary also referred to this principle's being an existing erga omnes obligation.[18]

Slovakia argued that no development of environmental law had given rise to environmental norms of jus cogens that could override the agreement.[19]

The ICJ referred in its decision, amongst other things, to the previously cited passage from *Legality of the Threat or Use of Nuclear Weapons*, where the ICJ argued that the environment provides the living conditions needed for human existence, including for future generations. All States should therefore ensure that such damaging pollution does not occur (prohibit it), including as regards transboundary pollution that has an impact outside the State's territory.[20] The ICJ elaborated on this statement in the *Gabčíkovo-Nagymaros project case* by underlining that this expressed a fundamental prohibition, not just for all States but 'also for the whole of mankind'.[21]

This may already provide an indication of the ICJ's progress in seeing this prohibition become a (peremptory) norm that is recognised by the majority of the international community, as the ICJ extended the prohibition to include 'the whole of mankind'.

The ICJ also went a decisive step further in the *Gabčíkovo-Nagymaros project* as, in the same paragraph, the Court chose to acknowledge the principles behind the statements made by Hungary pertaining to essential environmental norms potentially causing an agreement to be void due to a state of necessity. The ICJ stated:

> The Court has no difficulty in acknowledging that the concerns expressed by Hungary for its natural environment in the region affected by the Gabtikovo-Nagymaros Project related to an 'essential interest' of that State, within the meaning given to that expression in Article 33 of the Draft of the International Law Commission.[22]

[17] ibid para 49.
[18] ibid para 97.
[19] ibid.
[20] ibid para 53.
[21] ibid.
[22] ibid para 53.

The ICJ then went on to highlight those parts of the ILC's commentaries that stated that a grave danger to ecological preservation could be considered a state of necessity and that this was an *essential interest*, which it had become within the last two decades.[23]

Thus, already in 1997 the ICJ recognised that it is possible for environmental norms to be of such importance that they can be considered a state of necessity. A crucial step towards also deeming these to be jus cogens norms.

The ICJ further noted, obiter, that:

> Neither of the Parties contended that new peremptory norms of environmental law had emerged since the conclusion of the 1977 Treaty, and the Court will consequently not be required to examine the scope of Article 64 of the Vienna Convention on the Law of Treaties.[24]

This statement is not without importance, as the ICJ made it without being obliged to (it is obiter dictum) and it implicitly accepts that peremptory (jus cogens) norms of environmental law *could* exist, making the application of article 64 VCLT a possibility, but this was simply not contended by the parties.

This assumption is also based on the ICJ's use of the phrase 'new peremptory norms of environmental law', as this clearly implies that such peremptory norms already exist. Had it been a question of whether these norms even existed, the ICJ could have referred either to 'the possible existence of peremptory norms of environmental law' or merely to 'new peremptory norms of international law'.

The ICJ thus used the *Gabčíkovo-Nagymaros project case* to underline that the environmental norm established a year previously in the *Legality of the Threat or Use of Nuclear Weapons case* actually expressed a fundamental norm applicable to the whole of mankind, and that this was a norm affecting an *essential interest* that could give rise to a state of necessity.

Finally, the ICJ took it upon itself to implicitly state that peremptory (jus cogens) norms of environmental law *could* exist. Such a norm could, for example, prohibit transboundary pollution that may threaten the environment in ways that consequently may endanger the 'quality of life and the very health of human beings, including generations unborn'.

C. *Pulp Mills on the River Uruguay* (2010)

The ICJ *Pulp Mills on the River Uruguay case* from 2010, between Argentina and Uruguay, also entailed a point of (lesser) relevance regarding whether the Court would view an environmental norm as having obtained jus cogens status.[25]

[23] ibid para 53.
[24] ibid para 112.
[25] *Pulp Mills on the River Uruguay (Argentina v Uruguay* [2010] ICJ Rep 14.

The case involved an agreement entered into by Argentina and Uruguay in 1975, which sought to ensure that the River Uruguay, which runs between these two countries, was to be exploited optimally, rationally and jointly between them. The agreement also referred to the need for environmental circumstances to be taken into account when exercising commercial activities on the river. To ensure this, the parties established a joint authority, the Administrative Commission of the River Uruguay (CARU),[26] which had to be informed about such activities on the river to assess their environmental impact. However, Uruguay established two plants (pulp mills) by the river in 2006, Argentina contesting their legality. Argentina claimed that Uruguay had not informed CARU about it in advance, and that the establishment of these factories had a negative effect on the environment in and around the river.[27]

The ICJ did not rule in favour of Argentina, as the Court, after examining the environmental impact, did not find it established that the river had been contaminated or that the activities in question had adverse effects on human health.[28] This ruling is, in this context, of some relevance, as it shows the ICJ decided to make a substantive assessment of the environmental impact of the mills, including their potential adverse effects on human health.

The ICJ carried out detailed examination of whether the discharges from the plants had had an impact on the quality of the river water,[29] including focusing on the adverse effects of specific pollutants present such as phosphorus,[30] phenolic substances[31] and nonylphenols.[32] The Court also evaluated any damaging impact on animal life (fish, etc) in the river,[33] as well as on the biodiversity in general.[34] Finally, and very importantly, the ICJ assessed the air pollution emitted by the mills and the impact thereof.[35]

This in-depth assessment of polluting substances and their effects must lead to the presumption that the ICJ would also assess whether other forms of pollution, for example GHGs, are so harmful to the environment as to affect the 'quality of life and the very health of human beings, including generations unborn' – as required by the Court in *Legality of the Threat or Use of Nuclear Weapons* and in the *Gabčíkovo-Nagymaros project case*. This would allow the ICJ to determine whether an environmental norm that prohibits (limits) such pollution, for example as adopted by the IMO, could be considered a jus cogens norm.

[26] ibid 16; 'CARU' is the Spanish acronym for the Administrative Commission.
[27] ibid paras 94–111.
[28] ibid para 264 and paras 276–78.
[29] ibid paras 229–59.
[30] ibid paras 240–50.
[31] ibid paras 251–54.
[32] ibid paras 255–57.
[33] ibid paras 258–59.
[34] ibid paras 260–62.
[35] ibid paras 263–64.

D. Conclusions on the ICJ's Case Law

These three ICJ cases led to the following conclusions:

- The ICJ found in its Advisory Opinion in the *Legality of the Threat or Use of Nuclear Weapons* that a fundamental principle (obligation) exists for all States to prohibit transboundary pollution that can potentially threaten the environment in ways that consequently may endanger the 'quality of life and the very health of human beings, including generations unborn'. The ICJ also explicitly stated that this norm (prohibition) is 'now part of the corpus of international law relating to the environment'.
- The ICJ elaborated on this conclusion in the *Gabčíkovo-Nagymaros project case*, where it laid down that this norm applies not only to States but also to 'the whole of mankind'. The Court further found that environmental norms could be deemed norms affecting an *essential interest*, which can allow a State to invoke a 'state of necessity' in accordance with the ILC Articles on State Responsibility to preclude wrongfulness.
- The ICJ implicitly stated that peremptory norms of environmental law exist, although this was not argued by the parties.
- The ICJ will, given its detailed environmental assessment in *Pulp Mills on the River Uruguay*, consider itself capable of determining whether a specific polluting substance can result in such damaging contamination of the environment that it must be a threat to the 'quality of life and the very health of human beings, including generations unborn'.

It is thus assumed by this author that the ICJ will accept that an international environmental norm could be considered a jus cogens norm if it prohibits pollution that can endanger 'the quality of life and the very health of human beings, including generations unborn'. Such a peremptory norm would be considered a part of the corpus of international law relating to the environment, by which the whole of mankind must abide.

What forms this pollution might take is discussed in section VI of this chapter. Before that, relevant reports and commentaries from the ILC will be examined in section III, to determine whether there is any alignment between these ILC conclusions and the principles of the ICJ just described.

III. Reports and Commentaries of the ILC Accepting Environmental Jus Cogens Norms

The ILC was, as stated on its webpage, established by the UN in 1947 to 'initiate studies and make recommendations for the purpose of … encouraging the progressive development of international law and its codification'.[36] It is therefore common sense that the work of the ILC represents one of the most esteemed sources of international law, to which the ICJ often also refers in its decisions.

Statements and conclusions made by the ILC in its reports to the UN General Assembly and commentaries to conventions and other legislative works are therefore of significant value for determining whether an international environmental norm can be recognised by the ILC (representing the international community) as a jus cogens norm and under what conditions.[37] This is also emphasised in draft conclusion 9(4), pertaining to the recognition of jus cogens norms, which refers to the legal works of the ILC as a source for determining this.[38]

The ILC's work on creating the previously mentioned Articles on State Responsibility[39] in particular has given the Commission the opportunity to address many different aspects of international law, including the development and application of peremptory norms, which are explicitly mentioned in articles 40 and 41. The work was ongoing for several decades. The Articles were finally adopted in 2001, but they have not yet been transformed into a treaty. It is nonetheless believed that they and their commentaries reflect customary international law.[40]

That is also why the ICJ has applied *draft* versions of the Articles in different cases, for example in the *Rainbow Warrior* case between France and New Zealand, pertaining to circumstances precluding wrongfulness.[41] The ICJ also referred to these in the *Gabčíkovo-Nagymaros project case*.[42]

The Articles and commentaries have changed over the years, but many of the commentaries on previous drafts of the Articles are still relevant today, and still represent customary law, as the provisions (articles) to which they refer are still found in the adopted version, albeit with different wording and numbering.

The ILC commentaries to articles 40 and 41 of the adopted version and to article 19(2)–(3) of a previous 1976 draft version (preceding articles 40–41)[43] are therefore significant for establishing the ILC's position on accepting environmental norms as possible jus cogens norms, as all these articles relate to the definition and application of peremptory norms.

[36] See at http://legal.un.org/ilc/.

[37] The commentaries of the ILC are found in the *Yearbook of the International Law Commission*.

[38] *Report of the International Law Commission Sixty-ninth session*, n 1, ch VIII, 192–202.

[39] See n 2.

[40] A Henriksen, *International Law* (Oxford University Press, 2017) 128.

[41] *Rainbow Warrior (New Zealand v France)* [1990] ICJ Rep 215.

[42] *Gabčíkovo-Nagymaros project*, n 16, para 51.

[43] *Report of the Commission to the General Assembly on the work of its twenty-eighth session* (3 May–23 July) A/31/10 (1976) II (Part Two) *Yearbook of the International Law Commission* 96.

A. The ILC's Commentaries to Article 19 of the 1976 Draft Articles on State Responsibility

Article 19 of the previous Draft Articles related to the potential international liability of a State for 'international crimes and international delicts'.[44] Article 19(1) provided that if a State breached an international obligation, it was to be considered an *internationally wrongful act, regardless of the subject-matter of the obligation breached.* In other words, there was a clear liability for States violating international obligations. In accordance with article 19(4), any form of violation that was not an 'international crime' pursuant to article 19(2) and (3), was an international delict.

International crimes were defined as breaches of international obligations protecting fundamental interests of the international community, which the commentaries, inter alia, referred to as breaches of peremptory norms. Article 19(2) and (3), and the commentary thereto, thus contain important information, which is essential to the analysis in this chapter.

Article 19(2) read:

> An internationally wrongful act which results from the breach by a State of an international obligation so essential for the protection of fundamental interests of the international community that its breach is recognized as a crime by that community as a whole, constitutes an international crime.

Article 19(3) attempted to define the term 'international crime', and thereby the scope of article 19(2), by (non-exhaustively) listing breaches (violations) of four basic norms that were covered by the term and therefore by article 19(2). Article 19(3)(a)–(d) stated:

> Subject to paragraph 2, and on the basis of the rules of international law in force, an international crime may result, inter alia, from:
>
> (a) a serious breach of an international obligation of essential importance for the maintenance of international peace and security, such as that prohibiting aggression;
> (b) a serious breach of an international obligation of essential importance for safeguarding the right of self-determination of peoples, such as that prohibiting the establishment or maintenance by force of colonial domination;
> (c) a serious breach on a widespread scale of an international obligation of essential importance for safeguarding the human being, such as those prohibiting slavery, genocide and apartheid;
> (d) a serious breach of an international obligation of essential importance for the safeguarding and preservation of the human environment, such as those prohibiting massive pollution of the atmosphere or of the seas.

[44] ibid 95–96.

Accordingly, it is evident that the first three crimes listed ((a)–(c)) concerned violations of prohibitions that are deemed peremptory norms of a jus cogens character. Recalling that prohibitions on the unlawful use of force (acts of aggression), on refusing colonised peoples their right to self-determination, and on slavery, apartheid and genocide are all, as described in chapter 16, norms recognised by the international community as jus cogens norms, including explicitly by the ICJ and ILC. See, for example, the non-exhaustive list of jus cogens principles set out by the ILC in its 2019 Report, which includes all of these norms as fully recognised peremptory norms.[45]

Article 19(3)(d) is, for obvious reasons, the most interesting in this context, as it refers to 'a serious breach of an international obligation of essential importance for the safeguarding and preservation of the human environment, such as those prohibiting massive pollution of the atmosphere or of the seas' as (potentially) an international crime. The recognised jus cogens prohibitions (norms) listed in article 19(3)(a)–(c) make it logical to assume that the ILC viewed the prohibition of massive pollution of the atmosphere or sea ((d)) as being as important as the other prohibitions listed, thereby deeming it to be an equally important international obligation that is essential for the protection of fundamental interests of the international community in accordance with article 19(2).

Further, an 'international crime' was described in article 19(2) as the breach of an 'international obligation so essential for the protection of fundamental interests of the international community that its breach is recognized as a crime by that community as a whole', wording that is remarkably close to the requirements set out by the ILC itself in 2017[46] for establishing that an international obligation has jus cogens status (ie an international norm (obligation) recognised by the international community as a norm from which there can be no derogation).

The initial conclusion that may be reached is therefore that the ILC accepted that an international obligation (norm) prohibiting massive pollution of the atmosphere or of the seas could be deemed a peremptory norm. This deduction is supported by the ILC commentary to article 19(2)–(3).

The commentary to article 19(2), inter alia, stated that 'Paragraph 2 deals with the most serious breaches and introduces, with respect to them, the idea of "international crime"'.[47] This underlined that article 19(2) (and therefore also article 19(3)) covered the most serious breaches of international law, which is the general description also applied to peremptory (jus cogens) norms.

[45] See, eg, Special Rapporteur Dire Tladi, *Fourth Report on Peremptory Norms of General International Law (Jus Cogens)*, ILC, 71st session, A/CN.4/727 (April 2019) 63, para 137.

[46] *Report of the International Law Commission Sixty-ninth session*, n 1, ch VIII, 192–202.

[47] *Report of the Commission to the General Assembly on the work of its twenty-eighth session*, n 43, 118, para 59.

The link to article 19(2)'s encompassing jus cogens norms was also explicitly emphasised by the ILC in referring to article 53 VCLT as an *analogous text*:

> At first sight, the text of paragraph 2 [of article 19] may give an impression of tautology. In reality, what it says is no more tautological than the analogous text of article 53 of the Vienna Convention. What the latter article provides is that in order to be 'objectively' considered as 'peremptory', ie as not permitting of any derogation, a norm of international law must be 'subjectively' accepted and recognized as such by the international community as a whole. Similarly, paragraph 2 of the article under consideration provides that in order to be 'objectively' considered as an 'international crime', and as such liable to more severe legal consequences as a result of responsibility, an internationally wrongful act must be 'subjectively' recognized as a 'crime' by the international community as a whole.[48]

The assumption that article 19(3)(d) also represented a peremptory norm is strengthened by the ILC's affirming that the focus of that provision was the *survival of humanity* and not the protection of the environment itself. An approach very much in line with the normal application of the jus cogens principle. The ILC commentary on this stated:

> Lastly, within the equally wide framework of concern for the preservation of certain assets which are essential for the progress and survival of humanity, the Commission paid particular attention to recent developments in international law on the subject of the safeguarding and preservation of the human environment. In this sphere, it took as its example a breach of the obligations prohibiting massive pollution of the atmosphere or the seas. Some members expressed reservations regarding the choice of pollution as an example, because they thought that the notion of pollution was not defined as precisely as the other examples mentioned in paragraph 3. Nevertheless, they expressed full agreement with the general provision in sub-paragraph (d).[49]

The hypothesis that the prohibition of massive pollution of the atmosphere and seas, pursuant to article 19(3)(d), was a jus cogens norm, like the other recognised jus cogens norms codified in article 19(3)(a)–(c), is reinforced by the fact that the examples in article 19(3) are not set out in a hierarchal order. In fact, the ILC referred to *all* these examples (main categories) as rules that, within the international legal order, have been elevated to the rank of jus cogens:

> The rules of international law which are now of greater importance than others for safeguarding the fundamental interests of the international community are to a large extent those which give rise to the obligations comprised within the four main categories mentioned. It is mainly among them that are to be found the rules which the contemporary international legal order has elevated to the rank of jus cogens.[50]

Finally, the ILC asserted, in its conclusion on article 19, that the article sought to make a distinction between violations of international law in general (article 19(1) and (4))

[48] ibid 119, para 61.
[49] ibid 121, para 71.
[50] ibid 121, para 67.

and violations of jus cogens norms (article 19(2)–(3)) as seen within the law of treaties:

> In conclusion, the Commission wishes to emphasize that it is aware of the exceptional importance of the subject dealt with in this article. In the codification of the law of international responsibility, the adoption of a formulation which expressly recognizes the distinction between international crimes and international delicts is a step comparable to that achieved by the explicit recognition of the category of rules of jus cogens in the codification of the law of treaties.[51]

Thus, it can be concluded that the ILC in the 1976 Draft Articles on State Responsibility accepted that a violation of an international norm that is of essential importance for the safeguarding and preservation of the human environment by prohibiting massive pollution of the atmosphere or of the seas is a violation of a jus cogens norm.

B. The ILC's Commentaries to the 2001 Articles on State Responsibility

It is, following the conclusion on the 1976 Draft Articles referred to at the end of the previous section, relevant to examine whether there still is a link between draft article 19(2)–(3) and the commentary thereto and the Articles adopted in 2001.[52] The answer, in short, is 'yes'. The 2001 Articles have, to a large extent, codified the ILC's work on the previous Draft Articles, including the principles in article 19(2)–(3).

The provisions (paragraphs) in draft article 19 were redrafted in articles 40–41, the term 'international crimes' being abandoned in favour of directly referencing serious violations of 'peremptory norms'. Article 40(2) does, however, state that in order for a violation of a peremptory norm to be deemed serious, the violation must be attributable to a State's *gross or systematic* failure to fulfil the obligation. This requirement was, as described in chapter 16, also laid down by the ILC in its summation of the principles established by the ICJ's case law, pertaining to the Court's having indicated that the prohibition on certain illegal acts, such as torture, etc, has the character of jus cogens if it prohibits the perpetration of that act on a *widespread* or *systematic* basis.[53]

It should be noted that neither article 40 nor article 41 includes a list of examples in the same way as article 19(3). This is, however, merely for reasons of

[51] ibid 122, para 73.

[52] *Report of the Commission to the General Assembly on the work of its fifty-third session* (23 April–1 June and 2 July–10 August) A/56/10 (2001) II (Part Two) *Yearbook of the International Law Commission* 113.

[53] *Report of the International Law Commission Sixty-ninth session*, n 1, 23.

consistency with how the VCLT refers to jus cogens norms, as the ILC elaborates in the commentary to article 40:

> It is not appropriate to set out examples of the peremptory norms referred to in the text of article 40 itself, any more than it was in the text of article 53 of the 1969 Vienna Convention.[54]

The ILC has, however, unequivocally specified in both the 2001 commentary to article 40 and a 2012 report[55] that the examples listed in article 19(3) of the 1976 Drafts Articles are encompassed by and codified in article 40 as adopted.

The ILC commentary to article 40 states inter alia (footnote also cited):

> It must also be borne in mind that some of the peremptory norms in question, most notably the prohibitions of aggression and genocide, by their very nature require an intentional violation on a large scale.[651][56]

Footnote [651] stated:

> [651] At its twenty-second session, the Commission proposed the following examples as cases denominated as 'international crimes':
>
> '(a) a serious breach of an international obligation of essential importance for the maintenance of international peace and security, such as that prohibiting aggression;
> '(b) a serious breach of an international obligation of essential importance for safeguarding the right of self-determination of peoples, such as that prohibiting the establishment or maintenance by force of colonial domination;
> '(c) a serious breach on a widespread scale of an international obligation of essential importance for safeguarding the human being, such as those prohibiting slavery, genocide and apartheid;
> '(d) a serious breach of an international obligation of essential importance for the safeguarding and preservation of the human environment, such as those prohibiting massive pollution of the atmosphere or of the seas.'
>
> Yearbook … 1976, vol II (Part Two), pp 95–96.

C. The ILC's 2017 Report: Protection of the Atmosphere is a Basic Human Right

The previous ILC conclusions – that prohibitions on, inter alia, massive (ie widespread or systematic) pollution of the atmosphere can be jus cogens norms – are aligned with the considerations and conclusions of the ILC in a

[54] *Report of the Commission to the General Assembly on the work of its fifty-third session*, n 52, commentary to art 40, para 3, 112.

[55] UN, *Materials on the Responsibility of States for Internationally Wrongful Acts*, ST/LEG/SER B/25 (2012) 266, including fn 897.

[56] *Report of the Commission to the General Assembly on the work of its fifty-third session*, n 52, commentary to art 40, para 8, 113, including fn 651.

Report dated 2017, which includes a chapter (chapter VI) on the 'Protection of the atmosphere'. The ILC there underlined that it (still) considered protection of the atmosphere a subject deserving of special attention within international law.[57]

The Report stated that this topic had been part of the ILC's work programme since the sixty-fifth session in 2013. The Special Rapporteur assigned to this subject presented nine guidelines in the 2017 Report, to which States should adhere to protect the atmosphere, with accompanying commentaries.

The Report stated that some aspects of international law already referred to the protection of the atmosphere, including *the law of the sea*. This is, inter alia, a reference to articles 212 and 222 of UNCLOS.[58] It will be recalled, however, from the analysis in Part II of this book (especially in chapter 7) that the geographical and (extraterritorial) jurisdictional scopes of those articles are limited and somewhat vague, so that they should be used in conjunction with the other relevant articles of part XII of UNCLOS, including articles, 217, 218, 220 and 228.[59]

The ILC Report also considered that pollution of the atmosphere could be deemed a violation of *international human rights*. This assessment is extremely relevant to the hypothesis of this chapter, as international human rights encompass many of the previously mentioned jus cogens-covered rights by, inter alia, safeguarding 'the right to life', which includes prohibitions on torture, genocide, etc.

The ILC Report set out certain conditions for asserting that air pollution might be regarded as being a threat to, and thereby a violation of, basic human rights:

> In this regard, relevant human rights are 'the right to life', 'the right to private and family life' and 'the right to property'. Where a specific right to environment exists in human rights conventions, the relevant courts and treaty bodies apply them, including the right to health. In order for international human rights law to contribute to the protection of the atmosphere, however, certain core requirements must be fulfilled. First, as international human rights law remains 'a personal-injury-based legal system', a direct link between atmospheric pollution or degradation that impairs the protected right and an impairment of a protected right must be established. Second, the adverse effects of atmospheric pollution or degradation must attain a certain threshold if they are to fall within the scope of international human rights law. The assessment of such minimum standards is relative and depends on the content of the right to be invoked and all the relevant circumstances of the case, such as the intensity and duration of the nuisance and its physical or mental effects. Third, and most importantly, it is necessary to establish the causal link between an action or omission of a State, on the one hand, and atmospheric pollution or degradation, on the other hand.[60]

[57] *Report of the International Law Commission Sixty-ninth session*, n 1, ch VI, 147–62.

[58] ibid 158, para 9.

[59] Perhaps it could be envisioned that such future references to *protection of the atmosphere within the law of the sea* might encompass the conclusions in Part II, ie that port States have extraterritorial jurisdiction through art 218 to take measures against emissions violations.

[60] *Report of the International Law Commission Sixty-ninth session*, n 1, 160, para 11.

Three conditions may thus be extrapolated from the ILC text for establishing that a violation of an environmental norm can be considered a violation of a human right:

(a) It must be established that there is a *direct link* between a violation of an environmental obligation (norm/regulation) and the damage the environment (atmosphere) sustains. In other words, the damage must be attributable to violations of international regulations.
(b) The pollution must exceed a certain 'threshold', for example by being massive, widespread or systematic.
(c) There must be a causal link between a State's action or omission (resulting in the violation) and the pollution sustained.

If all three conditions are met, a violation can be considered a violation of a *basic human right.*

It is obvious that these conditions are very much in line with the conditions set out in the commentaries to the 1976 Draft Article and the 2001 Articles on State Responsibility, regarding a State's being held liable for violating an environmental peremptory norm protecting the atmosphere from massive pollution – pollution that results in damage to the human environment and human health.

Some human rights are considered peremptory norms. This conclusion therefore supports the argument that some environmental norms (if the potential pollution of the atmosphere is massive and damaging enough) can be considered norms recognised as jus cogens norms as well as human rights.

D. Conclusion on the ILC's Acceptance of Environmental Jus Cogens Norms

It is thus the conclusion of this author that the ILC recognises that an international norm (obligation), which is of *essential importance for the safeguarding and preservation of the human environment by prohibiting massive (widespread or systematic) pollution of the atmosphere* (or of the seas), can be considered a peremptory norm covered by the jus cogens principle.

IV. The International Criminal Court: Destruction of the Environment and Universal Jurisdiction

As mentioned previously, in the Articles adopted in 2001 the ILC abandoned the term 'international crimes' when describing violations of peremptory norms. The Prosecutor appointed at the ICC has, however, in a policy paper from 2016 on the prioritisation and selection of case proceedings before the ICC,[61] clearly indicated that *destruction of the environment* can be regarded as a *serious crime*, on a par with terrorism, human trafficking, illegal occupation of the territory, etc. The policy paper states:

> The Office will also seek to cooperate and provide assistance to States, upon request, with respect to conduct which constitutes a serious crime under national law, such as the illegal exploitation of natural resources, arms trafficking, human trafficking, terrorism, financial crimes, land grabbing or the destruction of the environment.[62]

It serves as an important signal when the ICC's Prosecutor directly states that the Prosecutor's Office will consider destruction of the environment a serious crime when selecting and prioritising cases to be brought before the Court. Especially considering that the ICC also can have jurisdiction over violations of jus cogens norms committed by individuals, as described in chapter 16.

Whether an environmental violation can result in 'destruction of the environment' should, according to the ICC Prosecutor, be assessed in light of, inter alia, the 'environmental damage inflicted on the affected communities'.[63] The ICC Prosecutor can ex officio (of its own accord) initiate investigations of such alleged infringements and, if the evidence so permits, institute prosecutions of such violations (see article 15(1) of the ICC Statute).[64] This would also apply to violations that result in destruction of the environment.

There is thus a connection between the ICC Prosecutor's viewing destruction of the environment as a serious crime and environmental norms' being deemed jus cogens norms, as this would, through the erga omnes principle and application of universal jurisdiction, allow all States to bring such violations before a court (the ICC or a national court), for them to be stopped and penalised.

The application of universal jurisdiction to bring environmental cases before a court is also directly mentioned as a possibility by the United Nations Environmental Programme (UNEP) in the UN Report on universal jurisdiction, which also refers to the ICC Prosecutor's prioritisation of these crimes:

> UNEP noted that, for greater effect and outcomes, these efforts could be supplemented by the application of universal jurisdiction in order to enhance the enforcement and adjudication capacities in the area of environment-related crime. UNEP drew attention to recent efforts by the Office of the Prosecutor of the International Criminal Court in its policy paper on case selection and prioritization as providing an encouraging avenue. In the policy paper, the Office of the Prosecutor sought to cooperate and provide assistance to States, upon request, with respect to conduct that constitutes a serious crime under national law, such as the illegal exploitation of natural resources,

[61] Prosecutor's Office at the International Criminal Court, *Policy Paper on Case Selection and Prioritisation* (September 2016).

[62] ibid 5, para 7.

[63] ibid 14, para 41.

[64] The Rome Statute of the International Criminal Court (adopted 17 July 1998, entered into force 1 July 2002) 2187 UNTS 3.

arms trafficking, human trafficking, terrorism, financial crimes, land grabbing and the destruction of the environment.[65]

V. Interim Conclusions on Acceptance of Environmental Jus Cogens Norms by International Bodies

The ILC's acceptance of international norms prohibiting massive and widespread pollution of the atmosphere (or sea) as peremptory norms is in alignment with the conclusion previously reached pertaining to the ICJ's case law, which found that a fundamental international norm (obligation) exists for all States – and the whole of mankind – to prohibit transboundary pollution that can potentially threaten the environment in ways that consequently may endanger the 'quality of life and the very health of human beings, including generations unborn'.

This also coincides with the ICC Prosecutor's deeming crimes that *destroy the environment* to be serious crimes of such a grave nature that they should be prioritised for presentation before the Court, in the same way as crimes such as terrorism, human trafficking and illegal occupation of territory. The UNEP also links the application of universal jurisdiction to enforcement and adjudication of these crimes.

When the conclusions of the ILC and the ICJ (and the ICC) are read together, several common criteria emerge for determining when an international environmental norm can be regarded as a jus cogens norm. These criteria, established by these executive sources of international law, clearly representing the majority of all States, will, when met, inevitably fulfil the *second condition* of draft conclusion 4 ((b)), described in chapter 16 and at the beginning of this chapter, pertaining to a jus cogens norm's being 'accepted and recognized by the international community of States as a whole'.

These common ILC/ICJ criteria for establishing that an international environmental norm can be considered a jus cogens norm are:

(a) It must be an international norm, for example an obligation or prohibition codified in an international convention.
(b) This norm must prohibit massive (gross, widespread and/or systematic) transboundary pollution of the atmosphere or of the seas.
(c) The type of pollution prohibited must potentially be able to threaten the preservation of the human environment by polluting the environment (atmosphere or seas) in ways that that can endanger communities and the quality of life and the very health of human beings, including for generations unborn.

[65] Report of the Secretary-General, *The scope and application of the principle of universal jurisdiction*, A/72/112, 72nd session, item 86 (2017) 10–11, para 46.

The next logical step is to examine what types of pollution of the atmosphere (or sea) could be deemed so massive, transboundary and harmful that they endanger human survival, including of future generations, thereby meeting the third criterion. This is done in section VI of this chapter.

VI. Types of Pollution that Can Threaten Human Existence

First, a distinction must be made between massive pollution of the *atmosphere* and massive pollution of the *seas* that can threaten the quality of life and the very health of human beings, including generations unborn.

When applying the jus cogens principle on environmental norms, it seems to be the very logical consensus of both the ILC and the ICJ (and the ICC) that not all environmental norms could be covered by the principle, merely those that are the most dangerous and damaging. Discussion of which forms of pollution (of the atmosphere or seas) might meet the criteria has been somewhat lacking in the aforementioned reports, case law and literature in general. The following subsections seek to answer the question.

A. Massive Pollution of the Seas

It is evident that pollution of the seas can have extreme and severe consequences for human existence. It can, for example, result in communities' losing sources of food and income from destruction of marine ecology, fishing and wildlife. It can also result in loss of income from tourism due to damage sustained by coastlines, etc. The clean-up costs can also be a heavy burden for such coastal communities to shoulder.

Violations of international norms that protect the seas, such as the regulations found within MARPOL Annexes I–V,[66] the BWM Convention[67] and other IMO Conventions, can result in *massive pollution* fulfilling the first two established requirements for deeming a norm a jus cogens norm. It is an international (codified) norm that can prohibit massive (transboundary, widespread) pollution of the seas.

However, it might be more difficult to establish that such massive pollution of the marine environment will also endanger the quality of life and health of human beings, including of 'generations unborn'. There is absolutely no doubt that the

[66] International Convention for the Prevention of Pollution from Ships (adopted 11 February 1973, as modified by the Protocol of 17 February 1978, entered into force 2 October 1983) 1340 UNTS 61 (MARPOL), Annexes I–V, IMO Publication: IMO-520E.

[67] The International Convention for the control and management of ship's ballast water and sediments 2004 (adopted 13 February 2004, entered into force 8 September 2017) ('the BWM Convention').

consequences of massive pollution of the seas, for example following an oil spill, can be widespread. But it must be deemed unlikely that the violation will present an immediate threat to human existence. The same applies to a spill of chemicals, sewage or ballast water containing highly invasive species. It is therefore considered unlikely that any IMO regulations, or other international regulations, protecting the seas and marine environment could fulfil this requirement for deeming the obligation a norm covered by the jus cogens principle.

It is imperative to ensure that these discussions on deeming environmental norms to be peremptory norms do not dilute the jus cogens principle. The jus cogens principle is reserved for prohibiting the most atrocious violations and crimes against mankind (genocide, illegal use of force, etc), and should therefore potentially only cover the most dangerous and damaging forms of pollution, that is, those that endanger the quality of life and health of human beings, including future generations. Massive pollution of the seas does not, in general, meet this criterion.

One could perhaps imagine an extremely dangerous chemical substance (perhaps not yet developed), covered by MARPOL Annexes II or III, which might interact with water and be potentially lethal to all people that come into contact with it. Prohibitions of such pollution could perhaps be deemed jus cogens norms, but would have to be determined on a case-by-case basis.

B. Massive Pollution of the Atmosphere

The other form of pollution mentioned in article 19(3)(d) of the 1976 ILC Draft Articles on State Responsibility, which is still encompassed by articles 40–41 of the 2001 Articles, is potential *massive pollution of the atmosphere*. The term 'pollution of the atmosphere' is in the following replaced with 'air pollution', to be consistent with the terminology used throughout this book.

It may be recalled that international prohibitions or limitations (regulations) on air pollution could be deemed jus cogens norms if they fulfilled the other ICJ/ILC requirements pertaining to the pollution's being *transboundary* (widespread), *massive* and threatening the preservation of the *human environment*, including the quality of life and health of human beings, now and for generations to come. An initial examination of what forms of air pollution, if any, can meet these criteria in this context must be carried out to determine if any regulation prohibiting or limiting such pollutants can be considered a norm (regulation) of a jus cogens character.

First, air pollution is nothing if not transboundary and widespread,[68] especially given that much air pollution, for example by carbon dioxide (CO_2)

[68] This is, for example, highlighted by the title (and content) of the Convention on Long-range Transboundary Air Pollution (adopted 13 November 1979, entered into force 16 March 1983) 1302 UNTS 217 (LRTAP Convention), mentioned in ch 1.

and ozone-depleting substances (ODSs), can remain in the atmosphere for decades.[69]

Second, some air pollution, such as by sulphur (SOx), nitrogen oxide (NOx), volatile organic compounds (VOCs) and GHGs (including CO_2), has, in very different ways, an enormous impact on the environment and human health, so that such pollution could be considered *massive*.[70]

The final (and most strict) requirement – that the pollution may threaten the preservation of the human environment, including the quality of life and health of human beings, including for future generations – narrows the field of prohibitions and regulations on air pollution that could be considered jus cogens. Nonetheless, regulations covering certain types of air pollution could perhaps meet this demand.

An example might be regulations covering SOx pollution, bearing in mind the damaging effects of sulphur and that the reduction of the sulphur content in marine fuels, from 3.5% to 0.5%, may (as described in chapter 1) prevent 137,000 early deaths and 7.6 million children from developing asthma each year.[71] It could, however, be difficult to establish whether sulphur pollution can have an impact on the human environment for generations to come. It thus cannot be ruled out that regulation 14.1.3 of MARPOL Annex VI might be considered a regulation that protects an environmental jus cogens norm, in the form of a prohibition on the use of marine fuel with a sulphur content above 0.5%.

However, as this work – establishing that environmental norms can be jus cogens norms – set out in part IV of this book is somewhat uncharted legal territory, it would seem most prudent to focus on the application of the jus cogens principle to the greatest pollution threat to human existence, particularly for future generations – *greenhouse gases*.

Greenhouse gas emissions, including CO_2 emissions, represent a source of massive pollution of the atmosphere that is more harmful than, for example, SOx or NOx pollution, if seen from an overall, global point of view, including when assessing the future impact on human existence. This assumption is supported by the conclusions reached by the ILC in the previously referenced Report from 2017, which includes a chapter on the protection of the atmosphere.[72] The ILC singles out the

[69] CO_2 can stay in the atmosphere for up to 200 years; see at https://archive.ipcc.ch/ipccreports/tar/wg1/016.htm.

[70] The adverse effects of SOx are described in ch 1, the effects of ODS, NOx, and VOCs in ch 14, and the effects of GHGs in the following paragraphs.

[71] M Sofiev et al, 'Cleaner fuels for ships provide public health benefits with climate tradeoffs' (2018) 9 *Nature Communications* article number 406, 4–5, available at https://www.nature.com/articles/s41467-017-02774-9#ref-link-section-d1456e583.

[72] *Report of the International Law Commission Sixty-ninth session*, n 1, ch VI, 147–62. This Report refers directly to the general IMO regulations on emissions from ships, such as those covering the regulation of SOx and NOx in MARPOL Annex VI. It states that these regulations can be enforced though the law of the sea (ie UNCLOS), as has also been established in Parts II and III of this book (ibid 158–59, para 9 and fn 727): 'IMO has sought to regulate vessel-source pollution in its efforts to supplement the provisions of the Convention and to combat climate change. The effective implementation of the applicable rules of the law of the sea could help to protect the atmosphere. Similarly, the effective implementation of the rules on the protection of the environment could protect the oceans.'

deterioration of the environment through *climate change*, and the adverse impacts thereof on human life, as being so severe that it could violate basic *human rights*:

> As for international human rights law, environmental degradation, including air pollution, climate change and ozone layer depletion, 'has the potential to affect the realization of human rights'.[73]

The reference to 'climate change', including *depletion of the ozone layer*, is a direct reference to the damage sustained by the release of GHGs. The ILC thereby indicates that this form of air pollution (GHGs causing climate change) is a much more serious and damaging form, as it can be deemed a violation of a human right. This also brings it closer to recognising such pollution as being serious enough to be covered by the jus cogens norm, given the often overlapping nature of basic human rights and jus cogens norms.

This author will therefore follow this premise set out by the ILC and focus on GHG (including CO_2) pollution as the form of air pollution that fulfils the requirements listed previously by the ILC and ICJ, relating to transboundary (widespread) and massive pollution of the atmosphere that has the potential to threaten the environment in ways that subsequently can endanger the quality of life and the very health of human beings, including of 'generations unborn'.

This conclusion does not rest only on the 'legal' assessments made by the ILC, but also on scientific descriptions of GHG pollution and its adverse impacts on the environment relating to human existence, including studies depicting the negative projections for the continued rise in global warming. Some of these impacts, scientific descriptions and studies are briefly referred to in the following subsections, to establish that GHG pollution meets the ILC and ICJ criteria.

i. Greenhouse Gas Pollution (Global Warming) Can Threaten Human Existence

It should initially be noted that the generic term 'greenhouse gases' covers, besides CO_2, other substances (gases), such as methane and ODSs, including CFC-11 (trichlorofluoromethane). 'Greenhouse gases' (or 'GHGs') is therefore used in the following as a term that covers all such sources, although the primary source of GHG pollution stems from CO_2.[74]

Greenhouse gases are, amongst other things, characterised by being gases that can absorb the heat radiation emitted from the Earth's surface.[75] This is, in principle, a natural part of the Earth's ecosystem in relation to retaining heat and generating temperatures that offer optimal conditions for sustaining life, including human life. However, the excessive release of manmade GHGs into the atmosphere

[73] ibid 159, para 10.
[74] See ch 14 for more on ODSs (including CFC-11) and CO_2.
[75] See at https://scied.ucar.edu/longcontent/greenhouse-effect.

creates a problem, as this results in too much of the radiated heat being absorbed, leading to *global warming*.

Global warming, caused by manmade GHG pollution, could therefore be covered by the jus cogens principle as defined by the ILC and ICJ, as this form of pollution has become a challenge for this and future generations due to its adverse effects.

Some of the most noteworthy and damaging consequences of global warming, according to the World Health Organization (WHO),[76] relate to the major climate changes caused by this, including:

- *Melting* of the polar ice caps, which can lead to a rise in sea levels, threatening low-lying island nations, coastal regions and human habitation near the sea.[77]
- *Droughts*, which can destroy vegetation and crops used for food, leading to famine, thus also impacting on the ability to feed and keep livestock for milk, food, etc.[78]
- *Droughts* can also lead to a shortage of drinking water, which probably is the most immediate threat to human survival, not just for future generations but also for the present.[79]
- The *lack of (clean) water*, besides causing death by dehydration, can also lead to other life-threatening circumstances, such as outbreaks of infectious diseases (cholera, diphtheria, etc) in already severely afflicted areas. It may also create more sustainable environments for animals spreading diseases, such as mosquitoes spreading malaria.[80]
- *Droughts* can also lead to forest fires, which can threaten lives, homes, crops, etc.[81] (Further, such fires release enormous amounts of CO_2 into the atmosphere, exacerbating the problem.)
- *Extreme weather conditions* in the form of floods and storms can threaten people and destroy homes and critical services, such as electricity, sanitation, clean water, heat, etc.[82]
- *Increases in UV rays* can cause skin cancers, which can be attributed to damage to the ozone layer caused by ODSs.[83]

Climate change can also, due to the abovementioned droughts, storms, floods, fires, etc, destroy trees and other plants that normally photosynthesise, converting CO_2

[76] WHO, *Protecting Health from Climate Change: Connecting Science, Policy and People* (World Health Organization, 2009), available at http://www.who.int/iris/handle/10665/44246.
[77] ibid 2.
[78] ibid 6.
[79] ibid 12–14.
[80] ibid 2, 4 and 12.
[81] ibid 8.
[82] ibid 6–7.
[83] See ch 14, section I.A for more on ODSs and their harmful effects.

in the atmosphere into breathable oxygen. Such destruction can lead to a vicious circle, speeding up a process that results in less conversion of CO_2 to oxygen, leading to more CO_2 build-up in the atmosphere, leading to more global warming, leading to more plants and trees being destroyed, and so on.[84]

ii. Global Warming – The Negative Projections

The application of the jus cogens principle to international regulations (norms) prohibiting/limiting the release of GHGs is of particular relevance, as many scientific studies and projections indicate that the impact of GHG pollution is not lessening and that the goals set out by the Paris Climate Agreement, including pursuing limitation of the increase in global warming to 1.5°C compared to pre-industrial levels, are not being met. This underscores the importance of strengthening the protection of the environment against such pollution by invoking the principles of jus cogens and erga omnes, and through the exercise of universal jurisdiction. This would apply in particular to enforcement of any future GHG regulations by the IMO, as shipping accounts for approximately 2.2% of all CO_2 emissions worldwide (ships of 5,000 GT or more are responsible for 85% of this release[85]) and the IMO Strategy has set out reduction goals that are aligned with the goals of the Paris Climate Agreement.[86]

This continued negative development of worsening climate change due to increased global warming, attributable to manmade GHG pollution (inter alia from shipping) has been described in several scientific studies, open letters, etc. The UN's Intergovernmental Panel on Climate Change (IPCC) published an alarming report in 2018 that advised that manmade GHG (CO_2) emissions should be drastically reduced and that a maximum 1.5°C increase in global warming should not just be pursued but met, as there could be vast consequences if the goal of 'only' a 2°C increase were to be achieved.[87] The report was, as a

[84] This short presentation does not consider the socio-economic costs associated with the above-mentioned environmental consequences as the jus cogens norm does not, in the view of this author, encompass such variables, despite having a great impact on societies and individuals. It is, however, obvious that these costs can be gigantic for society, for example in relation to hospital care, emergency responses to catastrophes, re-establishing infrastructure, etc. The economic costs for people and families affected by these climate changes can also be insurmountable as regards loss of jobs, loss of homes and medical costs, not to mention the psychological effects related to these consequences. See, eg, P Karfakis with L Lipper and M Smulders, 'The assessment of the socioeconomic impacts of climate change at household level and policy implications', paper for the Agricultural Development Economics Division (FAO, 2012) available at https://www.semanticscholar.org/paper/The-assessment-of-the-socio-economic-impacts-of-at-Karfakis-Lipper/79c506ec4ae363211ed84ba2be930bdac35f66ea.

[85] See points 1.3.3 and 1.6 of the IMO's GHG Strategy, MEPC Resolution MEPC 304(72), *Initial IMO Strategy on Reduction of GHG Emissions from Ships*. See also at http://www.imo.org/en/OurWork/Environment/PollutionPrevention/AirPollution/Pages/Default.aspx.

[86] Point 3.1 of the IMO's GHG Strategy sets the overall goal of reducing GHG emissions by a minimum 50% by 2050 compared to the GHG emissions level of 2008.

[87] O Hoegh-Guldberg et al, 'Global Warming of 1.5°C', *An IPCC special report on the impacts of global warming of 1.5°C above pre-industrial levels and related global greenhouse gas emission pathways, in the*

follow-up to the Paris Climate Agreement, prepared by 91 authors from 40 different countries.

The UN's World Meteorological Organization (WMO) published a report in 2017 that showed that the concentration of CO_2 in the atmosphere in 2016 rose to its highest level in 800,000 years.[88] The report, which was based on measurements from 51 countries, also showed that other GHGs, such as methane (CH_4) and nitrous oxide (N_2O), were also on the rise. A new study made available in May 2019 showed that this disheartening trend continues, as a new record amount of CO_2 was measured in the atmosphere.[89]

A NASA study from 2018 has proved that when global warming causes the polar ice caps to melt, it releases methane imbedded in the permafrost into the atmosphere, thereby dramatically intensifying the effect of global warming and climate change.[90]

These alarming indications that global warming is not slowing down, and the immense consequences it could have for the environment and for people, were emphasised by 15,364 researchers from 184 countries, who as early as 2017 signed a joint declaration (a *letter of warning to humanity*) to clarify the negative climate development and the impacts thereof, in order to prompt people and policymakers to address this.[91]

Finally, it may be recalled from chapter 14 that a study from 2018, made by the Canadian Research Institute NOAA, has shown a disturbing global rise in emissions of the harmful CFC substance CFC-11 (trichlorofluoromethane).[92]

Some of the most devastating predictions take into account the accumulated effects of global warming, including droughts destroying forests, the release of methane imbedded in the permafrost of the polar ice caps, etc, leading to accelerated deterioration of the environment, often described as the so-called 'hot-house effect'.[93] In 2019 these troubling forecasts were underlined by a new study, raising the argument that some places on Earth where people previously have lived, for example certain areas in India, are becoming too hot to sustain human life.[94]

A quick summary of these official UN reports and other studies indicates that the release of dangerous levels of manmade GHGs is predicted, by some, to

context of strengthening the global response to the threat of climate change, sustainable development, and efforts to eradicate poverty (IPPC, 2018) 253, available at http://www.ipcc.ch/report/sr15/.

[88] WMO Greenhouse Gas Bulletin (GHG Bulletin) no 13, 'The State of Greenhouse Gases in the Atmosphere Based on Global Observations through 2016' (WMO, 2017).

[89] See at https://scripps.ucsd.edu/programs/keelingcurve/2019/06/04/carbon-dioxide-levels-hit-record-peak-in-may/.

[90] See at https://climate.nasa.gov/news/2785/unexpected-future-boost-of-methane-possible-from-arctic-permafrost/.

[91] 'World Scientists' Warning to Humanity: A Second Notice' (2017) 67(12) *BioScience* 1026, available at https://academic.oup.com/bioscience/article/67/12/1026/4605229.

[92] See at https://www.eurekalert.org/pub_releases/2018-05/uoca-nfr051618.php.

[93] See at https://www.stockholmresilience.org/research/research-news/2018-08-06-planet-at-risk-of-heading-towards-hothouse-earth-state.html.

[94] See at https://edition.cnn.com/2019/07/03/asia/india-heat-wave-survival-hnk-intl/index.html.

continue, which could result in the temperature reduction goals of the Paris Climate Agreement and the IMO GHG Strategy not being reached, with catastrophic consequences for the environment and thereby for the conditions for sustaining human life, especially for future generations. This unfortunate projection supports deeming some environmental norms prohibiting/limiting the release of GHGs, for example a future IMO rule, jus cogens norms, as adherence to adopted goals and regulations must be an absolute minimum requirement to protect human existence.

VII. Conclusion

The aim of this chapter was to establish *whether* any environmental norms could be regarded as jus cogens norms and, if so, *which* environmental norms could be encompassed by this.

This chapter initially recalled the findings of chapter 16 pertaining to the ILC's having drafted a set of conclusions for establishing whether an international norm has obtained the status of jus cogens.[95] The fourth draft conclusion sets out the conditions for recognising a jus cogens norm, in terms of its being (i) a norm of general international law and (ii) accepted by the international community as a norm from which no derogation is possible.

The first condition (being an international norm) can be met if a norm concerns international environmental legislation, for instance IMO regulations.

The second condition (the norm is recognised by the international community as a jus cogens norm) is more difficult to meet.

The initial challenge was to clarify who the *international community* is. The ILC draft conclusions (conclusion 7.3) refer in this respect to acceptance by a 'large majority of States'. This is, in the view of this author, achieved by examining the conclusions on this of the ICJ and the ILC, as these UN bodies enjoy acceptance by a very large majority of States. Also, the ICJ applies and interprets official sources of international law in accordance with article 38(1) of the ICJ Statute, and the ILC is a UN Commission tasked with identifying and codifying international law.

The conclusions of the ICJ and ILC are also referenced in the draft conclusions (9.3 and 9.4) as valuable sources of international law for making such assessments.

It was therefore determined that the findings of the ICJ and ILC would represent the best sources within international law for establishing whether international environmental norms could achieve jus cogens status and under what conditions.

[95] *Report of the International Law Commission Sixty-ninth session*, n 1, 192–92, fn 808.

The ICJ case law that was examined comprised the Court's decisions in the *Legality of the Threat or Use of Nuclear Weapons*, the *Gabčíkovo-Nagymaros project case* and *Pulp Mills on the River Uruguay*.

In its Advisory Opinion in the *Legality of the Threat or Use of Nuclear Weapons*, the ICJ found that a fundamental principle exists within international law to prohibit transboundary pollution that can result in the *quality of life and the very health of human beings, including that of generations unborn*, being endangered. Further, in the *Gabčíkovo-Nagymaros project case* the Court found that compliance with this principle is owed to the *whole of mankind*. (Wording that also indicates application of the erga omnes principle, as seen in the *Barcelona Traction* case.[96])

The ICJ also stated in the *Gabčíkovo-Nagymaros project case* that environmental norms could be deemed of such *essential interest* that they would allow a State to invoke a state of necessity in accordance with the ILC Draft Articles on State Responsibility. The ICJ noted obiter in this context that peremptory norms could exist within environmental law, as it referred to article 64 VCLT's not having been invoked by the parties.

Given the ICJ's assessment in *Pulp Mills on the River Uruguay*, the Court would also see itself capable of determining, in great detail, what forms of pollution could be considered so damaging to the environment that they could pose a threat to the quality of life and the health of humans, including of generations unborn. The ILC stated in its 1976 commentaries on its Draft Articles on State Responsibility (article 19(2)–(3)) that *massive pollution of the atmosphere or of the seas* could be deemed an international crime alongside other crimes that are violations of recognised jus cogens norms.[97] These commentaries are implemented in those to the articles adopted in 2001 (articles 40–41).

This assumption is repeated to some extent by the ILC in its 2017 Report, where it stated that violations of environmental norms, which can lead to widespread or systematic pollution that endangers people and communities, could be considered violations of basic human rights – rights that often also encompass other recognised jus cogens norms.

The ICC Prosecutor has also indicated that serious violations of environmental legislation should be viewed as serious crimes that should be prioritised before the Court.

It was thus concluded that an international environmental norm prohibiting pollution of the atmosphere or seas could be considered *recognised by the 'international community'* (ie by the ICJ and the ILC) if the norm prohibited massive, widespread, transboundary pollution that could threaten human health and the quality of life, including for future generations.

[96] *Barcelona Traction case (Belgium v Spain)* [1970] ICJ Rep 3, para 33.

[97] It may be recalled that the ILC concluded, despite discussions in the commentaries among different States, that 'Nevertheless, they expressed full agreement with the provisions in sub-paragraph (d).' See extract reproduced in section III.A and cited in n 49.

It was then established that (massive) pollution of the *atmosphere* can more easily meet these conditions than pollution of the seas, as the latter form of pollution does not necessarily endanger human lives in the same way as the former.[98]

The next step was to establish what forms of pollution of the atmosphere (air pollution) could meet the ICJ and ILC requirements of endangering human lives and existence, including for generations unborn. Although many air polluting substances, some of which are regulated in MARPOL Annex VI, can present a threat to human health, especially sulphur pollution pursuant to regulation 14.1.3 of the Annex, the most damaging form of pollution of the atmosphere was deemed to involve GHGs, as it causes global warming and thus severe climate change.

This assumption was based on global warming's having several well-established harmful effects on the environment and climate, which either directly or indirectly can pose a threat to human existence, for example in the form of rising sea levels, droughts, floods, storms, diseases, etc. This conclusion is also supported by a 2017 ILC report, which directly refers to the adverse effects of GHGs.

Finally, it was stated that several scientific studies, including in official UN reports, have shown a clear need for protection of the environment, and therefore of human life, from GHG pollution, due both to the adverse effects of global warming and to the studies and their projections' indicating that existing regulatory measures and goals will not curb global warming to the extent required and needed. In fact, many studies showed the continued rise of levels of CO_2 in the atmosphere.

Pollution by GHGs therefore meets all the requirements set out by the ILC and ICJ, in terms of being transboundary, massive and threatening human existence, including for future generations. International rules (norms) regulating GHG emissions could therefore, in principle, be considered peremptory norms of a jus cogens character.

The impact of this conclusion on a potential future IMO regulation on prohibiting the use of fossil fuel (discussed in chapter 15, section III), and on the enforcement of this – through application of the erga omnes principle and the invoking of universal jurisdiction – is clarified in chapter 18.

[98] It was however emphasised that the prohibition of a substance so toxic that it could result in massive pollution of the seas and that *also* presented a clear and immediate threat to human life, could perhaps also meet these ICJ and ILC conditions.

18

Future IMO Regulation on Fossil Fuels as a Jus Cogens Norm

Chapter 16 explained what the jus cogens and erga omnes principles are, while chapter 17 established that there is a legal basis for assuming that the international community, through the International Court of Justice (ICJ) and the International Law Commission (ILC) (and the International Criminal Court (ICC)), would recognise international regulations prohibiting or limiting the release of greenhouse gases (GHGs) as international norms considered to be jus cogens norms, as well as regulating a basic human right.

This concluding chapter of Part IV will seek to establish if a regulation adopted by the International Maritime Organization (IMO) that, as discussed in chapter 15, section III, prohibited the use of fossil fuel, could be deemed a norm (regulation) of a jus cogens character. It will also be examined how this legislation could be enforced through application of the erga omnes principle and by invoking universal jurisdiction.[1]

Finally, section IV of this chapter will emphasise the jus cogens implications relating to the European Union (EU), should it choose to adopt its own regional GHG legislation.

I. IMO Regulation Prohibiting Use of Fossil-based Marine Fuel is a Jus Cogens Norm

It will be recalled that the ICJ and ILC have recognised environmental jus cogens norms if they prohibit massive, widespread, transboundary pollution of the atmosphere that can threaten human health, existence and the quality of life, including for future generations.

[1] Some will undoubtedly wonder whether the conclusions in this chapter could be applied to other international agreements and conventions on the reduction of air pollution from land-based sources, etc. This could, in principle, be a possibility, but since the focus of this book is on ships and their discharge of harmful substances into the atmosphere, it will be left to others to draw any inferences as to the applicability to other international climate agreements, conventions, etc of the conclusions reached in this part of the book.

Pollution by GHGs is such pollution, as it can cause global warming and climate change, which can result in damage to the environment in ways that endanger human health, existence and quality of life. (See the discussion in chapter 17.) This includes GHGs emitted from ships when they combust fossil fuel for propulsion, which is also directly stated by the ILC in its 2017 Report:

> Among the various human activities that have an impact on the oceans, greenhouse gas emissions from ships contribute to global warming and climate change. The 2009 study by the International Maritime Organization (IMO) on greenhouse gas emissions classified such emissions from ships into four categories, namely: emissions of exhaust gases, cargo emissions, emissions of refrigerants and other emissions. Research indicates that excessive greenhouse gas emissions from ships change the composition of the atmosphere and climate, and cause a negative impact on the marine environment and human health.
>
> The General Assembly has continued to emphasize the urgency of addressing the effects of atmospheric degradation, such as increases in global temperatures, sea-level rise, ocean acidification and the impact of other climate changes that are seriously affecting coastal areas and low-lying coastal countries, including many least developed countries and small island developing States, and threatening the survival of many societies.[2]

It can be assumed, in the view of this author, that if the IMO were to adopt a regulation that prohibited the use of fossil fuel, it could be recognised as a norm (regulation) of a jus cogens character, as it would meet the criteria set out in the ILC's draft conclusions 4(a)–(b), also discussed in chapter 17.

It would, as an IMO regulation, be a 'norm of general international law' (draft conclusion 4(a)) and a norm that was 'accepted and recognized by the international community of States as a whole from which no derogation is permitted' (draft conclusion 4(b)). The necessary acceptance and recognition by the international community would come from this prohibition's meeting the aforementioned criteria established by the ILC and ICJ, as the norm (the regulation) would prohibit transboundary, widespread and massive pollution of the atmosphere that can threaten human health, existence and quality of life, including for future generations.

A regulatory IMO prohibition on the use of fossil fuel would therefore represent an environmental peremptory norm of a jus cogens character.

II. Legal Implications of an IMO Ban on Fossil Fuel as a Jus Cogens Norm

This section will seek to elaborate on the legal implications for a State if it – or a ship, company, port, fuel supplier, etc under its jurisdiction – violates an IMO

[2] *Report of the International Law Commission Sixty-ninth session* (1 May–2 June and 3 July–4 August), A/72/10 (2017) 152, paras 3–4.

regulation deemed a jus cogens norm, both in respect of the *obligations* for all States to ensure compliance with the regulation (and the legal liability for failing to do so) and in respect of the *rights* for all States to take action in response to violations through the principles of erga omnes and universal jurisdiction.

Some of the legal consequences for a State associated with not complying with (environmental) jus cogens norms are explicitly codified in the Vienna Convention on the Law of Treaties (VCLT)[3] and in the ILC Articles on State Responsibility for Internationally Wrongful Acts ('Articles on State Responsibility').[4]

A. The IMO Prohibition Would Bind All States

The first consequence of an IMO (jus cogens) ban on the use of fossil-based marine fuel would be that all States would have to observe this prohibition in accordance with the IMO regulation, irrespective of whether the State in question was party to the legislation (which codifies the prohibition) or not. As provided in article 53 (VCLT), 'a peremptory norm of general international law is a norm accepted and recognized by the international community of States as a whole as a norm from which no derogation is permitted'.

All States would thus be responsible for ensuring that individuals and legal entities under their jurisdiction, including ships, shipowners and ports, comply with this peremptory norm.[5] Further, if such a jus cogens norm is deemed an erga omnes obligation, the right to take measures of enforcement against such violations can be invoked by other States. As previously concluded, there is sometimes an overlap, as jus cogens norms are often also perceived to be erga omnes obligations.[6] (See the discussions in chapters 6 and 16.) This is clearly stated by Karl Zemanek, who writes that 'all peremptory norms create obligations *erga omnes*, but not all *erga omnes* obligations derive from peremptory norms'.[7]

[3] Vienna Convention on the Law of Treaties 1969 (adopted 22 May 1969, entered into force 27 January 1980) 1155 UNTS 331 (VCLT).

[4] International Law Commission, *Draft Articles on Responsibility of States for Internationally Wrongful Acts*, November 2001, Supplement No 10 (A/56/10), chp IV.E.1, available at https://www.refworld.org/docid/3ddb8f804.html.

[5] The discussion in ch 15, section III, on detecting GHGs (CO_2), should be recalled, as it would be possible for States to detect violations of a ban on the use fossil fuels, eg by fuel sampling during a PSC, but also by using Continuous Emission Monitoring Systems (CEMSs), satellites and sniffers. The discussion in the same section on the economic incentive for shipowners not to comply with such an IMO ban on fossil fuels, despite having the right to choose their own means of compliance, should also be recalled.

[6] A Henriksen, *International Law* (Oxford University Press, 2017) 36.

[7] K Zemanek, 'New Trends in the Enforcement of erga omnes Obligations' (2000) 4(1) *Max Planck Yearbook of United Nations Law* 6.

B. Treaties and Conventions May Not Conflict with Jus Cogens Norms

Article 53 VCLT provides that:

> A treaty is void if, at the time of its conclusion, it conflicts with a peremptory norm of general international law.

Article 64 of the Convention states:

> If a new peremptory norm of general international law emerges, any existing treaty which is in conflict with that norm becomes void and terminates.

The scope of article 53 is therefore new treaties, conventions, etc that violate existing jus cogens norms, while article 64 pertains to existing treaties that violate new jus cogens norms.

The consequence of violating either of these provisions is that the treaty in question becomes *void*. The implications thereof are described in article 71(1)(a)–(b) VCLT for treaties that are void according to article 53 and in article 71(2)(a)–(b) for treaties that are void according to article 64.[8]

In terms of an IMO ban on fossil fuels representing jus cogens norms, this would mean that if two States entered into an agreement that violated this principle, it would become void – either according to article 53 if the environmental peremptory norms preceded the new treaty, or according to article 64 if the treaty was adopted before the new norms were recognised.

Such a violating treaty or agreement could attempt to circumvent this IMO GHG regulation. For example, two (or more) States might enter into a bilateral or multilateral agreement allowing ships under their flags to call into each other's ports without having to meet the IMO prohibition, perhaps by allowing the ships to bunker and sail using fossil fuels. Such a treaty would be void in accordance with article 53 or article 64, depending on the time of adoption. With this in mind, it should be recalled that article 66(1)(a) VCLT provides that any party to a dispute concerning the application or the interpretation of articles 53 or 64 may, by a written application, bring such dispute before the ICJ.

C. State Responsibility and Liability When Violating a Jus Cogens Norm

If a State violates a jus cogens norm, not just by adopting an international treaty but also by its actions, decisions, omissions, etc, it can incur State responsibility.

[8] ILC, *Fragmentation of international law: difficulties arising from the diversification and expansion of international law*, prepared by Martti Koskenniemi, Report A/CN.4/L.682 (presented at the 58th session in Geneva, 1 May–9 June and 3 July–11 August 2006) 184–85.

This would also extend to the obligations of all States to ensure that ships, companies, ports, fuel suppliers, etc under their jurisdiction comply with an IMO regulation prohibiting the use of fossil fuel, which would also include (bind) companies supplying marine fuel.

Any failure by a State to ensure such compliance with this regulatory measure (considered a jus cogens norm) could therefore lead to the State's becoming liable in accordance with the principles stipulated in the ILC's Articles on State Responsibility (also referred to as 'the Articles').

Article 40 and 41, found in chapter III of part two of the Articles, are of great relevance in this context as they relate to 'Serious breaches of obligations under peremptory norms of general international law'. Those articles were also examined in chapter 17, as the ILC, in its commentary to article 40,[9] makes explicit reference to article 19(3) of the previous Draft Articles,[10] which listed massive pollution of the atmosphere (or seas) as a serious crime alongside other already recognised jus cogens norms.

Article 40(1) sets out the scope of the chapter by stating:

> This chapter applies to the international responsibility which is entailed by a serious breach by a State of an obligation arising under a peremptory norm of general international law.

Unlike the previous draft article 19, this provision makes explicit reference to serious breaches of peremptory (jus cogens) norms.

Article 40(2) clarifies that:

> A breach of such an obligation is serious if it involves a gross or systematic failure by the responsible State to fulfil the obligation.

Thus, any gross or systematic breach of a jus cogens norm will always be considered 'serious'.

Article 41 requires all States to cooperate to stop such serious breaches of peremptory norms (article 41(1)), not to recognise or assist acts causing such serious breaches (article 41(2)) and that other consequences within international law may be invoked (article 41(3)).

If a State grossly or systematically violates, or allows violation of, a peremptory norm pursuant to articles 40 and 41, it can result in the State's becoming liable for these actions. This liability cannot, according to article 26 of the Articles, be precluded in any circumstances that would normally preclude wrongfulness, such as *consent* (article 20), *self-defence* (article 21), *countermeasures* (article 22), *force majeure* (article 23), *distress* (article 24) and *necessity* (article 25).

[9] *Report of the Commission to the General Assembly on the work of its fifty-third session* (23 April–1 June and 2 July–10 August) A/56/10 (2001) II (Part Two) *Yearbook of the International Law Commission* 113.

[10] *Report of the Commission to the General Assembly on the work of its twenty-eighth session* (1976) II (Part Two) *Yearbook of the International Law Commission* 95–122.

Article 26 stipulates that nothing precludes 'the wrongfulness of any act of a State which is not in conformity with an obligation arising under a peremptory norm of general international law'. A State's gross or systematic violations of a peremptory norm pursuant to articles 40 and 41 can therefore result in that State's becoming liable for these actions (or omissions), which means that it can be met with different 'legal consequences of an internationally wrongful act' in accordance with articles 29–31 of the Articles. These legal consequences include the '*duty to continue to perform the breached obligation*' (article 29), or to '*cease a violation*' and offer '*guarantees of non-repetition*' (article 30) and '*reparation*' for any material or moral damage (article 31).

Article 32 states, similar to article 27 VCLT, that national law cannot be used as a justification for not complying with the Articles.

It is primarily an injured State that, in accordance with article 42, has the right to invoke the legal responsibility of another State for that State's wrongful actions or omissions. An exception to this is found in article 48(1)(a)–(b), which allow any State to invoke the responsibility (infringements) of another State if:

(a) the obligation breached is owed to a group of States including that State, and is established for the protection of a collective interest of the group; or
(b) the obligation breached is owed to the international community as a whole.

This means that, pursuant to article 48(1)(b), any State has the right to invoke another State's violation of a jus cogens norm that is a codification of the erga omnes principle, as this principle, as described in chapters 6 and 16, allows all States to invoke any breach of an obligation (prohibition) owed to the international community as a whole.[11]

This is also emphasised by the ILC in the commentary to article 48(1)(b), as it explicitly refers to the erga omnes principle and the ICJ's acceptance of it in the *Barcelona Traction case*:

> States other than the injured State may invoke responsibility if the obligation in question was owed 'to the international community as a whole'. The provision intends to give effect to the statement by ICJ in the *Barcelona Traction* case, where the Court drew 'an essential distinction' between obligations owed to particular States and those owed 'towards the international community as a whole'. With regard to the latter, the Court went on to state that '[i]n view of the importance of the rights involved, all States can be held to have a legal interest in their protection; they are obligations erga omnes'.[12]

The erga omnes nature of article 48(1)(b) sets it apart from other relevant provisions regarding admissibility before the ICJ, for example article 66 VCLT or article 287 of UNCLOS,[13] as these articles refer to the right of the relevant *parties* to a treaty or dispute to bring a case before the court.

[11] *Report of the Commission to the General Assembly on the work of its fifty-third session*, n 9, 126, para 6.
[12] ibid 127, para 8, citing *Barcelona Traction (Belgium v Spain)* [1970] Rep 3.
[13] United Nations Convention on the Law of the Sea (adopted 10 December 1982, entered into force 16 November 1994) 1833 UNTS 3.

It should be noted that the ILC recognises that environmental norms (regulations) can be considered erga omnes norms (obligations), as in the commentary to this provision the Commission uses the example of a coastal State specially affected by 'pollution in breach of an obligation aimed at protection of the marine environment in the collective interest'.[14]

All States could therefore, by virtue of the erga omnes principle codified in article 48(1)(b), invoke another State's responsibility for grossly or systematically violating a peremptory (jus cogens) norm pursuant to article 40.

i. Flag States' Failure to Ensure Compliance by Ships under their Flag

A repeated (systematic[15]) disregard by the *flag State* to enforce an IMO ban on fossil fuels would thus not only result in the flag State's losing its primary jurisdiction in accordance with the second exception in article 228(1) of UNCLOS,[16] but it would also mean that the flag State could be liable in accordance with article 40(1) of the Articles on State Responsibility, as it would be a *serious breach of a peremptory norm*, given the reference in article 40(2) to such systematic failure always being of a serious nature.

All States would be allowed to invoke this systematic failure (repeated disregard) of peremptory norms pursuant to article 48(1)(b) of those Articles.

ii. States' Failure to Ensure Compliance by Ports and Fuel Suppliers under their Jurisdiction

All States could also be held liable in accordance with the Articles on State Responsibility if they failed to ensure that persons or legal entities under their jurisdiction complied with the IMO's prohibition on fossil fuels. A State that systematically failed to ensure that ports located under its jurisdiction acted in accordance with this IMO regulation could be an example of this.

The responsibility to ensure compliance falls on all States, irrespective of whether the violating port(s) under their jurisdiction are private or government-owned entities. Evidently, the same responsibility would also require States to ensure that fuel suppliers operating under their jurisdiction meet any requirements

[14] *Report of the Commission to the General Assembly on the work of its fifty-third session*, n 9, 127, para 10.

[15] It may be recalled from ch 16 that the ILC, in its 2017 report on 'Draft articles on crimes against humanity' refers to the ICJ's deeming crimes against humanity to be violations that are perpetrated on a 'widespread or systematic basis'. Special Rapporteur Sean D Murphy, *Third Report on Crimes Against Humanity*, ILC, 69th session (1 May–2 June and 3 July–4 August) A/CN.4/704 (2017) 152.

[16] The possible implications of accepting IMO GHG regulations as jus cogens could also affect the application of the *second exception* in art 228(1) of UNCLOS.

placed on them, including ensuring that they only supply compliant (zero-carbon or fossil-free) fuel. Notably, this also applies to the issuing of required documentation (a Bunker Delivery Note or its equivalent) and to ensuring the quality and quantity of such fuel, despite the fuel suppliers' being privately owned companies.

States that fail to ensure such compliance by ports or fuel suppliers within their jurisdiction can be held liable in accordance with the Articles on State Responsibility, particularly if there is a systematic (recurrent) failure to ensure compliance.

III. The Implications for Enforcement if an IMO Fossil Fuel Ban is a Jus Cogens Norm: Invoking the Principles of Erga Omnes and Universal Jurisdiction

If the IMO adopts a regulation prohibiting the use of fossil fuel, and this is considered to represent a peremptory norm, it could, in the view of this author, also have implications for how these rules can be enforced by non-flag States, including by States not party to UNCLOS, as it pertains to breaches of erga omnes obligations owed to the community as whole. To this end, the speculation in chapters 6 and 16 on the overlap between the jus cogens and erga omnes principles should be recalled, as well as the discussion of the ties between the erga omnes principle and universal jurisdiction.

A State that invokes the breach of an international environmental erga omnes obligation does so without necessarily having any connection to the offence or being impacted by it, yet the principle allows the State to seek to have the offence stopped, for example by bringing the matter before the ICJ or ICC.

Universal jurisdiction allows States to enforce and adjucate in the same way, by allowing such violations to be brought before a national court.[17]

It will also be recalled from chapter 16 that the United Nations Environment Programme (UNEP) stated in 2017 that '[u]niversal jurisdiction can play a significant role in bridging the gaps in the enforcement of international environmental law'[18] – something that was echoed by Matthias Reuß and Jakob Pichon.[19] It can

[17] Universal jurisdiction also grants the jurisdictional basis for a State to apprehend and detain a person suspected of such violations, and to deliver that person for prosecution before the ICC or to another State, despite none of the violating actions having an impact on the State. Such enforcement will nonetheless today often be subject to international multi- or bilateral agreements between States on extradition, etc.

[18] UN General Assembly, 'The scope and application of the principle of universal jurisdiction' A/72/112 (2017) 10.

[19] M Reuß and J Pichon, 'The European Union's Exercise of Jurisdiction Over Classification Societies – An International Law Perspective on the Amendment of the EC Directive on Common Rules and Standards for Ship Inspection and Survey Organisations and for the Relevant Activities of Maritime Administrations' (2007) 67 *Zeitschrift für ausländisches öffentliches Recht und Völkerrecht* 119, 136, available at https://www.zaoerv.de/67_2007/67_2007_1_a_119_144.pdf.

thus be assumed that the principles of erga omnes and universal jurisdiction would apply if an IMO prohibition on fossil fuels were to be accepted as a peremptory (jus cogens) norm.

This would, in the view of this author, have several implications for the enforcement of the prohibition, including in terms of enforcement through UNCLOS, as it would affect some of the conclusions made previously in Parts II and III regarding enforcement of IMO regulations on air pollution.

It may be recalled that if an IMO obligation banning the use of fossil fuels is deemed to be a jus cogens norm, it will bind all States. This would mean that all States would have to ensure that they, and all entities under their jurisdiction (persons, ships, companies, etc), *comply* with this prohibition, irrespective of whether the State itself is party to (having signed and/or ratified) the IMO regulation (convention) imposing the ban. Similarly, it would fall upon all States to *enforce* this IMO regulation, irrespective of whether the State is party to it or not.

A. All States Have a Right to Invoke the Erga Omnes Principle

If any State were to violate the IMO (jus cogens) regulation prohibiting the use of fossil fuels, it would give all other States the right to invoke the erga omnes principle and seek to have this violation stopped.

As discussed in section II of this chapter, a State can be liable for such violations if, for example:

(a) the State enters into an agreement (treaty) with another State allowing the use of fossil fuels; or
(b) the State, in its capacity as a flag State, neglects to effectively enforce the prohibition on ships flying its flag; or
(c) the State, in its capacity as a port State, neglects to effectively enforce the prohibition on ports, fuel suppliers, etc located within its jurisdiction.

All States would, in accordance with the erga omnes principle and the previously examined article 48(1)(b) of the ILC Articles on State Responsibility (which codifies the principle), be able to bring this matter before the ICJ[20] and claim that any violating State must fulfil its 'duty to continue to perform the breached obligation' (ie enforce the prohibition effectively) in accordance with article 29 of the ILC Articles, and also demand 'guarantees of non-repetition' from the violating State following article 30 of the Articles.

[20] Such a case could perhaps also be brought before the International Tribunal for the Law of the Sea (ITLOS), given that it concerns enforcement of maritime (IMO) legislation.

B. All States Have the Right and Obligation to Enforce the IMO Prohibition by Exercising Universal Jurisdiction

All States have the right to invoke their erga omnes right to stop a violation of a jus cogens norm (here, the IMO ban on the use of fossil-based marine fuels), including the right to extraterritorially enforce (sanction) and adjudicate upon this IMO regulation, in accordance with the principle of universal jurisdiction.

This is relevant when looking at enforcing such an IMO jus cogens regulation against private legal entities, such as ships and shipowners, when those ships violate the IMO ban on using these fuels.[21]

Evidently such enforcement would overlap with the measures of enforcement under UNCLOS already described in Parts II and III of this book. The implications for the normal enforcement paradigm found in part XII of UNCLOS of deeming an IMO prohibition on fossil fuels a jus cogens norm (and thus allowing enforcement through universal jurisdiction) will nonetheless be described separately in the next subsection, as this will focus on the basic legal implications of universal jurisdiction.

Exercising universal jurisdiction over a violation of an IMO ban on fossil fuels would allow all non-flag States to detain a ship and its crew,[22] either to prosecute the ship(owner) and the crew for the offence or to extradite them for prosecution in another State, following the principle of *aut dedere aut judicare*.[23]

This is supported by the ILC's having established that any State where a person is present who has violated basic human rights – which, as concluded in the previous chapter, can encompass environmental rights – should detain that person and either prosecute him or her for the offence, or extradite him or her following the above-mentioned principle of *aut dedere aut judicare*.[24] This is irrespective of that State's not having any territorial, national, passive personality or protective jurisdiction over the person:

> In such a situation, even if the crime was not committed in its territory, the alleged offender is not its national and the victims of the crime are not its nationals, the State nevertheless is obligated to establish jurisdiction given the presence of the

[21] The discussions in ch 15, section III and n 5 on detecting violations of an IMO prohibition on fossil fuels and the economic incentive for shipowners not to comply with this regulation are recalled.

[22] It would primarily be the master (captain), chief engineer and other senior personnel, to whom the crime can be attributed, who would be prosecuted for such a violation, given that they are in charge of the ship and its functions, including which type of fuel is used. An exception could be, of course, if other members of the crew also had knowledge of the violation and actively collaborated in the non-compliance.

[23] See at http://legal.un.org/ilc/texts/instruments/english/reports/7_6_2014.pdf.

[24] *Report of the International Law Commission, Sixty-ninth session*, n 2, 76–83.

> alleged offender in territory under its jurisdiction. This obligation helps to prevent an alleged offender from seeking refuge in a State that otherwise has no connection with the offence.[25]

This is a clear reference to the application of universal jurisdiction to ensure that all States have an accepted jurisdictional legal basis that enables them to detain and either prosecute or extradite such perpetrators.

In fact, it would, according to the ILC, not only be a *right* of every State, but also the *obligation* of every State to ensure that such violations are prosecuted. The ILC refers in this context to the Separate Opinion of ICJ Judge Guillaume in the *Arrest Warrant case*,[26] which describes the obligation for all the States to prosecute such violations:

> Whenever the perpetrator of any of the offences covered by these conventions is found in the territory of a State, that State is under an obligation to arrest him, and then extradite or prosecute. It must have first conferred jurisdiction on its courts to try him if he is not extradited. Thus, universal punishment of all the offences in question is assured, as the perpetrators are denied refuge in all States.[27]

Both the ILC and Judge Guillaume refer to detention and prosecution (or extradition) as obligations of the State in which the person is present.

Both the ILC Report and the Separate Opinion refer to the violation of basic human rights, but in the view of this author this can also encompass violations of jus cogens norms, given the previous discussion of the ties between these concepts. The reference to 'perpetrators' can also cover legal persons (entities), such as ships (shipowners) and crew members that knowingly violate an IMO ban on the use of fossil fuels.

Setting aside UNCLOS for now, this would mean that all flag States would be required to enforce an IMO prohibition on the use of fossil fuels over ships flying their flag. However, the obligation to enforce would fall upon all States, including non-flag States, such as coastal States and port States. This would mean, in short, that:

(a) all coastal States must enforce this prohibition in waters under their jurisdiction (ie in internal water, the territorial sea and in the EEZ), irrespective of whether the coastal State is party to this IMO regulation or not; and
(b) all port States must enforce this prohibition within their ports, irrespective of whether the particular port State is party to this IMO regulation or not.

The implications of this, including how it would affect a coastal State's right to enforce over ships making innocent passage or transit passage through an

[25] ibid 79, para 9.
[26] ibid 79.
[27] See the Separate Opinion of Judge Guillaume in *The Arrest Warrant Case (Democratic Republic of Congo v Belgium)* [2002] ICJ Rep 3, para 9.

international strait, and how it affects a port State's enforcement on the high seas, are discussed in the next subsection.

C. Enforcement of IMO Jus Cogens Regulation through Universal Jurisdiction Compared to Enforcement through UNCLOS

All States would, as mentioned, be required to take measures against violations of the IMO (jus cogens) regulation prohibiting the use of fossil fuels, due these also being violations of an erga omnes obligation, allowing for the application of universal jurisdiction. This is irrespective of whether these States are parties to the IMO regulation or UNCLOS. This would not be a new obligation for flag States, as it corresponds with the flag State obligations already found in articles 217 and 222 of UNCLOS.[28] The obligation for coastal States and port States to take enforcement action would, on the other hand, differ from the provisions set out in UNCLOS part XII, as these merely provide these States with different jurisdictional *rights* to enforce, not *obligations*.

The following paragraphs will focus on how non-flag States that wish to enforce can do so, by comparing the normal jurisdictions under UNCLOS, discussed in Parts II and III of this book, with the extended right (and obligation) to enforce through universal jurisdiction. This will enable us to determine where the jurisdictions overlap and where they differ.

i. The Rights of Coastal States to Enforce a Jus Cogens Prohibition on Fossil Fuels

If a ship calls into a port or at an off-shore terminal in a coastal State after it has been detected (by satellite, sniffer, CEMS, etc) violating a ban on using fossil fuels while sailing within the State's waters (internal or territorial waters, or in the EEZ), the coastal State can sanction the violation pursuant to article 220(1) of UNCLOS, as described in chapter 15, section III. The fact that the regulation that was infringed has the status of a jus cogens norm, and so universal jurisdiction can be exercised, provides no wider legal basis for sanctioning the violation than that found in UNCLOS.

However, if a ship were to make an innocent passage through the territorial sea or transit passage through an international strait located in this area, UNCLOS would not allow the coastal State to take any enforcement measures, as it relates to enforcement of regulations on air pollution, thereby allowing for enforcement

[28] See ch 8.

pursuant to article 220(2) and article 233 in accordance with article 42(b) of UNCLOS.[29] Evidently, the same conclusion would apply to the question of stopping a ship making a transit passage of the EEZ, as article 220(5)–(6) would not allow for this. See the discussions of these limitations in chapter 9 and chapter 15.

The universal jurisdiction that a coastal State could invoke in this instance, due to the violation of a jus cogens norm and thereby an erga omnes obligation, would, in the view of this author, change these conclusions, so that the coastal State could take action against any violation committed within its waters. This would allow the coastal State to stop, investigate, detain and sanction foreign ships en route through its waters, including detaining these ships until financial security has been posted, regardless of whether this is innocent passage through the territorial sea or a transit passage through an international strait or the EEZ.

This author bases this conclusion on a *lex specialis* assumption, as there can be no derogation from a jus cogens norm, bearing in mind the description of jus cogens found in article 53 VCLT (see section II.A). Universal jurisdiction makes it possible to respond to violations of such norms, as these are considered serious crimes and, to some extent, violations of basic human rights. They also constitute breaches of an erga omnes obligation owed to the international community as a whole. The universal jurisdiction that can be invoked in response to the violation of a jus cogens norm, for example a ban on the use of fossil fuels, must thus trump the exceptions found within the law of the sea (codified in UNCLOS), which 'merely' seek to allow ships to sail unhindered through foreign waters.

ii. The Right of Port States to Enforce a Jus Cogens Prohibition on Fossil Fuels

If a ship calls into a port or at an off-shore terminal in a port State and a violation is detected during a PSC (ie incorrect documentation or fuel sampling shows non-compliance) then the port State can sanction this under article 226 of UNCLOS. The possible application of universal jurisdiction does not change this outcome.

Further, article 218(1) of UNCLOS allows the port State to sanction any discharge violation (encompassing emissions) taking place outside its territory, including on the high seas (see chapter 10 and chapter 15). This right is also not affected by the universal jurisdiction principle.

A plausible difference in the enforcement measures a port State can utilise by virtue of using universal jurisdiction rather than UNCLOS, when acting in response to a discharge violation (ie a violation of a prohibition on fossil fuels), could be the need to meet the requirement in article 218(1) for the foreign ship to

[29] It will be recalled from ch 9 that a costal State can, in accordance with art 21 of UNCLOS, implement such IMO regulations and enforce them over ships making an innocent passage in accordance with art 220(2). This is not a requirement if the enforcement is based on the exercise of universal jurisdiction to stop the violation of a jus cogens norm.

voluntarily call into a port or at an off-shore terminal in the port State, before the State can invoke the extraterritorial jurisdiction of that provision.

If foreign ship has violated the ban on fossil fuel while sailing on the high seas and subsequently sails through the waters of a port State, including in the EEZ, that port State should have the right to stop the ship during its passage and investigate, detain and sanction the ship, irrespective of the ship's passage being innocent or a transit passage through an international strait or an EEZ. This possibility obviously exists if the ship continues its high-seas violation when sailing into the waters of the port State, as the (port) State would be able to sanction this violation by virtue of its being a coastal State, in accordance with the discussion in section III.C.i.

The same right for a port State should, in the view of this author, also exist if the ship does not continue its violation when it enters into the waters of the port State. Even if a foreign ship's violation is confined to the high seas, and the ship afterwards sails (using compliant fuel) directly through the waters of the port State, that State should still, by invoking universal jurisdiction to sanction a violation of a jus cogens norm, be able to stop, investigate, detain and sanction the ship for the high-seas violation while it is sailing in its waters. This is a clear expansion of the port State's extraterritorial jurisdiction under article 218(1) of UNCLOS.

iii. Overlapping Jurisdictions When Enforcing a Jus Cogens Prohibition on Fossil Fuels

The overlapping jurisdiction between a flag State and a coastal State or a port State is, as discussed in chapter 11, resolved in accordance with article 228(1) of UNCLOS.

But if a State sanctions a violation of an IMO prohibition on fossil fuels by invoking universal jurisdiction, due to its being the violation of a jus cogens norm, this would set aside the procedures under article 228(1), such as meeting the conditions for invoking the main rule or one of the two exceptions. On the other hand, the previously mentioned principle of *aut dedere aut judicare* would apply, resulting in the prosecuting (coastal or port) State being obligated to prosecute the ship (shipowner), and perhaps members of the crew, or 'extradite' them for prosecution in the flag State or in the State where the violation occurred.

IV. Implications for the EU's Adoption of GHG Regulations on a Regional Basis

On a final note, it might be asked whether it would have any impact on the possibilities for the EU to adopt regional regulatory GHG measures for ships, in the form of a regional (EU) ban on the use of fossil fuels within EU waters, if an IMO regulation on this were considered to represent a jus cogens norm. It should be recalled that the EU has previously created its own legislation modelled on IMO conventions, but with important differences: the Sulphur Directive,[30] the Regulation on Ship Recycling,[31] and the MRV Regulation[32] are examples.

It must initially be underlined that any EU measures regulating GHG emissions from ships would not achieve the status of jus cogens norms, despite the opposite applying to the IMO regulation. This is based on the fact that the EU's legislation is only *regional* in its application, while the IMO's regulations represent international (UN) rules that are applicable worldwide. Unlike IMO regulations, therefore, EU legislation cannot be considered to comprise *international rules*, thus being unable to fulfil the first condition laid down by the ILC in its draft article 4 for establishing jus cogens norms: that jus cogens norms must pertain to *international norms*.

Deeming an IMO regulation on GHG to be a jus cogens norm would hinder the EU in adopting legislation that contradicts or undermines this IMO regulation, as any such 'treaty' (directive or regulation) would be void in accordance with articles 53 or 64 VCLT.

However, the EU is not barred from adopting regulations with a higher level of protection against GHG pollution compared to the IMO's regulatory measure (a ban on fossil fuels), as the jus cogens norm (and erga omnes obligation) relates to the protection of human life and rights. A higher degree of protection against GHG pollution would therefore not violate but support the jus cogens principle. It would presumably also be possible for the EU to adopt rules specifying how the IMO ban on fossil fuels should be enforced by the EU Member States, provided this enforcement does not lessen the protection of the jus cogens norm or contradict the previously established rights and obligations within international law for all States to enforce the regulation and take action against violations in accordance with the principles of erga omnes and universal jurisdiction.

Recognising the IMO's prohibition on the use of fossil fuels as a peremptory norm would also have a direct impact on the possibilities for the ECJ to consider and apply this IMO regulation, if a question of the lawfulness of any EU GHG legislation were to be brought before the Court, despite the EU's not being an independent party to the IMO legislation (convention) that codifies the prohibition.

It may be recalled from the analysis in chapter 5 that the ECJ, in both the *Intertanko case*[33] and in the *Manzi case*,[34] refrained from answering questions

[30] Directive 2016/802 of the European Parliament and of the Council of 11 May 2016 relating to a reduction in the sulphur content of certain liquid fuels [2016] OJ L132/58. See the examination of the Sulphur Directive in ch 5.

[31] Regulation (EU) No 1257/2013 of the European Parliament and of the Council of 20 November 2013 on ship recycling and amending Regulation (EC) No 1013/2006 and Directive 2009/16/EC [2013] OJ L330/1.

[32] Regulation (EU) 2015/757 of the European Parliament and of the Council of 29 April 2015 on the monitoring, reporting and verification of carbon dioxide emissions from maritime transport, and amending Directive 2009/16/EC [2015] OJ L123/55.

[33] Case C-308/06 *Intertanko, Intercargo, Greek Shipping Co-operation Committee, Lloyd's Register, International Salvage Union v Secretary of State for Transport*, ECLI:EU:C:2008:312.

[34] Case C-537/11 *Manzi and Compagnia Naviera Orchestra*, ECLI:EU:C:2014:19.

brought before the Court regarding the legality of EU directives relating to IMO legislation (MARPOL Annexes I–II and VI), by placing great emphasis on that fact that the EU was not an independent party to these Annexes[35] and that these IMO regulations did not represent international customary law.[36] The ECJ stated in the *Intertanko case* that it was not 'incumbent upon the Court to review the directive's legality in the light of the Convention'.[37] This conclusion was also reiterated in Advocate General Wahl's Advisory Opinion in the *Bosphorus Queen*.[38]

A reverse conclusion establishes that if an IMO fossil fuel ban were to be considered a jus cogens norm representing customary law, this would have an effect on the ECJ's assessment in such cases, despite the EU's not being an independent party to the IMO regulations.

Legal questions as to the lawfulness of the EU's legislation relating to the implementation[39] and/or enforcement of the IMO's fossil fuel regulation could be brought before the ECJ. The Court would be required to adjudicate upon such questions, as the IMO legislation represents customary international law.

In conclusion, EU legislation regulating GHG emissions from ships, including a ban on fossil fuels, could not, unlike the IMO's regulation, be considered as representing international norms with recognised jus cogens status, as it has limited global applicability due to its regional scope. Deeming an IMO regulation to be a jus cogens norm would have an impact on the EU regulatory process, as the EU would have to ensure that any regional rules adopted on this matter (or on the enforcement of IMO's regulation) respect these jus cogens norms and the enforcement thereof, in accordance with recognised principles of international law.

The ECJ is competent to examine and determine whether the EU has fulfilled the obligations, despite the EU's not being an independent party to the IMO regulation. This is based on the IMO prohibition on fossil fuels representing international customary law (ie peremptory norms).

V. Conclusion

This chapter has sought to explore the possible consequences of the conclusions drawn in the previous chapters of Part IV.

Chapter 16 described what the principle of peremptory (jus cogens) norms entails, whilst also introducing the erga omnes principle and how this relates to universal jurisdiction.

[35] *Intertanko*, n 33, paras 47 and 49; *Manzi*, n 34, para 37.

[36] *Intertanko*, n 33, para 51; *Manzi*, n 34, para 39.

[37] *Intertanko*, n 33, para 50.

[38] Advocate General Wahl's Advisory Opinion of 28 February 2018 in Case C-15/17 *Bosphorus Queen Shipping Ltd Corp v Rajavartiolaitos*, ECLI:EU:C:2018:557, paras 46–47.

[39] The term 'implementation' is used in this context to describe the EU's independent adoption of a regional regulation that is based on IMO regulations.

Chapter 17 identified the conditions established by the ICJ and ILC for recognising an international environmental norm as a jus cogens norm. It must be a 'norm of general international law', which is 'accepted and recognized by the international community of States as a whole from which no derogation is permitted'. Recognition by the 'international community of States' (eg the ICJ and ILC) is achieved if the international norm (regulation) prohibits massive, widespread, transboundary pollution of the atmosphere that can threaten human health and the quality of life, including for future generations. Any gross or systematic violation could also be considered a serious crime that violates basic human rights, which all States are obligated to protect.

This chapter has concluded that an IMO regulation that prohibits the use of fossil fuel could fulfil these conditions. The legal implications (not exhaustive) were examined, and it was found that States would be unable to enter into bilateral or multilateral agreements that undermine or circumvent this IMO regulation. Such agreements would be void in accordance with articles 53 or 64 VCLT, depending on whether the specific IMO regulation was deemed a jus cogens norm before or after the agreement was made.

States could also encounter liability, as stipulated in the ILC Articles on State Responsibility (especially article 40), if they systematically fail to ensure compliance with the regulation. This would apply to flag States that, for example, repeatedly fail to ensure that ships under their flag adhere to the IMO regulation, or to port States that fail to ensure that persons or legal entities (companies, ports, fuel suppliers, etc) within their jurisdiction comply with the fossil fuel ban. The liability of a State under article 40 of the Articles on State Responsibility could be invoked by all States, pursuant to article 48(1)(b).

Violations of jus cogens norms are also, as concluded in chapters 6 and 16, deemed violations of erga omnes obligations, allowing all States to bring such violations before a court. However, as any such violation also could constitute the violation of a basic human right, it could lead to application of the principle of universal jurisdiction, as stated by the UNEP. This would bestow not only a *right* on all States to take measures against such violations within their own judicial systems, but also an *obligation* to do so, conditional on the ship's being within the territory (internal or territorial waters or in the EEZ) of a State, even though the violation may have occurred outside the territory. This would also apply to States that are not party to UNCLOS.

This chapter has therefore also elaborated on how all States, by virtue of the erga omnes principle, could invoke the application of universal jurisdiction to enforce violations on an IMO jus cogens regulation prohibiting the use of fossil fuels, and how this enforcement differs from enforcement through UNCLOS. One difference between enforcement through UNCLOS and enforcement by universal jurisdiction is that *coastal States* would be able to stop, investigate, detain and prosecute foreign ships that use non-compliant fossil fuel in their waters, including in the EEZ, and when foreign ships are making an innocent passage or a transit passage in an international strait. This sets the enforcement apart from the coastal States' normal UNCLOS jurisdiction pursuant to articles 220(2), 220(5)-(6) and 233.

Port States could also exercise a wider universal jurisdiction, as they should (also) be allowed to stop, investigate, detain and prosecute a ship transiting the port State's waters without calling into port, to sanction a violation that occurred on the high seas.

All States must also penalise the violation, or ensure that another does so, following the principle of *aut dedere aut judicare.*

Finally, it was concluded that deeming an IMO ban on fossil fuels to be a jus cogens norm could also have implications for the ability of the EU to adopt regional GHG regulations and how the ECJ could adjudicate on such matters. Due to its being regional in nature, EU legislation on GHGs cannot be considered as comprising peremptory norms. The EU must respect the jus cogens IMO regulation, but can, in its regional rules, offer a higher degree of protection of the same peremptory norms, that is, protection of the atmosphere against GHG pollution from ships.

The ECJ can apply the IMO legislation when deciding upon the legality of any EU-adopted GHG legislation, even though the EU is not an independent party to the IMO regulations. This conclusion is based on the decisions of the ECJ in the *Intertanko* and *Manzi cases* (affirmed by Advocate General Wahl in the *Bosphorus Queen*), which explicitly placed emphasis on whether or not IMO rules were considered to represent international customary law. Rules prohibiting the use of fossil fuels that are deemed to be jus cogens represent such customary law, thus having a binding effect on the EU and the ECJ.

A. Epilogue on Part IV

There have previously been discussions at international level about recognising environmental norms as jus cogens norms, especially pertaining to pollution of the atmosphere. Those discussions have often led to a legal impasse, however, as the theory of environmental jus cogens norms can be difficult to utilise and apply in practice.

This was eloquently stated by the ILC's Special Rapporteur, Dire Tladi, in the ILC's report on peremptory norms of general international law, where Mr Tladi summarised the status of the discussions on recognising environmental jus cogens as follows:

> By virtue of the importance of the subject matter and the catastrophic consequence that could result from the destruction of the environment, it might seem obvious that norms that aim at protecting the environment (at least some of them) would have the status of jus cogens. Yet, there seems to be little evidence of the required 'acceptance and recognition of the international community of States as a whole' that the environmental norms (or some of them) have acquired peremptory status, notwithstanding this empirical fact of the importance of environmental rules for the very survival of

> humanity and the planet. The paradox was noted by Krista Singleton-Cambage, who noted that, at the time (1995), 'environmental rights and responsibilities are not recognized as having' the status of jus cogens 'despite the fact that global environmental preservation represents an essential interest of all individuals within the entire international society'. In the Commission's own work, the importance of the atmosphere, as an empirical fact, has been acknowledged, yet there has been no recognition of the peremptory status of protecting the atmosphere – a resource on which life on earth depends. It is the case that there are many declarations and treaties on the environment, yet none of them provide strong evidence of non-derogability. Orakhelashvili does make a spirited argument for the jus cogens status of specific norms related to the environment, yet even he accepts that there is 'lack of evidence'. Although not referring to norms of jus cogens, John Dugard has described particular rules relating to the protection of the environment as establishing obligations erga omnes. It may well be that that some rules, like some relating to the environment, have the status of jus cogens which has yet to be accepted and recognized by the international community of States as a whole, with the result that the effects in law of jus cogens do not yet flow from such.[40]

It has, however, because of the legal basis established by the ICJ and ILC, been found in this part of the book (Part IV) that an IMO regulation that prohibits the use of fossil fuels could obtain the status of a peremptory norm by being recognised as such by the international community, due to the direct and immediate impact of this on prohibiting a massive source of atmospheric pollution (ie by GHGs and specially CO_2) that damages human life and quality, including for future generations.

As stated in the Preface to this book, my structure, language of argument, etc may be somewhat … unpolished, but this does not change the fact that a legal basis does exist for asserting that a future IMO regulation banning the use of fossil fuels could be considered a peremptory norm. If nothing else, it is hoped that this part of the book will inspire other writers to continue this work, to help protect the environment and thereby our future generations.

This concern was perfectly expressed by the ICJ in 1996 in its Advisory Opinion on the *Legality of the Threat or Use of Nuclear Weapons*, which statement shall therefore end this chapter and this part of the book:

> The environment is not an abstraction but represents the living space, the quality of life and the very health of human beings, including generations unborn.[41]

[40] See, eg, Special Rapporteur Dire Tladi, *Fourth Report on Peremptory Norms of General International Law (Jus Cogens)*, ILC 71st session, A/CN.4/727 (April 2019) 61–61, para 136.

[41] *Legality of the Threat or Use of Nuclear Weapons*, Advisory Opinion [1996] ICJ Rep 226, para 29.

PART V

Conclusion

19

Conclusion

This book has sought, throughout its four preceding parts, to establish how the strengthened sulphur limits of regulation 14.1 of MARPOL Annex VI[1] and a future IMO regulation banning the use of fossil fuels might be effectively enforced.

The first chapter of Part I focused on providing a basic introduction to the historic development of MARPOL and UNCLOS,[2] and the impact of sulphur pollution on human health, including descriptions of the economic incentives for shipowners not to comply and the challenges involved in detecting violations.

The following chapters of Part I included reiteration and examination of the key regulations and provisions of MARPOL Annex VI, UNCLOS and the EU Sulphur Directive.[3] Port State Control (PSC) enforcement, encompassing the principle of 'no more favourable treatment' (NMFT) and international PSC collaboration through the different Memorandums of Understanding (MoUs), was also discussed, as well as the basic jurisdictional principles of international law, some of which were established in the *Lotus case*.[4]

The seven chapters in Part II attempted to present a thorough analysis of the relevant provisions of part XII of UNCLOS, including the jurisdictions of coastal States and port States pursuant to articles 220 and 218, as well as laying out the obligations of flag States in accordance with article 217. The subsidiary nature of articles 212 and 222 pertaining to enforcement of air pollution regulations was also defined. The extremely important article 228(1), for determining primary jurisdiction, comprising a main rule and two exceptions, was also subject to in-depth study, especially pertaining to the three detailed requirements for invoking the main rule and to the second exception, which applies to flag States that repeatedly disregard their obligations to effectively enforce. Finally, Part II offered some insights into the safeguarding provisions of part XII of UNCLOS (section 7) and the provisions of part XV of UNCLOS for dispute settlement, and their relevance for enforcing the sulphur regulations of MARPOL Annex VI.

[1] International Convention for the Prevention of Pollution from Ships (adopted on 11 February 1973, as modified by the Protocol of 17 February 1978, entered into force 2 October 1983) 1340 UNTS 61 (MARPOL), Annex VI, IMO Publication: IMO-520E.

[2] United Nations Convention on the Law of the Sea (adopted 10 December 1982, entered into force 16 November 1994) 1833 UNTS 3.

[3] Directive 2016/802 of the European Parliament and of the Council of 11 May 2016 relating to a reduction in the sulphur content of certain liquid fuels [2016] OJ L132/58.

[4] *SS Lotus (France v Turkey)* PCIJ Rep Ser A No 10.

The conclusions in Part II established as follows:

- *Flag States* are, pursuant to article 217 of UNCLOS, required to take effective enforcement measures against violations of environmental regulations committed by ships under their flag. This includes violations of MARPOL Annex VI, irrespective of where these occur, including on the high seas. Effective enforcement of regulation 14 of MARPOL Annex VI must, at a minimum, involve confiscating all savings achieved *and* deter future violations by imposing a dissuasive punitive element that increases in the event of aggravating circumstances, such as repeated violations – a principle also set out in article 18 of the Sulphur Directive. The flag State must subsequently inform all States and the International Maritime Organization (IMO) of its effective enforcement.
- *Coastal States* are able to take action against all violations of the sulphur limits of MARPOL Annex VI within their territory, including in the EEZ, pursuant to article 220(1) of UNCLOS, provided the violating ship subsequently and voluntarily calls into a port or at an off-shore terminal in the coastal State. Certain exceptions apply to the coastal State's right to enforce in territorial waters, pertaining to foreign ships that are making transit passage through an international strait in accordance with article 38, or innocent passage in accordance with articles 17–19 of UNCLOS. A coastal State can, however, adopt environmental rules pertaining to innocent passage in accordance with article 21(1)(f) and enforce these pursuant to article 220(2). A coastal State's jurisdiction could, in principle, be extended in accordance with the principle of *hot pursuit* pursuant to article 111, for example by the use of drones with sniffers and visual uplinks attached. Article 220(5)–(6) are not applicable to enforcement of MARPOL Annex VI in the EEZ, as a single emissions violation, for example of the sulphur limits, cannot result in tangible and significant or major damage.
- *Port States* are provided with an extraterritorial jurisdictional basis under article 218(1) of UNCLOS, for taking measures in response to discharge violations on the high seas or in the waters of another State, conditional upon the ship's afterwards calling into a port or at an off-shore terminal in the port State. The term 'discharge', as used in article 218(1), can, by interpreting it in accordance with article 31(1) of the Vienna Convention on the Law of Treaties (VCLT),[5] be applied to cover *emissions*, a conclusion supported by the definition of 'discharge' in article 2(3)(a) of the MARPOL Convention and how the term is already applied in regulation 12.7.3 of MARPOL Annex VI relating to the discharge of ozone-depleting substances into the atmosphere. Article 218(2)–(4) enable coastal States to empower a port State to take measures against a violation on their behalf. This allows port States to carry

[5] Vienna Convention on the Law of Treaties 1969 (adopted 22 May 1969, entered into force 27 January 1980) 1155 UNTS 331 (VCLT).

out end-to-end enforcement in response to a foreign ship's violation of regulation 14 of MARPOL Annex VI.

- Any *overlapping jurisdiction* between a flag State and a port State or a coastal State is resolved in accordance with article 228(1) of UNCLOS, which consists of a main rule and two exceptions. Applying the main rule is conditional upon the flag State's meeting three requirements: (i) reacting within *six months* by (ii) bringing *corresponding charges* and afterwards (iii) *proving* that it brought corresponding charges that resulted in effective enforcement. However, a port State or coastal State can deny a flag State's request for primary jurisdiction if the *first exception* (coastal State suffering *major damage*) or the *second exception* (flag State *repeatedly disregarding its obligations to effectively enforce*) in article 228(1) applies. The second exception is, unlike the first, of considerable relevance to the enforcement of regulation 14 of MARPOL Annex VI.

These conclusions were applied in Part III of this book, to assess how existing IMO regulations, including the other provisions of MARPOL Annex VI, could be enforced in accordance with part XII of UNCLOS. Part III also sought to clarify the framework of the IMO's GHG Strategy and its reduction goals, especially pertaining to enforcement of the short-, mid- and long-term candidate measures proposed in the Strategy through UNCLOS. These measures include revised Energy Efficiency Design Index (EEDI) phases, speed optimisation, JIT Arrival, cold ironing and, last but not least, provisions on the use of alternative (zero-carbon/fossil-free) fuels. It was also discussed whether the IMO regulations on GHG could result in the IMO's adopting a regulation that totally prohibits ships from using fossil fuel, including obligations for fuel suppliers and ports not to assist in the procurement of fossil fuel. It was concluded that such a prohibition could easily be envisioned as being adopted by the IMO, and that it would allow shipowners to decide how to comply with the prohibition, for example by using alternative fuel as stipulated in the GHG Strategy, or by other means, such as solar power, wind, electricity (batteries), etc.

Part IV of the book attempted to outline the possibilities in relation to whether this potential future IMO regulatory measure on GHGs, prohibiting the use of fossil fuel, could be considered a peremptory norm of a jus cogens character, that is, a norm that would coincide with erga omnes obligations owed to the international community as a whole. After an introduction to these principles, the book sought to establish what conditions an international environmental norm (regulation) must meet to be recognised as a jus cogens norm. Conditions laid down by the International Court of Justice (ICJ) and the International Law Commission (ILC) state that if an international norm (regulation) prohibits massive transboundary pollution of the atmosphere that can threaten human health and quality of life, including for future generations, it could be considered a peremptory norm.

It was subsequently concluded – due to the harmful effects of GHG pollution on the environment and on the conditions needed to sustain human life and habitation – that GHG pollution could meet the ICJ/ILC requirements for massive, transboundary pollution of the atmosphere that can threaten human existence, including for future generations. It could therefore be envisioned that an IMO regulation prohibiting the use of fossil fuel might be considered a jus cogens norm. Finally, this led to analysis in the concluding chapter of Part IV of what the legal consequences of this conclusion might be. As such, it was established as follows:

- A new IMO regulation banning the use of fossil fuel could be considered a peremptory (jus cogens) norm. Other treaties therefore may not, according to articles 53 and 64 VCLT, infringe upon such a regulation, and any derogation by a State, be it a flag State or a State with ports or fuel suppliers under its jurisdiction, could lead to the State's becoming liable in accordance with articles 40–41 of the Articles on State Responsibility – a liability that all States can invoke pursuant to article 48(1)(b), relating to jus cogens norms also representing erga omnes obligations.
- A violation of a jus cogens norm, as mentioned, can also be regarded as a violation of an erga omnes obligation, but it could also constitute a violation of a basic human right, allowing the application of universal jurisdiction when the ship is present within the territory of a State, even though the violation occurred outside the territory. This would give any State not only a *right*, but also an *obligation* to take enforcement measures against such violations within its own judicial system, with the principle of *aut dedere aut judicare* applying in this respect.
- It could also influence enforcement by non-flag States, compared to the jurisdiction granted by UNCLOS, if this IMO regulation were to be deemed a jus cogens norm, as coastal States could, by virtue of the erga omnes principle and by invoking universal jurisdiction, stop, investigate, detain and prosecute foreign ships for using non-compliant (fossil-based) fuels, despite these ships' making innocent passage through the territorial sea or transit passage through the EEZ or an international strait.
- Lastly, IMO GHG regulations of a jus cogens character could also influence the ability of the EU to adopt regional GHG regulations and how the ECJ could adjudicate on such matters.

International law is constantly evolving to adapt to the developments and needs of the international community. It is my belief that UNCLOS provides a legal basis for port States to take enforcement measures in response to violations of the 0.5% sulphur limit in MARPOL Annex VI, including on the high seas and, under certain conditions, in spite of flag State objections.

It is also my profound hope that a future IMO regulation prohibiting the use of fossil fuel may, in time, be considered a peremptory (jus cogens) norm.

INDEX

Introductory Note

References such as '178–79' indicate (not necessarily continuous) discussion of a topic across a range of pages. Wherever possible in the case of topics with many references, these have either been divided into sub-topics or only the most significant discussions of the topic are listed. Because the entire work is about 'air pollution' and 'UNCLOS', the use of these terms (and certain others which occur constantly throughout the book) as entry points has been restricted. Information will be found under the corresponding detailed topics.

www.ingramcontent.com/pod-product-compliance
Ingram Content Group UK Ltd.
Pitfield, Milton Keynes, MK11 3LW, UK
UKHW021911150926
475560UK00003B/314

9 781509 927760